▼ **Graphic Communications Today**

SECOND ▼ EDITION

Graphic Communications Today

Theodore E. Conover
University of Nevada, Reno

West Publishing Company
St. Paul   New York   Los Angeles   San Francisco

Copyediting: Elaine Levin
Illustrations: Brenda Booth, Rolin Graphics
Composition: Parkwood Composition
Index & Glossary: Lavina Miller

COPYRIGHT ©1985   By WEST PUBLISHING COMPANY
COPYRIGHT ©1990   By WEST PUBLISHING COMPANY
                   50 W. Kellogg Boulevard
                   P.O. Box 64526
                   St. Paul, MN 55164-1003

Printed in the United States of America

97 96 95 94 93 92 91   8 7 6 5 4 3 2 1

Library of Congress Cataloging-in-Publication Data

Conover, Theodore E.
   Graphic communications today / Theodore E. Conover. — 2nd ed.
      p.   cm.
   Includes bibliographical references.
   ISBN 0-314-66570-6
   1. Printing, Practical—Layout.   2. Newspaper layout and typography.   3. Newsletters—Design.   4. Magazine design.   5. Graphic
arts.   I. Title.
Z246.C58   1990
686.2'24—dc20   89-70567
                                                         CIP

▼ **To Edna**

Contents

C H A P T E R   T H R E E

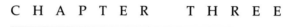

▼ **Type: The Basic Ingredient**   **43**

C H A P T E R   F O U R

▼ **Creative Typography**   **65**

Contents

C H A P T E R   F I V E

▼ Art and Illustrations

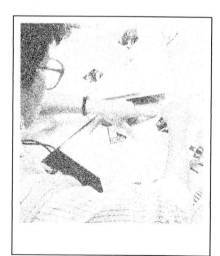

C H A P T E R   N I N E

▼ Preparing for Production   **179**

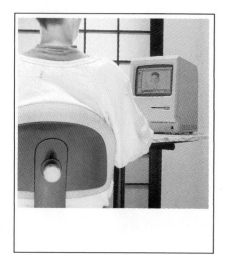

C H A P T E R   T W E L V E

▼ Advertising Design   263

C H A P T E R   T H I R T E E N

▼ Designing the Magazine   285

C H A P T E R   F O U R T E E N

▼ Inside and Outside the Magazine   303

C H A P T E R   F I F T E E N

▼ **The Newspaper and the Designer**   **329**

C H A P T E R   S I X T E E N

▼ **Designing and Redesigning Newspapers**   **355**

C H A P T E R   S E V E N T E E N

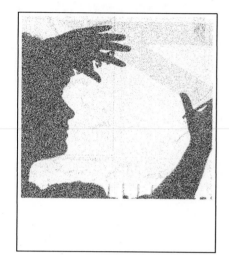

▼ **Newsletters**   **395**

C H A P T E R   E I G H T E E N

▼ **Designing in the Twenty-First Century**   **423**

APPENDIX   A

▼ A Basic Communications Design Library   **439**

APPENDIX   B

▼ Type Specimens   **445**

Foreword

We've come full circle in American print communication. And this book is a celebration of that fact.

The first newspaper editor was a printer who established a weekly to make sure there was always something "on the hook" for the craft personnel to work on. The first advertising designer was a printer. The first direct-mail layout was done by a printer. So was the first magazine and the first annual report and the first decorative label on a ketchup bottle or a box of matches. Even the first "graphic" artist—who showed the position of the pallbearers at Washington's funeral—was a printer who fashioned the newspaper diagram with ordinary printing rules and extraordinary ingenuity.

But the industry expanded so rapidly that it divided like growing cells. Magazines were separated from newspapers, advertising from news matter, public relations mailings from billboards. And within each division, individual jobs were specialized. Newspeople and ad staffers left production to compositors, engravers, and those who ran the presses.

But today the editor again becomes a printer. Typesetting, page makeup, advertising design—whether for newspaper, magazine, print ads or direct mail—are not only controlled by, but are actually produced by, what we can collectively term "editorial people."

Today's journalism student—and, indeed, the professional communications practitioner—must know what happens after his or her copy goes "to the backshop." This book tells that. As well as what, today, must happen even before that stage.

The graphic arts are both a craft and an art. They are more than a commerical enterprise; they are warp and woof of our culture, the flowering of and the mirroring of our history. So this book acknowledges and honors the past of our craft and profession even as it acknowledges and anticipates the exciting future that lies ahead of our exciting present.

Communicators who lived in the watershed years of Johann Gutenberg must have felt an electric excitement when they realized the momentous change from handwritten manuscript books to those printed by movable type.

We are living in an era just as revolutionary (a word used advisedly and precisely, not as in a "revolutionary new toothpaste"). We tend to be a bit blase about our vocations, though, perhaps a bit jaded by space walkers and artificial hearts. But we have every reason for excitement and pride and anticipation. All of us—students and professionals alike—have many new tools to master and to communicate with.

Ted Conover is the right man to write a book like this. His distinguished career has spanned both the hot-metal era (actually little changed from Gutenburg's methods) and today's electronic wizardry. He is more, even, than a rare combination of craftsman and artist. He is a fine teacher who has insisted that his students be more than tradeschool graduates, that they be university graduates and, as such, the Universal Man and Woman. He knows that the printed word—be it printed on paper or on a television screen—is the thread that weaves together our civilization and culture. What's more, he knows that this craft, this art, this pursuit—label it what you will—is fun, lots of fun. All together, this is a universal book that a respected and cherished colleague has written. I gladly share it with you, Gentle Reader.

Dr. Edmund C. Arnold
Virginia Commonwealth University

Preface

My hope is that this book will provide useful information for those who desire to produce more effective and attractive communications.

I have other goals for this book as well. First I hope this book will help designers understand the communications philosophies of editors. Second, I hope this book will help editors understand the philosophies of designers so they can work together in reaching common communications goals.

Third, there are those who aspire to careers in communications and find themselves putting together publications. Suddenly they have to know layout, typography, and graphics—and know them fast. They have to put out a newsletter, company magazine, financial report, or brochure. This book, hopefully, will help them, too. It is written for the writer who now must be the editor and layout person and for people who shift from the electronic to the print media.

One of the basic rules of effective communications is "keep it simple." Every attempt has been made to keep it simple. At the same time, every attempt has been made to include what is needed to plan and produce effective communications.

This book is intended to be a beginning. A starting point. The student or professional can use it to enter the fascinating, fast-changing world of graphic communications. I hope that the book will open a door and the reader will go on from there to continuing and more complex studies of one of humanity's oldest, and at the same time newest, arts—the art of communication.

I also hope that practicing professionals will find this book worthwhile. Even though busy communicators may know typography and graphics thoroughly, a review of the field from a different perspective may trigger new inspiration and enthusiasm. This book provides a chance for professionals to stand back and reexamine the way things are being done, to break out of daily routines to consider improvements so that the messages produced are more readable, more attractive, and more effective.

▼ Acknowledgments

It would be impossible to list everyone who has had a part in making this second edition of *Graphic Communications Today* a reality. As in the first edition, I have contacted magazine art directors, newspaper designers and executives, designers, equipment manufacturers, and educators by the score for advice and permission to use materials. The response has been overwhelming and all have my everlasting gratitude.

Several people, though, must be citied for their special help. Those who helped with the first edition include the following:

George Weiss, author and director of planning, Metro Associated Services, Inc.; John L. Rush, president, Dynamic Graphics, Inc.; Hal Metzger, Graphic Products Corporation; William Marken, editor, *Sunset;* Mark A. Williams, editorial art manager, the *Orlando Sentinel;* Dan Vaccaro, general manager, Printers Shopper; Donald H. Duffy, corporate art director, *Reader's Digest;* Cliff Kolovson, Atex, Inc.; Rebecca Marrs, Texet Corporation; Don Watkins, author and advertising designer; and Jean Stoess, who prepared the manuscript and helped with revisions.

The following people read the manuscript and offered many valuable suggestions for its improvement:

Laura Berthelot
Louisiana State University

Daniel Boyarski
Carniegie-Mellon University, Pennsylvania

Alan Dennis
University of Alabama

Robert J. Fields
Virginia Polytechnic Institute and State University

Marie Freckleton
Rochester Institute of Technology, New York

Charlotte R. Hatfield
Ball State University, Indiana

Robert H. Hawlk
Ohio University

Kenneth F. Hird
California State University, Los Angeles

Tom Knights
Northern Arizona University

Sean Morrison
Boston University

W.S. Mott
California Polytechic State University, San Luis Obispo

James F. Paschal
University of Oklahoma

Bob Pike
Syracuse University, New York

Roger Remington
Rochester Institute of Technology, New York

Jerry Richardson
North Dakota State University

David W. Richter
Ohio State University

Bill Ryan
University of Oregon

Thomas E. Schildgen
Arizona State University

Sexton Stewart
Moorpark College, California

Terry Whistler
University of Houston, Texas

Robert Willett
University of Georgia, Athens

Harold W. Wilson
University of Minnesota

Karen F. Zuga
Kent State University, Ohio

Two who helped with the first edition and contributed much to this second edition were Professor Myrick Land, who gave the valuable advice and reviewed the manuscript, and Edna Conover, who read the manuscript, made suggestions, and kept records. In addition, I would like to thank the following for their help on the second edition: Michael Kennedy, designer; Susan Zucker, editor, *Plus Business*, Metro Creative Graphics; Richard A. Nenneman, editor-in-chief, *Christian Science Monitor*; Richard Stout, president, Reno Printing; Bob Felten, advertising account executive; Michael J. Parman, editor, *Press Democrat*; Bill Burns, S.D. Warren Company; J. Michael Pate, publisher, *Sun News*; H. Doyle Harvill, editor, *Tampa Tribune*; Richard Curtis, *USA Today*; Fred Liddle, Carpetbagger Press; Paul H. Conover; and my colleague, Phillip Padellford.

Edmund C. Arnold, who has contributed much to the world of graphics and who played a part in making this book possible, deserves my sincere thanks.

Once again, Clyde Perlee and his fine staff at West Educational Publishing made me realize many times how fortunate I have been to have them guiding this project to completion. A special thanks to Jan Lamar, developmental editor, and Ann Rudrud, production editor.

In addition, I would like to thank my associates for many years in the American Amateur Press Association who encouraged my interest in the printed word for the pleasure it can bring, rather than for its profit potential.

Theodore E. Conover

Introduction

Photograph: Mark Jenkinson

There are a number of reasons why people in the world of communications need a comprehensive understanding of graphics and typography today.

This need began to emerge in American society in 1956 when white-collar workers outnumbered blue-collar workers for the first time. This fact heralded, though it was mostly unnoticed at the time, the forthcoming *information age.*

John Naisbitt points out in *Megatrends,* that "in 1950 only about 17 percent of us worked in information jobs. Now more than 65 percent of us work in information."[1] He notes that most Americans spend their time creating, processing, or distributing information.

What does this have to do with graphics and typography? This flood of information threatens to bury us, making it increasingly difficult for pertinent and essential material to attract the attention of its intended audiences. It challenges communicators to find ways to present information dramatically and effectively.

Thus there is a clear need for a *graphic journalist*—a communicator who understands the effective use of words *and* who understands how to present information in a graphic and effective manner. In the world of communications, graphic journalists are now filling an important role in the editorial offices and newsrooms of both the print and electronic media. This role will increase in importance as time moves on.

Consider what an editor in the magazine division of McGraw-Hill told a journalism instructor; "If you do not do anything else beyond trying to make your students write well, be sure they understand the basics of graphics, typography, and printing."

The instructor was working in the magazine division of the publishing giant on a program sponsored by the American Business Press. Again and again staffers on *Engineering and Mining Journal, Business Week,* and *Chemical Week* emphasized the need for journalists who knew something about the mechanics of effective printed communication. They lamented that journalism school graduates did not know how to select and arrange types, how to lay out pages, or even how to use white space effectively.

At the same time, designers and photo editors complain that editors do not understand graphics, good design, and good visual display.

This need for understanding graphics has been accented in recent years for a number of reasons. One concerns the technological advances in the profession. Another has to do with the challenge to communications from competing forces such as changing lifestyles, opportunities for activities that infringe on time for reading, and the declining emphasis on reading in our education system.

A third concerns the inclination for people in the communications industry to switch from one area to another. Reporters shift to public relations, editors shift to advertising, newspaper people shift to television, and then they shift again.

Then there is the new technology. It has made printing and duplication of printed messages easily available to everyone involved in the distribution of information. As a result we are being flooded with an incredible amount of poorly designed and poorly executed communications. A lot of wastebaskets are overflowing with unread communications, brochures, and newsletters because they are visually unattractive.

Notes

[1] Naisbitt, John, *Megatrends*, (New York: Warren Books, Inc., 1984), pp. 4–5.

Why Typography and Graphics?

Fig. 1-1 A typical newsroom in 1969 before the electronic era. Layout people used manual typewriters, rulers, pencils, paste jars, and scissors to design the newspaper. (Photo by Howard Decker)

▼ Before the Big Change

It is midsummer 1962 in the editorial department of the *Daily Times.*

The *Daily Times* is a typical small-city American newspaper, and the staff is working hard because it is thirty minutes before deadline. About an hour from now, today's edition will roll off the press.

One of the reporters is tapping out a story on her battered manual typewriter. She has already banged out three pages of her account of last night's city council meeting, and the pages of copy paper are strewn across her desk.

In the corner, surrounded by his teletype machines, is the wire editor, rolling up strips of yellow paper covered with tiny holes as it comes off a perforating machine. This device punches ⅞-inch-wide tape in concert with the teletype that is typing out on paper a story sent by telephone wire from New York.

The wire editor will edit the printed story and write a headline in pencil on a strip of copy paper. Then he will send the three items—printed copy, headline, tape—to the composing room. The tape will activate typesetting machines to transform copy into type. The headline will go to a "floorman" who will set it into type on a Ludlow machine, which casts type from molten metal. The printout copy will be used by the proofreader to check for errors.

Over in the advertising department salespersons are turning in the copy they have collected from the businesses to whom they have sold space. The artist is sketching illustrations, advertising layout people are "marking up" advertisements. They are adding instructions for the workers who will prepare the advertisements for printing in the composing room. Others are making layouts from the salespersons' notes, clipping illustrations from mat service books—collections of generic artwork. Drawing boards, t-squares, pens and pencils, scissors, and many other tools are being used to create the layouts.

Others in the advertising department are marking the areas for advertisement placement on dummies, which are miniature sketches of the newspaper pages. These page dummies will be sent to the editorial department where space for stories, photographs, and features will be designated. The completed dummies will be sent to the composing room where they will be used much like blueprints to guide those workers who will prepare the pages for printing.

Back in the composing room, it's bedlam. Linotype machines seven feet high are clashing and clanging as they cast lines of type for the page forms from molten metal. People are rushing about making proofs of galleys of type, sawing strips of metal for spaces between lines of type, mitering corners on other strips of metal borders to make boxes for some stories and advertisements, casting the headline type, and using routers to shave unwanted metal off castings of illustrations.

Other people are placing all the metal pieces in proper order within chases, the metal frames that hold the pages together while they are put through a molding press to make mats, or molds, of the pages for casting in the foundry. The resulting castings, or plates, will be put on the press for actual printing.

Meanwhile, back in the editorial department, the reporter has finished the story. She pulls the last page of copy from her typewriter and hands the story to the city editor. He asks a few questions, makes a note or two on the copy, and takes it to the copy editor. ·

The copy editor is seated in the "slot." He is in the middle of a U-shaped desk, and four copy readers are on the "rim," or outside edge of the desk, busily making corrections and writing heads for stories.

Fig. 1-2 A composing room worker moves an assembled page on a table with wheels (called a turtle) toward the pressroom in the days of hot type (below, right). The ponderous process was replaced by the drawing board and cold type tools (left). Today the computer is replacing both.

The copy editor sees that one of the readers has just finished editing a story, so he tosses the latest arrival in that direction for a final going over before it is sent to the composing room.

Before handing over the story, the copy editor has diagrammed the area it will occupy on a page dummy. The makeup man will use this sketch to guide him in putting the page together for the molding press.

While all of this is going on at the newspaper office, farther down the street partners in a public relations consulting business are contemplating their future. They recently opened their office and business is slow. They had assumed that their job would be to provide publicity for the clients they could obtain in the community. They were discovering how wrong their idea of public relations was.

Both partners had some media experience. One had been a reporter for the *Daily Times* for three years. He had covered city politics. The other had been on the public affairs staff of the local television station. They thought they knew all they needed to handle the tasks of a public relations operation.

They were good writers and editors. They knew the media and what editors and news directors wanted in the way of usable copy. They had a lot of good contacts in the community.

But they soon found out that they would be expected to do a lot more besides turning out press releases. Their clients wanted them to give advice on policy questions, find out what employees thought about their organizations, and evaluate community opinion. And their potential clients expected them to produce brochures, information folders, and all sorts of other printed material.

Two potential clients had specific projects. One wanted a monthly employee magazine or newsletter. Could the partners produce one—from writing all copy to taking care of printing and distribution?

The other prospective client wanted a bimonthly newsletter for its members and patrons.

Fig. 1-3 The tools of the trade are changing rapidly. This is a typical newsroom today. Gone are the typewriters and disappearing are the pencils, rulers, and paste jars. Electronic equipment is now used for writing stories and designing pages. (Photo by Larry Brooks)

The partners had no trouble coming up with lively copy and illustration ideas for the magazine and newsletter, but when it came to putting the publications together and working with printers, they didn't know where to begin.

The printers asked questions like:

"Shall we set the body matter 8 on 10 or 9 on 11?"

"Would you like an Oxford rule to box the masthead?"

"Would you like us to print it on 60-pound machine finish book, or do you think 70-pound would be better?"

The partners had never worked with printers and they had never selected types and rules or designed attractive pages.

They lost both clients.

Back at the newspaper, the editorial and advertising staffs are not concerned about the printing of the *Daily Times*. They know that once they send copy to the composing room skilled craftspeople will space everything on the pages so the printed *Daily Times* will look attractive. They know that if they write a headline or advertising copy block that is slightly too long for the space allotted, the workers will adjust the spacing between words or lines to make things fit. Proofreaders will catch most errors.

▼ The Big Change

Now it is the decade of the nineties, and before we know it a new century will arrive.

At the *Daily Times* there have been a lot of changes. It is quiet. The clatter of typewriters and staccato beats of the teletypes and perforating unit are gone. Reporters are staring intently at video display screens as they type their stories on computers. Wire copy is spewing forth at a rapid rate from a printer, which quietly zips and buzzes, while the same copy is being stored in a computer.

Computer-enhanced wire service photographs are transmitted by satellite to give the editor a wide selection of better quality and more timely illustrations from around the world.

When the reporters finish their stories they press code keys and the stories disappear from the screens to be stored in a computer until an editor wants to call them up to process them for printing.

The U-shaped copy desk is gone. So are the copy readers. But editors are still there. They are seated in front of their computer work stations. When an editor wants a reporter's story she will type a code and the story will appear on her editing screen. She will scan the story and make changes—shift paragraphs, add words, delete words, make necessary corrections—simply by sending instructions to the computer.

She will also tap out a headline and then instruct the computer as to the size and style of the head. She will store the story with others in her computer. Then she will switch to a page makeup program, a grid will appear on the screen, and she will lay out the page—integrating the stories and illustrations—and then generate a hard copy of the page with the laser printer.

Fig. 1-4 The Editor's Page Planner is a personal computer-based layout and page design system. It automates the page design process, enabling layout specialists to make quick and easy layout revisions, modify page column and section grids, and design pages for newspapers or magazines.

The advertising department, too, has been moving rapidly into the computer age. The artist is creating most of her illustrations on a computer. She has access to a vast array of stock art through a software program. The advertisements created on computers are placed in position on pages which are transferred to the editorial department by modems that connect the computers in each department.

When a page is completed it will go to the "back shop"—or printing department—where a plate will be made to be clamped on the press for printing.

The *Daily Times* is somewhat ahead of its time in adopting computer technology, and there are still many publications that use photocomposition and pasteup to create pages for printing, but more are changing every year.

There have also been a lot of changes at the public relations firm. One partner is at his desk discussing page dummies for a company magazine with the company's personnel manager. The other partner is at work on the Macintosh computer. She is using a publishing program to integrate headings, copy, and illustrations on a four-page newsletter.

After the initial shock of losing those two promising clients, the partners rushed to learn all they could about printing, graphics, and typography. It was not easy, and they lost a lot of time and money during those first few months they were in business. But now, they are training two recent journalism school graduates to continue the business when they retire.

They've also made contact with a graphics design firm to do the art and design work on projects they cannot handle adequately with their in-house equipment.

"We thought we knew it all," they said many times. "How much easier it would have been if we had known something about printing, typography, and graphics before we struck out on our own in this business."

They, along with the editors at the *Daily Times,* also have added a word to their vocabulary that they keep in mind as they work on the computer terminals: WYSIWYG (wizzy-wig)—what you see is what you get!

The communicator has become not only the writer and editor but the typesetter, proofreader, typographer, and makeup person as well. For these tasks knowledge of typography and graphics is essential.

▼ A Revolution in Communications

In the past two decades a revolution has swept through editorial, public relations, and advertising offices across America. This revolution has engulfed the world of printed communications and has changed practices, some of them going back 400 years, and the way communicators do their jobs. It has been a great challenge, and it has created a great opportunity as well.

As with all change, there was considerable resistance for a while. There still are those who are reluctant to come to terms with the new technology. But many have found that the new technology has provided the tools to do a better and more satisfying job. They have also discovered that, in

Fig. 1-5 Public relations and advertising agency personnel, as well as others working in communications, are producing in their offices much of the material formerly turned over to commercial printers. In this case, the Ventura Publisher program is being used. (*Courtesy Xerox Corporation*)

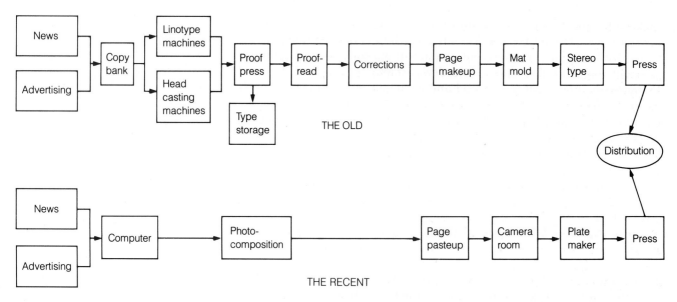

THE OLD

THE RECENT

Fig. 1-6 (Above) Not too long ago copy written and edited in the front office passed through as many as ten stations where people could catch errors and back up the editor as the copy made its way to the press. Today advanced technology and automation are eliminating these stations. The writer or editor has lost his backup and must make more and more of the graphics decisions.

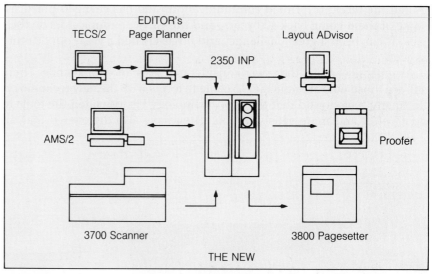

THE NEW

Fig. 1-7 Configuration of a paginated publication. This is full-spectrum publishing with all the operations (up to the making of printing plates) performed electronically.

The PC-based TECS/2 workstations process editorial copy and support classified advertisement creation. With a dedicated archive option, these stations may also be used to store, search, and merge information by writers, editors, and classified ad personnel. EDITOR'S Page Planner receives display ad geometry from Layout Advisor and—when tightly coupled with TECS/2 Editorial—enables editors to complete the page design process. In such instances, TECS/2 Editorial users learn the layout status of articles via a "story budget" from EDITOR'S Page Planner, and may then write copy to fit specific newsholes. Note that all three systems—TECS/2, EDITOR's Page Planner, and Layout Advisor—are PC-based.

The Sun-based AMS/2 terminal automates the prepress production of display advertisements, including all graphics. The 3700 InfoScan Laser Scanner crops, scales, and digitizes line and continuous tone art and prescreened material. The hub and manager of all these operations is the 2350 Image Network Processor (INP). It manages the publication database—consisting of all text, graphics, halftones, line art, and so on. Finally, fully-composed page images are dispatched from the INP to the 3800 Laser Pagesetter, which produces high-quality pages that include color and full graphic effects. The 3800 pagesetter generates pages ready for plate making at a rate of one per minute for a full-size newspaper page. (*Courtesy Information International, Inc.*)

making the job easier, the new technology is also capable of making the end product of their profession—words on paper—less effective.

Not too long ago, the principal method of producing printed material was with metal type, engravings, borders, ornaments, and the letterpress. Everything printed was composed and arranged to very precise dimensions. In the metal, or "hot type," method, rules and borders have to be cut by hand. Type, illustrations, and borders have to be fitted much as carpenters fit materials to produce a building that is straight and true. Tedious craftsmanship was needed to produce satisfactory results on the printed page.

With the advent of cold type composition and the increased use of offset lithography, the process was simplified. It was possible to arrange words and illustrations on paper simply and easily. It was possible to produce excellent printed pieces with comparative ease. On the other hand, it was also possible to produce poorly conceived and executed work. Since the composition and printing processes had been simplified, there was a temptation to take the easy way out and be satisfied with work that did not quite measure up to its potential.

Another dramatic change occurred with the arrival of the computer and "desktop publishing" capabilities in the editorial, advertising, and public relations offices, a change that further increased the potential for both good and poor communication. Many communicators now find that they are functioning as designers and compositors as well as writers. They, like the staff at the *Daily Times,* are having to make decisions about type style and size, spacing, placement of elements, and all the graphics decisions handled by production departments not too long ago. Communicators are becoming compositors as well as writers as they operate the computers and electronic gear that have moved the composing room into the editorial offices.

▼ Bombarded with Communications

The situation is further compounded by the nature of life today. The busy citizen may not have the time or inclination to decipher unattractive and difficult materials. Sloppy, poorly designed printed material will be tossed aside. There is too much that is interesting and attractive clamoring for attention.

We are bombarded with communications from the time our clock radios blare forth in the morning until we doze in front of the television set in the evening. In 1980 researchers estimated that the average person received between 1,500 and 1,800 messages a day. The competition for our attention is awesome and it is getting worse. Columnist George F. Will recently noted that the typical American is exposed to approximately 3,000 commercial messages a day. He pointed out that the average American spends one and a half years of his life watching television commercials.

Between 1967 and 1982 the number of messages transmitted by broadcast and print media doubled and may double again by 1997.

Try counting the number of messages that appeal for your attention during a prime-time hour on television. Counts can go as high as fifty. In the 1950s the sixty-second spot was the standard. Now there is a shift to fifteen-second formats and a further movement toward seven-and-a-half-second commercials.

A study by John C. Schweitzer, of Indiana University, reported in the American Newspaper Publishers Association's "News Research for Better Newspapers," that the great majority of American newspaper readers spend less than thirty minutes a day with their papers. Many people surveyed said they only glance at the newspaper.

Continuing research by the Roper Organization between 1959 and 1972 showed a growing reliance on television by many Americans. By 1963, television overtook newspapers as the source of most people's news about the world, and television became the most "believable" medium in 1961. In 1972, 56 percent of the people surveyed by Roper said if they could keep only one medium it would be television. Newspapers would be retained by 22 percent, and only 5 percent would give up the other media and keep magazines.[1]

And in spring 1981, a Gallup poll showed that 71 percent of the people believe that network television does a better job of providing accurate, unbiased news than anyone else. The poll revealed that local television rates much higher in public trust than the printed media (69 percent for local television, 66 percent for news magazines, and only 57 percent for newspapers).

Times have changed since the days when Will Rogers was quoted as saying, "All I know is what I read in the newspapers."

The point is, the person who works in communications has a challenge. Audiences are going to have to be lured away from other forms of communication. In addition, the communicator will have to strive harder to entice the audience to select one message from the scores of others that are demanding attention. It just is not possible for everyone to examine every message. Some will be selected and some will be rejected.

Much of the communications that threatens to suffocate people in a torrent of paper is hardly worth pursuing. It can be fired directly into the wastebasket. But, on the other hand, much significant information may be thrown out because it is poorly written and poorly presented.

Some communications miss their targets because they are not attractively packaged. They do not make initial contact or, if they do, the reader quickly loses interest because of poor content or poor typography.

Many communicators have turned to design specialists to help them produce presentable printed material. If the communicator can afford it, the help of a professional designer can be a valuable investment—a designer, that is, who understands how to communicate and does not design simply for design's sake.

Many communicators, however, especially those in entry-level jobs, do not have the luxury of access to a professional designer. They must produce printed communications on their own, unassisted.

Donna Valenti Weiss, publications manager for the National Rehabilitation Association, tells of arriving at work one day recently to find a Macintosh computer being installed in her workplace. She was expected to produce the association's monthly newsletter from that day on and take her camera-ready pages to the commercial printer, who had done the whole job in the past.

Donna, along with many others in similar positions, had to learn a lot about type and how to use it in a hurry. For example, she had to learn that some type styles present a psychological image or create a mood but are unreadable. Bold, heavy borders alongside lightface reading matter may create contrast but cause readers to rub their eyes with fatigue. Lines of all capital letters set in Black Letter (also called Text type and more commonly by the misnomer Old English) must be deciphered a letter at a time before they make sense. Full pages of reverse type (white letters on a black background) can be found in magazines. More often than not, readers will skip such messages rather than take the trouble to try to figure them out.

Improper spacing between letters and words turns headlines into globs of black rather than crisp, hard-hitting messages. Vertical lines of type make pleasing arrangements but require the reader to twist and turn to figure them out. Borders used for design effects sometimes become walls over which eyes must climb—but too often don't—to get at the message.

This list of typographic offenses could go on and on. However, rather than spending too much time belaboring the point, let's concentrate on creating the most effective communication possible.

Where to begin?

The starting point for any journey through the world of printed communication is an understanding of the complete process. Graphics and typography are only parts of the whole. Effective communicators must understand all of the elements that make up communication.

Consider, as an example, an auto mechanic. The mechanic cannot rebuild a carburetor without knowing how that part fits into the whole mechanics or working machinery of an automobile engine. That is, the

Fig. 1-8 Typographic devices such as lines set in all capital letters, lines too long for easy reading, lines with improper spacing, heavy rules, and reverses can hamper effective communication if not used with care and planning.

SIXTY ROWS OF VARIOUS GRASSES WILL PROVIDE A SITE FOR VO-AG TRAINING IN GRASS IDENTIFICATION. THE ROD-ROW PLANTING IS A COOPERATIVE PROJECT BETWEEN THE HIGH SCHOOL VO-AG PROGRAM AND THE SOIL CONSERVATION SERVICE. THE FFA CLUB PLOWED AND DISCED THE SCHOOL 924SITE. THE CONSERVATION DISTRICT PROVIDED FUNDING. THE PLANTING INCLUDES BLOCKS OF BROME GRASSES, FESCUES, WHEATGRASSES, AND NATIVE

Extension efforts are also directed toward educating growers in the use of integrated pest management techniques. According to the university program coordinator, some 98 percent of all growers are using these techniques as a result of a state-wide certification training program.

Pest Management

Extension efforts are also directed toward educating growers in the use of integrated pest management techniques. According

Now you can shop in your own home with the most complete, 100 pg. original fine art catalog ever offered to collectors — with discount prices normally made available only to dealers. Join the many other collectors who are enjoying huge savings on their purchases as

The planting includes blocks of brome grasses, fescues, wheatgrasses, and native range grasses. Irrigated and dryland hay or pasture grasses are included along with rarely seen special purpose grasses. This project has attracted attention from local ranchers who are interested in observing.

mechanic needs to know how all the parts of the motor fit together to make the car go before she or he can rebuild a carburetor that will do its job. The same is true with communications. A knowledge of how everything fits together is needed before a communication can be made to "go."

▼   Complete Communication

A complete communication consists of five parts: the sender, the message, the delivery system (medium), the audience, or receiver, and, finally, some way to indicate that the communication was received and understood. The communicator calls this latter part *feedback*. Leave out any of these and the communication might miss its mark.

The student of communications outlines this process like this:

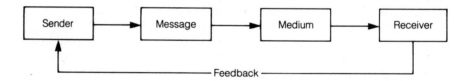

Fig. 1-9

With this model in mind, let's see how graphics and typography fit in. No matter what sort of printed communication is planned, it will not do its job unless each element in the complete communication model is doing its job.

So let's take a close look at that model and then zero in on our particular concern.

Consider the Sender

First of all, we must consider the sender. What does the sender need to know to put together effective printed communications? A number of things. For instance, the sender needs to know the reason for communicating. Is it to sell something? Is it to keep people informed? Is it to try to change their minds or rally their support behind a cause? Is it to create a particular image for an organization? Or is it to get action and get it fast?

Before you do anything else, brainstorm the reason *why* you are going to communicate. Then, write it down on paper. Know exactly what is to be accomplished. Know exactly what is expected of the person who reads the message. If the message has no specific purpose, you can waste time and money. No one will pay attention.

Then consider what words will cause the audience to stop and look and read.

Words are needed that the audience will understand. Graphics and typography cannot cover up inadequate copy. Sometimes a poorly conceived message is dressed up with attractive typography and impressive graphics. But it still will not work. If words cannot be put together in a way that will attract attention, arouse interest, create curiosity, and convince the audience that what is said is worthwhile, the effort to communicate has failed.

Robert E. Huchingson, owner of a public relations counseling firm in St. Louis and former vice president, public relations, of a major corporation, likes to relate how he approached the problem of writing advertising copy for chicken feed early in his career. He didn't know anything about chickens and chicken feed. What did he do? Before he wrote a line, he bought some chicks and raised them in his kitchen while researching chickens and feed.

The typographer and the designer, too, must know the subject of their concern to do as effective a job as possible with their part of the communication process.

Of course, this is a rather superficial examination of the first element of a complete communication. However, it does put our concern, typography and graphics, into perspective.

Media Deliver the Message

The media—newspapers, magazines, brochures, and so on—can be considered channels for delivering a message. Communicators need to know which channels are best for a particular message. And they need to know how to prepare the message so it will be in its most effective physical form for each channel.

Here it is time to clarify something about the audience. In the past the audience, or receiver, of the message was often referred to as "all those people out there." In the days when communicators emphasized the "mass media" and "mass communications," one philosophy of communication was that there was a vast audience of people "out there" and that communications were designed to reach them all. Messages were aimed at the "most common denominator" or the largest numbers of people through the mass media—or the media that reached the largest numbers of people.

Now communicators are becoming convinced that messages aimed at the largest numbers are not always the messages that are the most effective. Many mass circulation newspapers and magazines are on the decline. Media tailored to specialized audiences are on the rise. The *Wall Street Journal*, aimed at an audience with a specific interest, business, continues as the newspaper with the largest circulation in the United States. Even *USA Today*, which has moved into second place in the circulation race, was carefully tailored to a specific audience after considerable research.

National Geographic, another specialized publication, has the fourth largest circulation in the United States. *Modern Maturity*, also aimed at one specific group, attained the largest periodical circulation in the country in fall 1988 with more than 19,000,000 subscribers. It topped *TV Guide* and *Reader's Digest*, which had been the leaders for years, by more than 3,000,000. *Smithsonian*, a special interest magazine, is growing rapidly.

Creative Communication

Editor, art director, designer—all are communicators and all have a common goal—to communicate.

Some definitions are in order before we continue.

Communication to those who work with words and graphics is a process by which understanding is reached between people through the use of symbols. These symbols can be the letters of the alphabet or they can be graphics.

Graphics to the communicator are all the elements used to design a communication.

Effective communication requires that letters and graphics be combined in such a way that the message is attractive, interesting, and understandable. The composition and design of an effective communication can be enhanced if it is produced as a result of *creative* thinking and action.

The editor, art director, designer—communicator—who is creative as well as technically competent will produce newspapers, magazines, and brochures that break out of the ordinary.

Fig. 1-10 USA TODAY arrived on the scene in 1982 after extensive research of its potential target audience. Its extensive use of color and graphics has caused many newspaper designers and editors to reevaluate their layout philosophies. (© 1989 reprinted with permission)

Saturday Evening Post, Collier's, Life, Look, and *American,* circulation leaders of the past are gone. The *Post* and *Life* have been revived but on a very limited basis.[2]

So, from now on, when we discuss the audience, we mean a "target audience." This is an audience that has been carefully defined and its interests and concerns clearly identified.

Let us return now to the medium or channel. The channel is the means of transmitting the message to the target audience. It could be radio, television, postcards, billboards, newspapers, newsletters, handbills, magazines, and so on.

Communicators usually devise a *communications mix* that will use several media because repetition and reinforcement are vital if a message is to be seen and remembered. *Repetition* is a key word in successful communication. This includes sufficient repetition of the message in a single medium as well as repetition through more than one channel. The successful communicator does this all the time.

Typography can help here. For instance, the use of a certain type style to emphasize a key word or phrase in all printed communications can reinforce the impact of this word or increase its memorability. Using the same border designs can promote recognition of a theme. The use of the same typeface in subsequent messages can help the reader recognize the message.

The communicator must also deal with two kinds of "noise" that can affect the message. Noise is anything that interferes with the message as it moves toward its target. There are two kinds of noise in communication: semantic noise and channel noise.

Semantic noise refers to the words chosen for the message. When the communicator uses words that the target audience does not understand, semantic noise is created. When words are used that have different meanings for the target audience than for the communicator, semantic noise distorts the message. For instance, if a paper bag is called a "poke" in southern Ohio, most people would know what is meant. But if it is called a poke on the West Coast, communication would probably not occur. Semantic noise would interfere. And, to compound the problem, people in some parts of the country associate the word poke with vegetable "greens."

Channel noise occurs when there are problems with the medium itself. When there is static on the radio or a fuzzy picture on the television, channel noise is present. When there are typographical errors, borders that block a reader's eyes, a poorly printed page, or hard-to-read type, then print media channel noise is interfering with the communication. Channel noise occurs when a fine-screen plate of a photograph is used

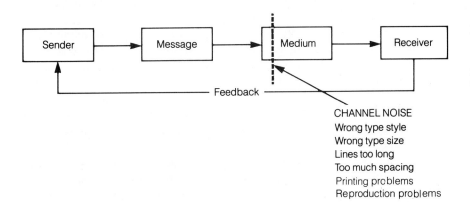

CHANNEL NOISE
Wrong type style
Wrong type size
Lines too long
Too much spacing
Printing problems
Reproduction problems

Fig. 1-11 Eliminating typographic channel noise so the message can flow with a minimum of interference is one of the goals of the designer or editor of printed communications.

on textured paper and the details do not print sharply. It occurs when the harmony of a layout is destroyed by a mixture of bold, sans serif type with old-style Roman types.

A knowledge of typography and the proper use of design elements will help communicators eliminate channel noise and thus communicate more effectively.

Consider the Receiver

One of the secrets of effective communication is the realization on the part of the message sender that we are a civilization of diverse interests, increased specialization, and groups brought together because of common interests.

Thus there is a lot we should know about the target audience. What are its interests and concerns? What are its demographic characteristics—age, income level, and the like? What are its physical and psychological characteristics? What is its life-style?

If the target audience, for instance, is the outdoor type, certain typefaces and illustrations can be chosen to say "this is for you." If the target audience is highly artistic and appreciates excellence in esthetics, type-faces, illustrations, and borders can be arranged to appeal to this audience and convince another audience that what is being said is not for them.

The communicator also needs to know what media or channels the target audience reads, views, and trusts. What channels it does not trust must be recognized as well. If the target audience does not trust a channel you must use, you may need to change the current typography and graphics to change the channel's image so that trust can be developed.

However, if the receiver does not trust the communicator's client, no amount of distinctive or impressive typography will make the message acceptable.

It is possible, though, that a careful choice of words and arrangement of type and art on pages, and even the proper choice of paper, backed by quality performance on the part of the source, can begin to turn a poor image around.

Feedback Is Vital

Finally, a complete communication must have a way of letting the sender know the word is out, that it was received and understood by the target audience, that the message hit the target. A complete communication must be read, understood, and acted on.

A newspaper publisher, for example, needs to know if the readers are actually reading. Feedback is a must. It can be obtained by watching circulation figures and those of the competition. It can be obtained by the communicator going out and talking to readers. And it can be obtained with scientific research.

Feedback is important for typography and design, too. For years it was an unwritten law that publications should be arranged so that all the advertisements are placed next to editorial matter. Or, at least, every effort should be made to do this. The theory was that people reading the content of the magazine or newspaper would more likely see the adver-tisements if they were immediately adjacent to articles. This led to many

Fig. 1-12 Art, type style, and the use of white space create a unified layout that produces a mood of elegance and "eloquence."

horribly unattractive inside pages of publications, especially newspapers.

Research, or feedback, then revealed that the proximity of editorial material to advertisements has little or no effect on the pulling power of advertisements.

There are many ways to obtain feedback, but obtain it we must. And it must be obtained each and every time communication is attempted if communications are going to work.

Most of the design and typographical techniques and suggestions you will find in the chapters of this book are based on years and years of feedback concerning the most attractive and effective ways of putting messages on paper. The techniques also apply to visual graphics for the electronic media as well.

Our goal in this book is to investigate ways we can eliminate as much channel noise as possible and create communications that do their jobs.

In the early 1800s, Thomas Codben-Sanderson, a British printer and typographer, explained what typography is all about. His explanation is as valuable today as it was some 175 years ago. In this "age of graphics" it can be applied to all the elements that are included in a visual communication. He wrote: "The whole duty of typography . . . is to communicate to the imagination without loss by the way the thought or image intended to be communicated by the author." [3]

Printing during Codben-Sanderson's time employed methods and equipment most of us would not recognize. However, whether you were a printer who set type by hand a hundred years ago, or you are sitting before a video display terminal today, the goal is still the same—to communicate.

If this goal is to be reached, communicators need to know something about type, printing processes, paper, illustrations, color, and design. Each of these elements is examined in subsequent chapters. Then we will see how to put them all together to produce better-read newspapers, lively and attractive magazines and newsletters, result-getting advertisements, and public relations communications that make people stop, read, and act.

▼ Graphics in Action

1. Examine a copy of a newspaper and see how many instances of channel noise you can find. Analyze them, try to determine the cause of each, and, if possible, tell what you would do to remedy the situation. Example: Narrow columns of type running around a photo, causing awkward spacing between words. Solution: Enlarge photo to eliminate need to set type in very narrow widths.

2. Examine a magazine, brochure, or direct mail piece and analyze it for channel noise. Discuss your findings in a small group and then report to the entire class (if you are in a classroom situation). If your group is made up of family or friends, just get their feedback on what they like or don't like about the printed piece. Jot the reactions down and later, as your graphic perceptions expand, decide how you would have designed the printed piece.

3. Plan a printed communication for a small target audience (such as an organization to which you belong). Analyze that audience. Prepare a profile of the audience's characteristics. File for further reference in planning printed communications.

4. Refer to the information you have collected in (3). What sort of printed communication might appeal to your target audience? What type style and art do you think would appeal to this audience? (Refer to the type styles in Appendix B.)

5. Obtain copies of the *World Almanac* or *Reader's Digest Almanac* for 1950, 1960, 1970, 1980, and 1990. List the ten magazines with the largest circulations in those years. What conclusions can you draw concerning the changes that might be occurring in magazine audiences? Can you draw any tentative conclusions about society in general from your findings?

Notes

[1] Earnest C. Hynds, *American Newspapers in the 1970s* (New York: Hastings House, 1975), pp 18–19.

[2] *Reader's Digest 1989 Almanac and Yearbook* (Pleasantville, N.Y.: Reader's Digest Association, Inc., 1989), p. 695.

[3] Daniel B. Updike, *Printing Types*, Vol. 2 (Cambridge, Mass.: Harvard University Press, 1937), p. 212.

From Gutenberg to Pagination

Art: Culver Pictures

Fig. 2-1 A modern composition and pagination system for producing books and publications. But even though technological advances are made constantly, the basic principles of letterpress, gravure, and lithography are still applied in printing most communications.

The goal of communication is to get a message across. When we achieve this we have effective communication. Those working in the advertising branch of the profession have put together a formula for effective communication that goes something like this: Our communication should be delivered by the right medium to reach the right audience at the right time at the most reasonable cost. As we noted in the previous chapter, to do this the professional must understand each step in the process.

Although most of the communicator's time will be devoted to writing, editing, designing, and preparing material for the printing press, it is important that he or she also know what happens in the back shop. *Back shop* is a term used in the industry to refer to the production end of the business.

Most of the printing you will be concerned with is produced by one of three methods—letterpress, offset, and gravure. The easiest way to understand how these methods work and what they can do for you is to take a look at how they developed. None of them is new. Even though most printing now is done by offset, a modification of the lithographic process, printing by this method actually started back in 1906. And lithography itself dates back to 1798! Gravure, or printing from a recessed surface, was developed in the 1400s!

Regardless of the method chosen and regardless of what methods may supersede these three methods in the future, all printing has one thing in common—the result is always a quantity of the same visual image. And printing will be one of the basic methods for obtaining duplicate images of messages as long as civilization exists. Even though the electronic media are surging ahead and some prognosticators see the day when printing will be relegated to displays in museums, there are important reasons why the printed image will always be an important part of the communications mix.

Communication by multiplied impressions—that's a good definition of printing—provides many advantages over other methods of communication.

With the printed message in hand, our audience can turn back and reread. The message is there in solid form. It is not a fleeting notice on a screen that may be missed when the receiver glances away or is distracted. The audience can speed up or slow down its reading pace. Readers can skip ahead to another section or topic without waiting through long-winded oratory as too often happens with the spoken word. And, the audience can stop at any word or statement that seems to call for thought, verification, or a trip to the dictionary before moving ahead.

The printed message is permanent; it is easily stored, transported, and filed for future reference. It can be taken wherever a person goes and it can be read in the most unlikely places—places where electronic equipment could not function.

The printed communication is still the basic medium of many communications programs. Communicators who devise long-range programs of messages aimed at target audiences usually build a plan, or communications mix, based on one of the printed media such as a newsletter or in-house magazine. This basic medium is then supported by other channels of communication.

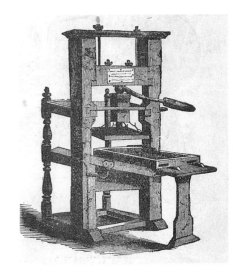

Fig. 2-2 This type of printing press, known as a wooden common press, was used for about 400 years. This model was used by Benjamin Franklin, and many similar models were made in Europe and the United States.

▼ Printing Before and After Gutenberg

When we think of printing we think of Gutenberg. For Johann Gensfleisch zum Gutenberg, of Mainz, Germany, brought it all together.

But Gutenberg did not invent printing. His contribution was to devise a method of casting individual letters and composing type and combining all of the necessary printing components—typecasting, ink manufacture,

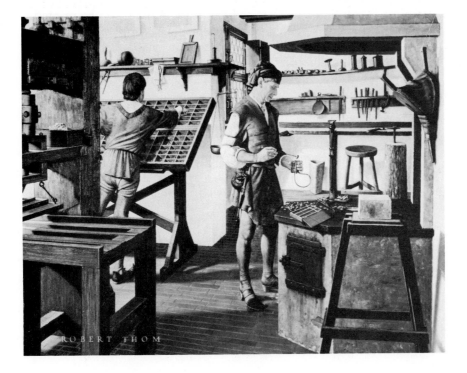

Fig. 2-3 Artist Robert Thom's version of Johann Gutenberg's shop in Mainz, Germany. One worker reaches into an upper case for a capital letter while another casts an individual type letter. (Courtesy Rochester Institute of Technology)

Fig. 2-4 The press room of a print shop in colonial America. (Courtesy Rochester Institute of Technology)

punch cutting, composing, a press, and paper—into a workable system.

Printing had existed long before Gutenberg's marvelous achievement. Printing from movable type was used in China and Korea in the eleventh century. However, because of the fragility of the materials used and the enormous number of individual pictographs used in the Chinese alphabet, it remained an obscure practice.

These early attempts at printing were based on the *letterpress* principle. Letterpress is printing from a raised surface. The image to be printed is raised above the base on which it rests. This image is inked and paper is pressed against it to transfer the image to the paper. If the image is a letter, then it is pressed to form a print; thus, letterpress.

Letterpress printing is also called *relief* printing. And letterpress printing today is but a refinement of the way it was done 500 years ago. Letterpress was the principal method of commercial printing until quite recent times.

If you were in the communications business 200 or so years ago and you wanted a broadside, pamphlet, or handbill printed, you might have taken it to Benjamin Franklin in Philadelphia or William Bradford in New York or one of the other printing shops in the thirteen colonies. All of these shops would produce your handbill by this most basic of printing methods—letterpress.

It would be a slow, laborious process, as all the type used would have to be set by hand, a letter at a time. The press would be a crude affair by today's standards. But this same press would have been used for more than 300 years, though it had been improved somewhat through the years. It would continue to be the basic printing press for 20 or 30 more years, and it would continue to see service in some parts of the world for another 100 years.

The first letterpress, used in Europe in the early 1400s, was modeled after presses used to make wine. It was known as a *platen* press. Essentially, it operated by lowering a heavy plate, the platen, under controlled pressure against the type form. After the type had been inked with a pair of ink balls, a sheet of paper was laid on the tympan* against guides. The frisket** was closed over it, leaving exposed only the section of the paper to be printed. The paper was dampened, sheet by sheet, to obtain a good print.

Two men working one of these wooden or iron presses could turn out about 500 or 600 impressions a day. Today a simple offset press can produce 5,000 to 9,000 impressions an hour.

▼ Three Types of Letterpress Printing Presses

The letterpress, which evolved from the simple wine press, can be found in commercial printing plants, but its use is limited. There are three types of letterpress printing presses: the platen, the flatbed cylinder, and the rotary.

The Platen Press

The platen press is known as the faithful "job press." Its main function is to produce miscellaneous items of printing—letterheads, envelopes, cards, tickets, handbills, and so on. It is used mainly for short runs such as imprinting Christmas cards and paper processing such as die cutting, creasing, and perforating.

The press is called a platen press because it operates by having a platen, or flat surface, on which the paper is placed, move against the stationary type, which is locked in a chase, which in turn is locked in the bed of a press. The operation reminds some observers of the opening and closing of a clamshell, so it has been called a *clamshell* press.

Some platen presses are operated by hand. The operator places paper, a sheet at a time, on the platen. Others have automatic feeders.

*The tympan is a sheet of heavy, oiled paper in a hinged frame the size of the platen.
**The frisket is usually made of cloth and it, too, is in a frame the size of the platen and hinged so it can be lowered in place between the tympan and the form to be printed.

Fig. 2-5 Three methods of letterpress (raised image) printing: platen, or flat surfaces (left); flatbed cylinder (middle); rotary (right); (Courtesy Eastman Kodak)

Fig. 2-6 An early steam-powered flatbed cylinder press. Steam power was first used to run printing presses in about 1860. This was a major breakthrough that led to greater speeds and much larger presses. (Courtesy Bettman Archive/The Mead Corporation)

The Flatbed Cylinder Press

In 1814, Friedrich Koenig sold the first power-driven press to *The Times* of London. Koenig, a German, solved the problem of pressing two stationary surfaces together to get an impression. He was able to devise a press operated by power, in this case steam, that continuously rotated a round impression cylinder.

The form to be printed was placed on a bed that had tracks along the sides (actually rows of teeth that meshed with teeth on the round impression cylinder). The flat platen was replaced with a round cylinder. The paper to be printed was held against the cylinder by metal fingers, or grippers. As the cylinder rotated, the bed slid back and forth, going under the rotating cylinder where the paper was pressed against the type form on the bed.

The first issue of *The Times* printed on the Koenig cylinder press appeared on November 29, 1814. *The Times* proclaimed the issue was the "result of the greatest improvement connected with printing since the discovery of the art itself." Koenig's press inaugurated a great revolution in the industry. It was the breakthrough that led to the giant presses of today. And the principle of the revolving cylinder was adapted to offset and gravure presses.

However, the flatbed cylinder press still had a serious shortcoming when it came to producing mass volumes of books, magazines, and newspapers. It still printed only one sheet at a time. When one side of a sheet was printed, it had to be turned and printed again if both sides were to receive an impression.

The solution became known as the *web perfecting press*. This press is fast and efficient and can be used for long runs of magazines, newspapers, advertising brochures, and other publications with many pages. It uses a rotary impression (or platen) cylinder and a rotary type bed as well, plus a continuous roll of paper, rather than single sheets. It prints both sides of the sheet as it travels through the press.

The Rotary Press

The rotary press uses forms cast into curved plates, a process known as *stereotyping*, which are locked onto the cylinder. As many duplicates of the form as desired can be made, and a number of presses can turn out the same product simultaneously.

The Times of London again scored a triumph in printing when in 1869 it was printed on a continuous roll of paper from curved stereotype plates on a rotary press. During the next ten years Robert Hoe, the American printing press manufacturer, improved and developed newspaper printing by making it possible to deliver a complete newspaper—folded, counted, and ready for sale—from a roll of blank paper.

Although we are in the middle of the age of offset printing, letterpress is not dead. It is still used for many magazines, newspapers, books, and brochures, though the number is dwindling.

Letterpresses are used for numbering, creasing, folding, embossing, die cutting, and other forms of printing and processing paper and card stock. One manufacturer, Heidelberg of West Germany, turns out a letterpress every fourteen minutes.

▼ Offset and Lithography

In recent years many newspaper publishers and printers have abandoned the letterpress and turned to *offset*. Offset is actually a modification of *lithography*, or printing from a flat surface. (Lithography is often also called *planographic* printing.)

This method of printing did not come to us through centuries of evolution. It was invented—almost by accident.

One day in Munich, Bavaria, a twenty-five-year-old artist who lived with his mother and who dabbled in the theater, playwriting, sketching, and drawing, was busy in his workshop. The year was 1796. He was Aloys Senefelder, bachelor son of an actor.[1]

Since Senefelder was a sometime artist he turned to engraving. He experimented with copper plates and practiced on limestone slabs. His idea was that he could etch his copy into the plates and then print them. This would be quicker and cheaper than the complicated and costly letterpress task of setting them into type by hand.

While Senefelder was busy trying to perfect his engraving skill, his mother came in and asked him to write a list of items to hand to the waiting laundress. He reached for whatever was handy and picked up a piece of limestone and some correction fluid he had concocted to fill in the errors he made on the engraving plates.

Senefelder wrote the laundry list with the correction fluid, which had a greasy base, on the flat limestone slab. Later, while experimenting with this list, he noticed that ink would stick to the words he had written and

---

Creative Communication

How can an editor or designer be creative—seek an approach that is different, that will make her or his efforts stand out and yet be effective in communicating the message? Many people in the communications industry and the arts maintain that creativity cannot be learned. Are we sure?

What is creativity?

Create (from the Latin *creatus*) means to bring into being, to cause to exist, to produce, to evolve from one's own thought or imagination, to make by investing with new character or functions.

Formal definitions of creativity include such phrases as—the ability to bring into existence—to produce through imaginative skill—to design.

Design is defined as a mental project or scheme resulting from deliberate planning to achieve a purpose, a sketch showing the main features of a communication, or the arrangement of the elements in a communication.

To put it more precisely, design is the organization of verbal and visual material (elements) on behalf of communication.

Creative design results from creative thinking. Psychologists define creative thinking as a method of directed thinking applied to the discovery of new solutions to problems, new techniques and devices, and new artistic expression.

Fig. 2-7 Aloys Senefelder at work in his lithographic shop in the late eighteenth century. (Courtesy Rochester Institute of Technology

water would wash off the ink on the blank areas of the stone.

Senefelder had hit upon the chemical fact that is the basis of the lithographic printing process—oil and water do not mix!

How do we know so much about Senefelder's achievement? Well, he wrote it all down in his *Invention of Lithography*, which was published in New York in 1911. In addition, he wrote an autobiography, *Complete Textbook of Stone Printing*, in 1818.[2]

Senefelder abandoned his efforts to become a playwright and devoted his time to perfecting the lithographic printing process. A friendly, outgoing person, he shared his invention with fellow artists, traveled extensively, and became famous.

As mentioned, the central principle of lithography is the fact that oil and water do not mix but repel each other. All procedures of lithography observe this basic tenet.

The stone is the carrier of the printing form. This stone is called a *lithostone*. It has the unique qualities necessary for lithographic printing. It is quarried only at Solenhofen, a village near Munich. It must be polished to a smooth, slightly grainy surface.

The design to be printed is applied to this surface by various techniques. Drawing pens, brushes, or grease crayons are used. Although lithographers have complete control over their creations, they must prepare the design so that it reads backwards or in a mirror image. Then, when a print is made from the design, the lefts and rights will be in their proper positions when a person views the object or reads the page.

After the design is completed, the stone is covered with a watery and slightly acidic gum arabic solution. This "seasons" the stone. There are some additional steps in preparing the stone, but for our purposes it is enough to say that the design on the stone is "prepared for printing."

The stone is dampened and rolled with an ink roller. The grease-receptive design areas will accept the ink while the areas that have been moistened will not pick up the ink. After a series of inkings and moistening, paper is pressed against the stone and the reverse or mirror image of the original sketch will be printed on the paper.

The press and method used by Senefelder are used today to make artistic lithographic prints.

Changes and improvements in lithography in the early 1800s made it practical for mass production. Better plate preparation and the use of power-operated presses plus the ability to make duplicate plates contributed to the development of lithography for commercial printing.

The development of lithography as a commercial printing method was hampered by the slow and cumbersome process of printing from the stone "plates." At the same time, letterpress was progressing at a rapid pace, and Senefelder had mentioned in his writings that he thought zinc could be used as well as heavy stone.

In 1889 the first lithographic press to print with zinc plates went on line. This greatly increased the printing speed.

Offset was discovered quite by accident, and we do not know for sure just when and how. If the lithographic press missed a sheet an impression would be made on the paper carrier (platen or cylinder). The next sheet through the press would pick up this impression in reverse (backwards) on its back. Someone noticed that the quality of the reverse was quite good.

We can visualize the principle of offset by thinking of the letter E carved in relief, inked, and pressed against a sheet of paper. The image will appear in reverse—Ǝ. But if another sheet of paper is pressed against this image before the ink dries, the image on the second sheet will be in the correct position—E.

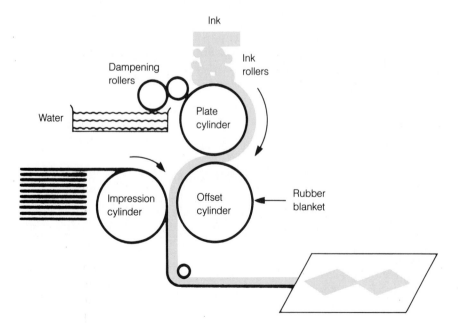

Fig. 2-8 A schematic drawing of the offset lithography printing process. The printing plate with the planographic image is clamped on the plate cylinder, the dampening rollers coat the plate with water, the ink rollers ink the plate (the ink adheres to the image area but not to the area not to be printed), the image is offset to the offset cylinder and then transferred by impressions onto the paper. (Courtesy National Association of Printing Ink Manufacturers)

Fig. 2-9 A New Jersey native, Ira Rubel, designed the offset press in which a rubber cover is used on the cylinder of a lithograhic press to produce a vivid image on paper. (Courtesy Rochester Institute of Technology)

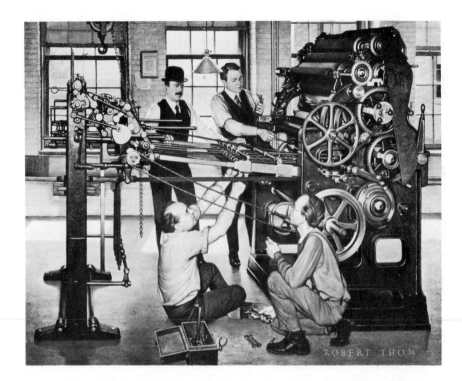

So, after experimentation, offset was born. The final printed product is obtained from an image of what is to be printed. This image is printed on a cylinder covered with a rubber blanket. The rubber blanket then becomes the printing plate and the image is printed (or offset) on the paper.

In 1906 the first offset press as we know it today began rolling out printed sheets in Nutley, New Jersey. Although some offset printing had been done in Europe well before this, the press developed by Ira A. Rubel, a paper manufacturer, is considered the one that started the offset revolution. The press used by Rubel and his partner was built by the Potter Press Company, which later merged with the Harris Press Company. Today it is known as the Harris-Seybold Press Company and is a major manufacturer of offset presses.

▼ Gravure Printing

Gravure, or *intaglio* as it is also called, is simply printing from a recessed surface.

Just as letterpress printing evolved through the centuries until Gutenberg put the elements together in a manageable fashion, so gravure printing also has a long and sometimes murky history. Its beginnings can be traced back to the 1400s.

A finely detailed print of the Madonna enthroned with eight angels is considered the earliest known engraving. It was produced by an unknown German artist with the initials E. S. and is dated 1467.

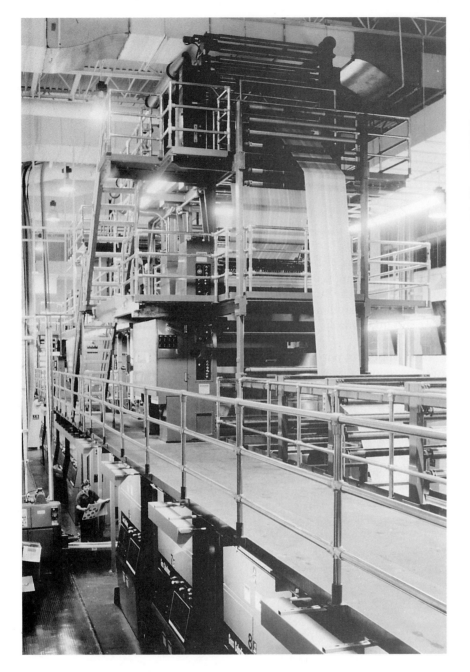

Fig. 2-10 (Above) A printer attaches an offset plate to a cylinder of a Goss Colorliner press. (Courtesy Graphic Systems Division, Rockwell International Corporation)

Fig. 2-11 A giant Goss Colorliner offset press. Part of the color revolution in newspaper design, this press enables a newspaper to place process color on any and every page of up to 160 pages. Up to 1989 more than forty of these presses, costing more than three-quarters of a billion dollars each had been ordered. (Courtesy Graphic Systems Division, Rockwell International Corporation)

Many early books were printed by letterpress while the illustrations were engraved on copper plates and printed by the gravure method. Rubens, Rembrandt, and Van Dyke were among the artists who worked in gravure. Rembrandt produced many etchings that he printed by the gravure method in the middle 1600s.

Gravure is a simple method of printing. It is called intaglio from the Italian word *intagliare*, which means "to carve." And that is what the

Fig. 2-12 A schematic drawing of the gravure process. The image to be printed is engraved or carved into the impression surface. The entire area is inked, and the ink is removed from the nonprinting area. Pressure of paper on the plate pulls the ink out of the recessed areas. (Courtesy National Association of Printing Ink Manufacturers)

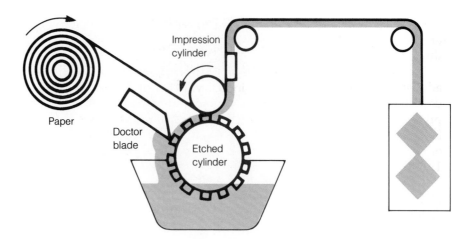

gravure printer does by engraving the design into the printing surface.

Gravure printing is done on a flat surface in which the letters or designs are sunk below the surface. The design is engraved into a smooth plate, the plate is covered with ink, and the ink is wiped off the surface. But look, the ink remains in the letters and designs that are sunk below the surface.

Then a sheet of paper or other material is pressed against the plate and an impression is made.

A good gravure print needs to be made under great pressure. The gravure press consists of a bed that travels between two steel rollers. A felt blanket is placed between the upper roller and the paper, and the paper is placed on the engraved plate. A roller beneath helps exert the forward pressure generated when the blanket, paper, and plate move between the rollers.

Fig. 2-13 Diagram of a rotogravure press that uses the gravure process. Many of the largest magazines in the United States are printed by this method. (Courtesy Eastman Kodak)

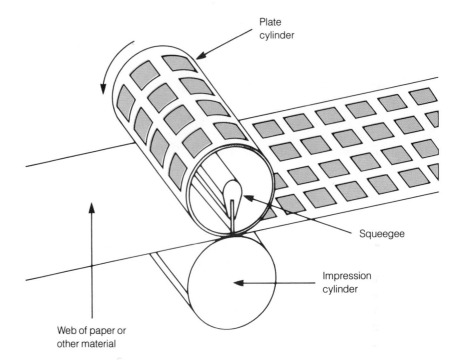

The felt blanket, under all that pressure, acts as a self-adjusting overlay and helps force the paper into the crevices so it contacts and withdraws the ink onto its surface.

The development of *rotogravure printing* has been one of the "hottest" advances in the communications industry. Many of the magazines, Sunday supplements, and catalogs we see today are the result of rotogravure printing. This method was invented by an Austrian, Karl Kleitsch, who lived in Vienna and later moved to England. He developed a rotary method of printing gravure plates in his workshop in Lancaster, England, in 1894. The *New York Times* installed a rotogravure plant in 1914.

Rotogravure requires a rather complicated photomechanical method of plate making. Today, however, it is done rather easily with an electronic stylus or laser. A rotogravure plate is capable of a million impressions, far more than plates made for letterpress or lithography.

Acceptable quality, speed, and the use of inexpensive ink and paper make rotogravure popular for medium- and large-volume jobs, where speed and economy are deciding factors.

Gravure printing is considered excellent for reproducing pictures. A distinct feature that makes it easy to recognize is that the entire image must be screened—type and line drawings as well as halftones of photographs.

As with the rotary letterpress and offset, gravure presses are manufactured for both sheets of paper (sheet-fed gravure) and rolls (rotogravure), but most gravure printing is done from rolls. Many of our largest magazines are printed by rotogravure. These include *National Geographic, Women's Day, Family Circle, Reader's Digest,* and *Redbook,* to name a few.

The *National Enquirer* is printed in Buffalo, New York, on rotogravure presses. Each press can reach speeds of 35,000 copies of a ninety-six-page newspaper per hour. A quarter of the ninety-six pages can be done in four colors.

▼ Other Printing Methods

Most books on printing methods have listed the advantages and disadvantages of each. It was safe to do this in the past. However, modern technology has erased many of the distinguishing features of the three basic printing methods.

Not long ago experienced communicators could distinguish among products of the letterpress, the offset press, and the gravure press. Now each method is capable of producing excellent quality, and the advantage of one over the others has largely disappeared. It is more the skill of the printer than the method used that should be the concern of the communicator.

Of course, most small communications operations will almost exclusively use offset and perhaps letterpress printers. But startling advances are being made each year in printing processes, and the communicator should be familiar with what is happening.

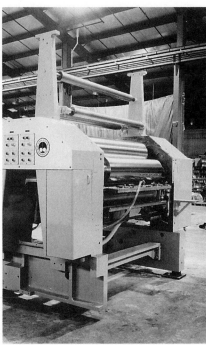

Fig. 2-14 (Above) Flexo press lines during assembly in one of Publishers Equipment Corporation's Rockford, Illinois, factories.

Fig. 2-15 (Above, right) A flexo unit ready for shipment from the factory. The ink chamber is in the "off" position to show the anilox roller. Ink is fed to the chamber through the transparent tube. (Courtesy Publishers Equipment Corporation, Dallas, Texas)

Flexographic Printing

The future of flexography (flexo) looks bright. During the past decade it has been the fastest-growing printing process in the world of communication and currently accounts for 25 percent of the total output of the printing industry. In the past the flexographic method was considered practical only for specialized uses such as printing paperback books, packaging materials, and plastic bags. Today flexography is used to print Sunday comics, newspaper inserts, television news magazines, business forms (the Internal Revenue Service tax forms are printed by flexography), telephone books, point-of-purchase material, and magazines.[3]

Newspapers, too, are installing flexographic presses. Among the newspapers that are using flexo are those in Portland and Bangor, Maine; Modesto and Monterey, California; Buffalo, New York; Knoxville, Tennessee; and Trenton and Atlantic City, New Jersey, to name a few.

Flexo printing is a comparatively simple process. It uses an ink that usually is water based. The ink comes in contact with an engraved "anilox" cylinder. This cylinder distributes the ink to another cylinder which carries a flexible letterpress-type plate that is made out of rubber or photopolymer (a plastic material). This inked plate makes the impression on the paper.

Anilox cylinders have a hard surface with millions of tiny cells of equal size and shape. The ink fills these cells and the excess is scraped off by doctor blades, much as the ink is scraped off a gravure printing plate. However, the function of the anilox cylinder is to distribute the ink on the printing plate. This gives an even, reliable distribution of ink.

Flexo advocates cite a number of advantages of the method. It can print on very light paper with no show-through. Full-color illustrations with fine definition can be printed with excellent results. Since water-based inks are used, the problem of rub-off is virtually eliminated. There is little waste of paper compared with offset. Vibrant colors are possible with flexo, and it can print large solids without "ghosting" (the appearance of light areas in large areas printed in color).

As more and more flexo presses move into the publication and commercial printing fields, advertisers, editors, and graphics people will need to become familiar with this method of printing.

Laser Printing

Laser printers are playing an increasing role in the world of printed communications and most people who enter public relations, advertising, or editorial positions will sooner or later use a laser printer.

Laser printers are used for creating mechanicals. These are completed pages of newsletters, flyers, magazine pages, and so on that can be used for making printing plates. Laser printers are also good for making proofs and for producing limited copies of printed materials.[4] Another important use of the laser printer is to produce copy directly from a computer. It can merge type and graphics.

The laser printer uses a tiny pinpoint of light that passes through a finely tuned and complex optical system and lands on a light-sensitive drum. As the drum revolves, a strong charge is applied to its surface. The laser beam pulses on and off at amazing speed as it scans the surface of the drum. Wherever it strikes, a different charge appears and dots are

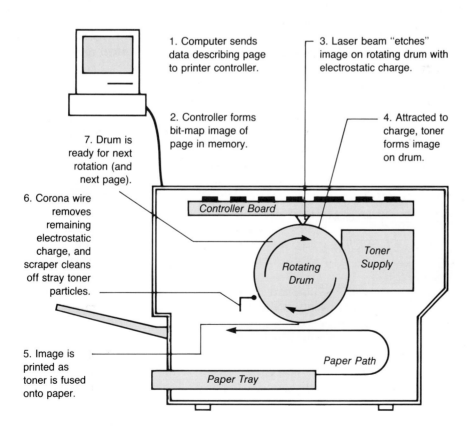

1. Computer sends data describing page to printer controller.

2. Controller forms bit-map image of page in memory.

3. Laser beam "etches" image on rotating drum with electrostatic charge.

4. Attracted to charge, toner forms image on drum.

7. Drum is ready for next rotation (and next page).

6. Corona wire removes remaining electrostatic charge, and scraper cleans off stray toner particles.

5. Image is printed as toner is fused onto paper.

Controller Board

Rotating Drum

Toner Supply

Paper Path

Paper Tray

Fig. 2-16 A side view of a laser printer. The printer receives a page that has been created on a computer and reproduces it to produce a mechanical of the page ready for plate making. Or, the laser printer can be used to produce a limited number of printed copies of the page. (Used by permission of *Publish! the How-to-Magazine of Desktop Publishing,* from Vol. 2, No. 4, May 1987, published at 501 Second St., San Francisco, California 94107)

formed at these points on the drum. A series of dots forms the letters or illustrations on the drum. When toner is applied, an image appears. As the drum rotates this image is attracted to an oppositely charged sheet of paper. Heat and pressure fuse the toner to the paper, creating a permanent image.

The quality of the printing produced by a laser printer is determined by the density of the dots. The more dots per inch, the greater the quality or clearness of the image to be printed. For instance, a printer that uses 200 to 300 dots per inch (dpi) is considered at the low end of the quality scale. It is adequate for making copies but not for acceptable-quality printing of flyers, newsletters, or brochures. A range of 200 to 600 dpi will produce fair quality suitable for inexpensive printing. Many jobs that require a limited number of copies are being produced by laser printers. If a printer has a 300 to 1,100 dpi range it can turn out high-quality press proofs and camera-ready mechanicals.

In 1973 Xerox introduced the first laser printer, making it possible to produce copy directly from the computer. Through the years the device was improved, and it can now be used to produce high-quality type.

The computer and the laser printer have moved everything but the printing plate and the printing press into the office. More and more businesses and organizations are producing their printing "in house." The operator can enter illustrations into the computer. This art can be enlarged, reduced, or cropped on the screen. Headlines, copy blocks, and illustrations can be moved electronically to form a page, and that page can be transferred to the laser printer and copies produced.

Screen Printing

Screen printing, also called screen process printing, uses a fine, porous screen made of silk, nylon, Dacron, or even stainless steel mounted on

Fig. 2-17 In the screen printing process, the ink is squeezed through a stencil and a screen onto the surface to be printed. (Courtesy National Association of Printing Ink Manufacturers)

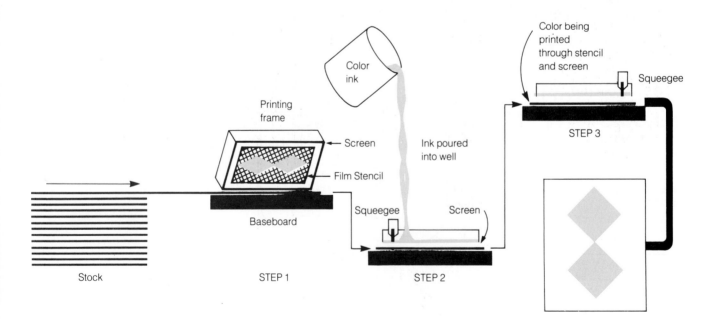

a frame. A stencil is cut either manually or photomechanically with the lettering or design to be reproduced and then placed over the screen. Printing is done on a press by feeding paper under the screen, applying paintlike ink to the screen, and spreading and forcing it through the fine mesh openings with a rubber squeegee.

The use of the stencil to produce images can be traced back to 1,000 B.C. in China, but screen printing did not become an important part of the printing industry until after World War II. Today automatic screen presses, four- and five-color screen presses, and rotary screen presses are used to turn out posters, sheets for billboards, menu covers, and bumper stickers.

▼ Advances in Putting Words on Paper

In recent years we have seen the introduction of ink-jet printers that produce excellent color work. In addition, ink-jet typewriters are coming on the market.

Universally used in the business world is xerography. *Xero* is a Greek word meaning "dry"; thus, dry printing. Xerography is used for copying text and graphic material and improvements are constantly being made in these copying devices. The Xerox uses reflective light to expose a photoconducting surface. A negatively charged toner on this surface is transferred to a positively charged sheet of paper and fused through the use of heat and a pressure roller. It produces an exact copy of the original.

Another printer that has come on the scene is the ion deposition printer. Ion means "atom" or "group of atoms." An ion cartridge creates an image on a rotating drum by shooting charged particles on it in dot matrix format. Then toner is employed to develop an image which is fused to a sheet of paper. Cold roll pressure is used to make the image stick, rather than heat and pressure as is used in a laser printer.

Ion printing produces a very high-quality product that can sustain long runs on a printing press.

Many new advances in science, such as the laser beam, the computer, and the cathode ray tube, are a part of the printing industry. However, it is safe to say that no matter how we place words on paper, the basic theme of this book will not become obsolete. That is, no matter how we print our message, no matter whether we arrange the elements of our printed message with a pagination device that puts those elements in place on a video screen or with a pencil on a piece of paper, we must still make basic typographic decisions.

Those decisions involve the choice of type styles and sizes; the width of the printed lines; the amount of spacing between letters, words, and lines of type; and the size and composition of pieces of art. All these decisions are still critical to the effectiveness of a message.

The starting place in creating effective printed communications is deciding on the basic ingredient of any recipe for effective printed communication. That is *type,* and type is what we consider next.

▼ Graphics in Action

1. The best way to acquire an understanding of the basic printing methods is to see the actual processes in action. Plan a trip to a local printing plant. Join a team of classmates to visit a letterpress, lithographic/offset, screen, gravure, or flexo facility. Obtain samples of printed work from the plant you visit. Have your team report to the class on your findings. (Note: It might be worthwhile to plan a visit to a printing plant after learning about layout and preparing for production [Chapters 8 and 9]. Perhaps the group can prepare a mechanical and have the printing plant tour conductor demonstrate from start to finish what happens in the printing process.)

2. Prepare an essay that updates trends in printing methods. Consult trade publications such as *American Printer* and *Graphic Arts Monthly* and interview printing plant managers, if possible.

3. Find out what printing methods your school or office uses. Learn all you can about their operation. Do you think these are best for the purposes? Explain.

Notes

[1] Alois Senefelder, *A Complete Course of Lithography* (New York: Da Capo Press, 1977, reprint of original published in London in 1819), p. 22.

[2] Warren A. Chappell, *A Short History of the Printed Word* (New York: Alfred A. Knopf, 1970), pp. 171–173.

[3] American Printer, "Bullish on Flexography," March 1989, p. 26.

[4] White Paper, *Lasers in Graphic Arts* (Wilmington, Mass.: Compugraphic Corporation, 1987), p. 6.

Type: The Basic Ingredient

Fig. 3-1 Is it an A or an a? Letters are available in all shapes and sizes. The designer selects the best for the situation.

The recipe for building an effective printed communication includes a number of ingredients. But the basic ingredient is type.

No matter what sort of communication is planned—newspaper, magazine, brochure, letterhead, or business card—it should be designed to accomplish five things.

1. *It should attract the reader.* If a printed piece does not attract the reader at the very beginning, it may not work. And it must attract a reader who is a member of the target audience—that particular segment of people out there we specifically want to reach.

2. *It should be easy to read.* People simply will pass up material that appears difficult, unless they know it contains information they really want so badly they are willing to overcome any barrier to get to it. Busy people today avoid unattractive and difficult type masses.

3. *It should emphasize important information.* One of the secrets of effective communication is to make sure the arrangement of type on the page and the size and styles of types employed make the heart of the message—the points we want to make—easy to recognize quickly as important and easy to absorb.

4. *It should be expressive.* Everything in the message—the paper used, type, art, and ink—should provide a unified whole to reinforce the message and make it clear to the target audience. The reader should never say of the communication or any element in it, "I wonder what that means?"

5. *It should create recognition* Printed messages can make a communications program more effective and can help build a permanent identity for our organization, our position, even our can of beans in the minds of the target audience, not just once, but time after time. Publications achieve this visual identity, for instance, by using the same type style for the nameplate, the masthead, or as a logo for house advertisements. The identity is extended by using that type style on trucks and T-shirts for any teams they sponsor, for instance.

The starting place in planning printing that will accomplish all this is with the basic ingredient. Proper selection and use of type can support the message, and improper selection and use can make the message less effective.

An effective printed message must combine good press work, good design, good placement of elements but, most of all, it must first be set in type that is legible, suitable, and readable. Of course, illegible type can be used sometimes to transmit a unique message, but in this discussion our concern is with messages that are to be read.

Effective use of type involves (1) the ease with which it can be read, (2) its grouping into words, lines, and masses, (3) the nature of the message, (4) the kind of reader, and (5) the printing process. Effective use also includes esthetics, for the printed piece must be attractive to be effective.

All communicators can increase the effectiveness of their efforts if they acquire a basic knowledge of type and how to use it. This understanding can make work easier and faster, and it can help cut printing costs. It will enable communicators to give explicit instructions to printers and to communicate with them so that everyone can work together to produce the best possible product.

Understanding type and its use involves some basic concepts.

First of all, communicators need to know how type is designed, the various typeface patterns available, and the suitability of each for specific messages.

Then, we must know how type is measured and how it is set and spaced. In addition, it is helpful to know basic typographic principles that have been established through practice and testing.

This chapter attempts to sort all this out, and the chapters that follow add the various ingredients needed to make sound decisions concerning printed communications.

▼ Some Essential Terms

We need to begin with an agreement on terms. These will be kept to a minimum, but the terms included in this chapter are considered by many professionals as essential for anyone who works with words, type, and printing. Our discussion starts off with some basic concepts concerning typefaces and how they came into existence.

When you look at a letter what do you see?

Is it an *a* or a *B*? This little squiggle might be any one of fifty-two different ones. A designer or typographer sees a squiggle that will create a sound that will create an image in someone's mind. The professional also notes the ascenders and descenders. Are they too long or too short, or just right? Do the serifs make the letter easier to read? Do the thick and thin letter strokes create an interesting contrast? Are the letters the right size on the body for maximum readability or legibility? Do the letters appear modern or old-fashioned?

In picking just the right type for a publication, the communicator needs to consider all these variations in letters and more. But before deciding on one of the more than 6,000 different designs of letters available in America today, the communicator needs to know what a serif is, what strokes are, and how big on the body a particular type might be.

So, in beginning to work with type, we need to know the anatomy of letter forms.

The basic elements of a letter are the *strokes,* or lines that are drawn to form the design. These can be *monotonal* (all the same width) or they can vary from *hairline* to quite *thick.*

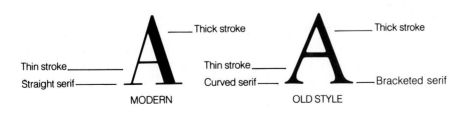

Thin stroke — Thick stroke
Straight serif — MODERN

Thin stroke — Thick stroke
Curved serif — OLD STYLE — Bracketed serif

Fig. 3-2 Characteristics that differentiate modern and old style Roman types.

Some designers have finished off the strokes that form the letters with rounded or straight lines called *serifs*. There are *rounded serifs* and *flat serifs*. Serifs that are filled in are called *bracketed serifs*.

Fig. 3-3 Serif treatments vary with type styles. Sans Serif typefaces have no serifs, old style Romans have bracketed serifs, and modern Romans have flat, straight-line serifs.

SANS (WITHOUT) SERIFS   BRACKETED SERIFS   FLAT SERIFS

An alphabet of a certain design will be drawn to a consistent size for all letters. This size is determined by the *x* height of the letters. This is the height of the lowercase *x*.

Fig. 3-4 The x height (the height of the lowercase x) is a critical basic measurement in the design of letters.

Some letters have a *stroke*, or stem, which extends above the *x* height. This stroke is known as an *ascender*. The letters *g, j, y, p,* and *q* all have a tail or *descender* that drops below the *x* height.

In the capital alphabet only the *Q* has a descender. There are no ascenders in capital letters.

Fig. 3-5 The letter stroke that extends above the x height is called the ascender. If the stroke extends below the x height, it is a descender.

Now that we are acquainted with a few terms, we can start our consideration of type and its use. As mentioned, there are more than 6,0000 type designs in the Western world. In the days when printing was done from metal type cast by foundries, the letter designs were fairly consistent. The Caslon design, for example, which first was cast by William Caslon in England in 1734, was and still is one of the basic types.

With the advent of what is known as *strike-on* (letters composed on a typewriter or other device that strikes the image of the letter on paper), *photocomposition,* and digital composition there has been a proliferation of type designs of various names. Many of these designs have characteristics of long-established and recognized styles but each manufacturer has made slight changes.

In addition, new faces are constantly being produced by the hundreds of new companies that have entered the printing market during the past two decades of the technological explosion.

Communicators can no longer rely solely on old family names for type styles. It is necessary to compare types on the specimen sheets of the various printing establishments with the styles that have become basic types through the years.

As far as learning about the correct use of types, we can start our discussion with the letter designs that are traditional and recognized by all good printers and graphic designers.

To create an effective communication, we need to decide which typeface would be easiest to read and give the best "image" to go with the message. We need to be acquainted with enough designs to understand the strengths and weaknesses of each.

▼ Typefaces

Type is classified in a way that makes differentiation among the designs easy. It is sorted rather like humans have been sorted by anthropologists. Just as there are various races of people, there are various *races* of types. Some communicators refer to these basic divisions as "species" of types.

Within these races are *families* of types. Each family has the basic characteristics of its race but in addition it has slight differences in letter design from the other families of the same race.

The Caslon family and the Bodoni family are both members of the *Roman* race of types. They have certain basic similarities, but they also have subtle differences.

An inspection of the Caslon and Bodoni types reveals their similarities and differences. They both have serifs, and they both have variations in the letter strokes. However, the Caslon serifs are rounded and bracketed, and the Bodoni serifs are straight-lined. The letter strokes of the Caslon letters do not have as pronounced a variation between thick and thin strokes. The Bodoni strokes are very thick and very thin. Note also the different treatment of the juncture of the thick and thin strokes at the top of the letters.

Once such differences and their effects on the printability, legibility, and suitability of a printed piece are mastered, the effective selection of letter styles becomes quite easy.

Many typographers and graphic artists recognize six races of typefaces. There isn't full agreement on this, but the majority usually classify all the thousands and thousands of letter styles within six categories. These races are the Romans, the Sans Serifs, the Square Serifs (or Egyptians), the Text types (or Black Letter), the Scripts and Cursives, and a catch-all category for those designs that defy specific identification called Miscellaneous (or Novelty).

No matter how the letters selected for a printed communication are imposed on the printing surface—by using the foundry types of the letterpress process, or letters produced from film by the photocompositor, or strike-on letters of an electric typewriter—they all will exhibit characteristics of the various races. They can all be discussed within the framework of this race-and-family system of classification.

Let's examine the characteristics of each of these six races.

ABCabc

ABCabc

Fig. 3-6 Caslon (top line) and Bodoni are both members of the Roman race, but there are differences in the letter structures. Caslon is an old style Roman and Bodoni is a modern Roman.

Fig. 3-7 The six races of type in lower-case letters for comparison.

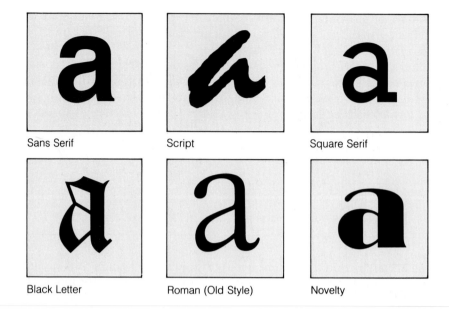

Sans Serif   Script   Square Serif

Black Letter   Roman (Old Style)   Novelty

Text or Black Letter

The first printing produced from movable type was, of course, the product of Johan Gutenberg's shop in Mainz, Germany. It is interesting to note that Gutenberg's most lasting achievement in printing was his famous forty-two-line Bible. Each page contained two columns, and each column was forty-two lines long. Nearly five hundred years later this Bible is still considered by typographers to be a superb example of printing and layout.

The letters that Gutenberg designed, cast, and printed and the letters used in the Gutenberg Bible were copies of the heavy black letters of the northern European hand. They are thought of today as "Old English." Actually Old English is a family of the Text or Black Letter race and is not a race itself. Black Letter types are ornate with a great variation in the strokes. They are ponderous and difficult to read.

Fig. 3-8 This Black Letter typeface is in the Wedding Text family designed by American Type Founders.

abcdefghijklmnopqrstuvwxyz

ABCDEFGHIJKLMNOPQRSTUVWXYZ

These Text or Black Letter types do have a place in modern graphic design, however. What better way to say "medieval banquet," or "ye olde gift shoppe"? Although these types may remind you of Gothic cathedrals, don't call them "Gothic." Gothic is an entirely different race of types.

Some more confusion can arise about this race of types. The name Text was attached to these designs because they were the "text" or reading matter types of the medieval northern Europeans. But today the term *text* is used to designate any reading matter. So from now on, to avoid this confusion, this race is referred to only as Black Letter.

Use Black Letter when you want to emphasize tradition and solemnity and when you want to give the image of strength and efficiency. Many

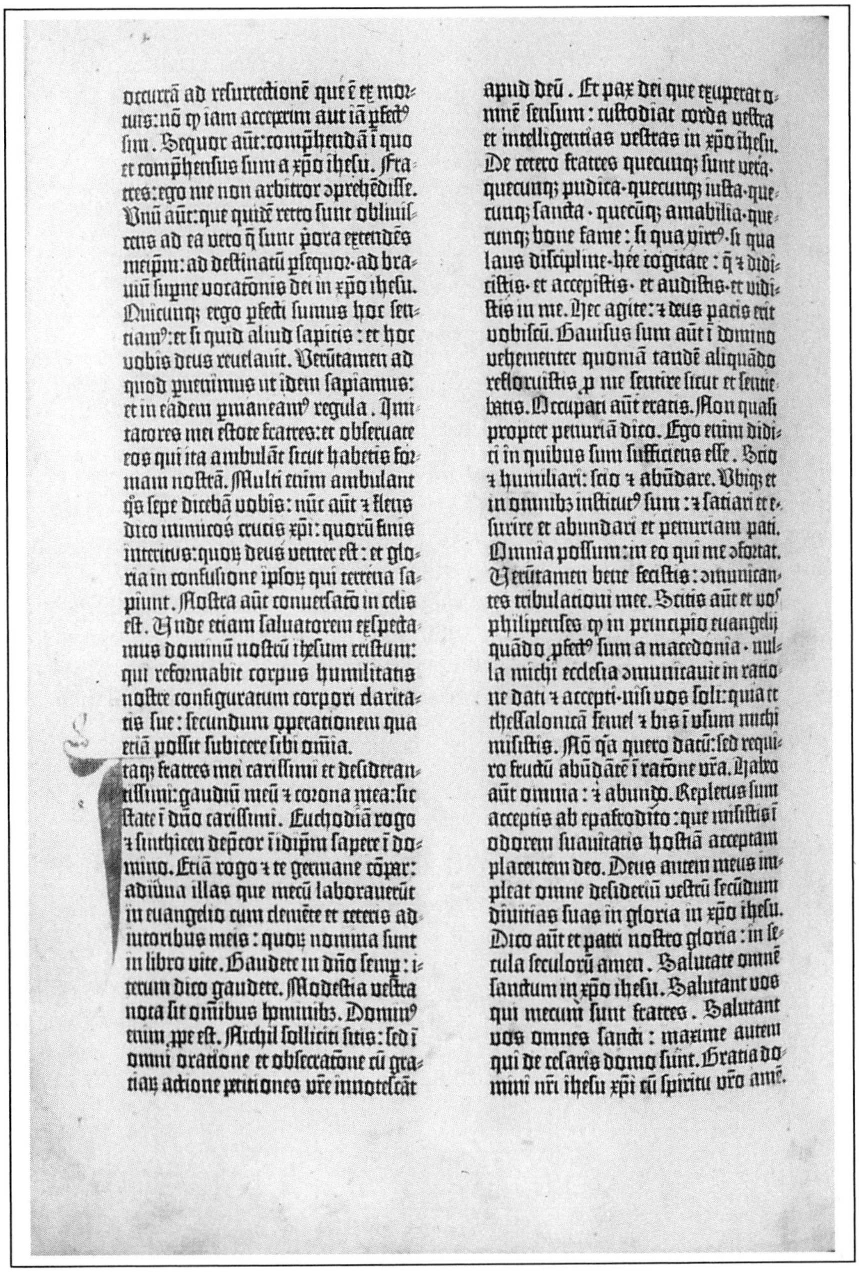

Fig. 3-9 A page from the famous Gutenberg 42-line Bible, considered an excellent example of typography, printed in 1455. (Courtesy Lilly Library, Indiana University, Bloomington, Indiana)

newspapers adopted Black Letter types for their nameplates for these reasons. Avoid using Black Letter in all-capital lines; the letters are difficult, if not impossible, to read when set in all caps. Use them sparingly for title lines, logos, and subheads, where they are suitable. They are useful for formal invitations and stationery.

The New York Times

Fig. 3-10 The New York Times adopted a Black Letter typeface for its nameplate many years ago and still uses it today. Black Letter gives the image of a long-established, serious publication.

Roman

The birthplace of printing in the Western world was northern Europe. From there it spread south and west. Venice became a center for printing in the late 1400s. However, the people in southern Europe did not like the heavy type used in the north. They preferred the simpler and more open letters of the Romans. So typefaces developed in Venice, Paris, and elsewhere followed Roman letter forms and became known as the Roman race.

Fig. 3-11 This type is Caslon, designed by International Typeface Corporation for use with electronic typesetting equipment. It is a member of the Roman race.

abcdefghijklmnopqrstuvwxyz
ABCDEFGHIJKLMNOPQRSTUVWXYZ

The ancient Romans developed their alphabet from the Greeks and the other early civilizations of the eastern Mediterranean area. In fact, some letters of the Roman alphabet and thus of our alphabet can be traced back to the picture writing, or hieroglyphics, of the Egyptians.

Historians believe that the Phoenicians adopted and modified Egyptian letters and then the Greeks took some letter forms from the Phoenicians, rearranged their constructions, and used them to form the Greek alphabet. These forms passed from the Greeks to the Romans.

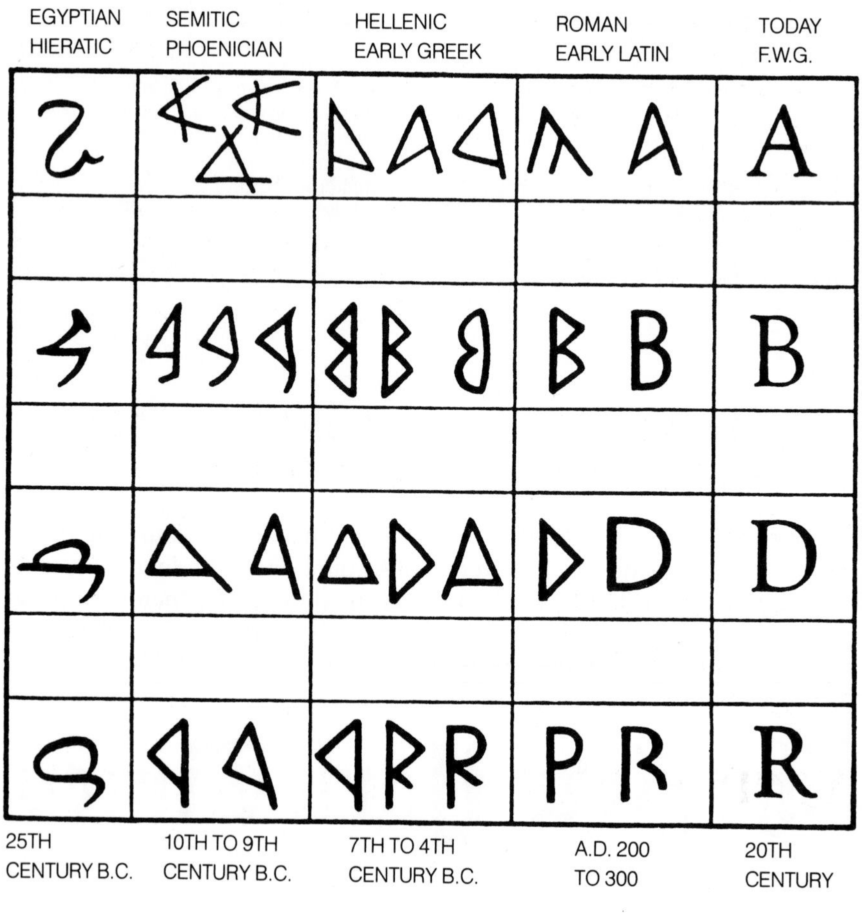

Fig. 3-12 The development of some modern letter forms can be traced back to the hieroglyphics of the Egyptians.

EGYPTIAN HIERATIC	SEMITIC PHOENICIAN	HELLENIC EARLY GREEK	ROMAN EARLY LATIN	TODAY F.W.G.
25TH CENTURY B.C.	10TH TO 9TH CENTURY B.C.	7TH TO 4TH CENTURY B.C.	A.D. 200 TO 300	20TH CENTURY

The serif angles out sharply from the stem and terminates in a crisp point.

There is a slight tilt to the angle of the swells of the round letters.

There is greater contrast between thick and thin strokes.

Fig. 3-13 Characteristics of old style Roman types. (Courtesy Westvaco Corporation)

The tool most used by the Romans in writing inscriptions on their buildings was the brush. They painted inscriptions on the masonry and then stonemasons carved over the painted outline. The Romans could turn a brush to make curves and continuous lines, and thus they developed letter strokes of varying widths for their letters. They also finished off long broad strokes with a narrow stroke using the side of the brush. These endings became the serifs on our letters and one of the distinguishing features of the Roman race of types. All Romans have serifs and letter strokes that vary in width from thick to thin.

Roman types are classified as old style, transitional, and modern.

A typeface is classified as *old style* if it has little contrast between the thick and thin lines, if the strokes are sloping or round, and if the serifs are slanting or curved and extended outward at the top of the capital T and the bottom of the capital E.

In 1773 Giambattista Bodoni, an Italian type designer and a contemporary and friend of Benjamin Franklin, introduced the so-called *modern* Roman types. These types are distinguished by strong contrast between thick and thin strokes and straight, thin, unbracketed serifs. Any type

Galliard

ABCDEFGHIJKLMNOPQRSTUVWXYZ
abcdefghijklmnopqrstuvwxyz

OLD STYLE

Baskerville

ABCDEFGHIJKLMNOPQRSTUVWXYZ
abcdefghijklmnopqrstuvwxyz

TRANSITIONAL

Bodoni

ABCDEFGHIJKLMNOPQRSTUVWXYZ
abcdefghijklmnopqrstuvwxyz

MODERN

Fig. 3-14 Examples of the three subgroups within the Roman race.

Fig. 3-15 Characteristics of transitional Roman types. Transitionals are basically old style with modifications that move them toward modern Romans. (Courtesy Westvaco Corporation)

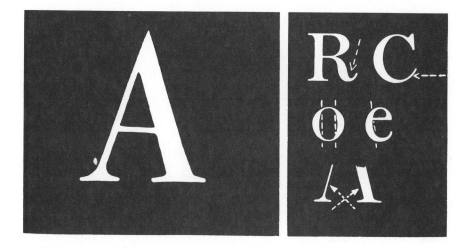

ITC
Fenice Bold

Fig. 3-16 Fenice Bold, a type designed by International Typeface Corporation, is modeled after Bodoni and is thus a modern Roman design. The contrast between thick and thin strokes is carried to the extreme and there is a vertical emphasis as shown by the thin strokes in the letter bowls, such as the top and bottom of the o above.

with these characteristics, even if a century and a half old, or if designed only yesterday, is classified as a modern Roman.

Type that falls between the sloping or rounded strokes, little contrast between thick and thin lines, and slanting or curved serifs of the old style Roman and the contrasting thick and thin strokes and straight thin and unbracketed serifs of the modern Roman is considered a *transitional* Roman face. It is said to be half-way out of the old style Roman design and half-way into the modern Roman configuration.

In the transitional types the angle of the thin strokes is not as pronounced as in the old style Roman, and the variation between thick and thin strokes is more pronounced.

Romans are considered the basic types. They come in all weights and sizes. There are Romans available to help communicate virtually any message esthetically and effectively.

Fig. 3-17 New type for a new age. Lucida is the first original typeface designed for use with a laser printer. The laser printer creates images in the form of "pixels," or tiny squares that create a mosaic of the letter or illustration. This typeface was developed by Kris Holmes, a free lance typeface designer, and Charles Bigelow, a Stanford University professor. (Courtesy Holmes & Bigelow, Palo Alto, California)

ABCDEFGHIJKLMNOPQRSTUVWXYZ
abcdefghijklmnopqrstuvwxyz &ÆŒØøæœßÇçÅåfi fl
0123456789$^1/_4$ $^3/_4$ $^1/_2$/-.,; :!?()[]{}/_|'' " " „¿¡— - -
†‡§¶*@©®™$¢£f#%‰˜ ¯ ˆ ˇ «»+-=±><'"°• □ ~

**ABCDEFGHIJKLMNOPQRSTUVWXYZ
abcdefghijklmnopqrstuvwxyz
&ÆŒØøæœßÇçÅåfi fl   0123456789$^1/_4$ $^3/_4$ $^1/_2$
/-.,; :!?()[]{}/_|'' " " „¿¡— - -
†‡§¶*@© ® ™$¢£f#%‰˜ ¯ ˆ ˇ «»+-=±><'"°• □ ~**

*ABCDEFGHIJKLMNOPQRSTUVWXYZ
abcdefghijklmnopqrstuvwxyz &ÆŒØøæœßÇçÅåfi fl
0123456789$^1/_4$ $^3/_4$ $^1/_2$/-.,; :!?()[]{}/_|'' " " „¿¡— - -
†‡§¶*@©®™$¢£f#%‰˜ ¯ ˆ ˇ «»+-=±><'"°• □ ~*

Sans Serif

The third broad category, Sans Serif, is comparatively new in name but it also has an ancient derivation, probably from the flat, even-bodied lines of the Greek and early Roman letters. Sans Serifs are called Gothics by some typographers, and in Europe they are known as the Grotesque types.

abcdefghijklmnopqrstuvwxyz
ABCDEFGHIJKLMNOPQRSTUVWXYZ

Fig. 3-18 An example of the Sans Serif race, this type is a member of the Avant Garde Gothic family designed for the International Typeface Corporation.

As the name implies (*sans* means "without" in French), Sans Serif types have no serifs. They are geometric, precise, and open. Some forms of Sans Serif types have slight variations in the letter strokes but most are monotonal.

Sans Serifs are excellent all-purpose types. Some of our newest designs, such as the Avante Garde Gothic, are renderings of the sans serif style of lettering. Sans Serifs are suitable for every purpose. They print well and have a modern look with good punch. And they probably will continue to be among the most popular faces for years to come. They are especially good for cold type production and offset printing methods as they stand up exceptionally well in photographic reproduction.

The Sans Serifs became especially popular in the 1920s when the Bauhaus movement in Germany, which emphasized functional design, made its influence felt in the world of typography. Sans Serifs may be rather monotonous when printed in mass. However, they are being used more and more, especially in newsletters and in-house magazines.

Square Serif

The fourth race of type is the Square Serifs. These have the general characteristics of the Sans Serifs except that a precise square or straight-line serif is added. Some typographers refer to the Square Serifs as Slab Serifs or Egyptians. They were called Egyptians when they first appeared.

In the early 1800s there was great interest in Egypt, most likely because of the discovery of the Rosetta stone in 1799. The Rosetta stone contained an inscription in three languages—that of the rulers, that of the common people, and Greek. The Greek was used to unlock the secrets of Egyptian hieroglyphics and of ancient Egyptian history.

abcdefghijklmnopqrstuvwxyz
ABCDEFGHIJKLMNOPQRSTUVWXYZ

Fig. 3-19 This example of the Square Serif race is a member of the Lubalin Graph family produced by the International Typeface Corporation.

About the same time, Vincent Figgins, a British type designer and founder, produced a typeface characterized by thick slab serifs and a general evenness of weight. The face had a certain quality reminiscent of Egyptian architecture in its capital letters. So it was only natural for it to be dubbed "Egyptian."

Today there are many versions of Square Serif types, many with Egyptian family names or with names that are descriptive of the type style. There are Memphis, Cairo, Karnak, and Luxor. And there are Stymies, Girders, and Towers.

Today Square Serifs are more logically associated with modern buildings than Egyptian architecture. They are sturdy and square. The capital letters look like steel girders, and the design gives the reader a feeling of strength, stability, and ruggedness.

The Square Serifs were not meant to be used in mass. They are excellent for headlines, headings in advertisements, and posters. They are monotonous and tiring when used in long columns of reading matter.

These types were quite popular about the turn of the century, and they are found in the headlines of that day. The "wanted" posters of the Old West were also often set in Square Serif.

Square Serifs faded from favor during the first decades of the twentieth century. They were considered too old-fashioned. Interest in these types revived in the late 1940s and 1950s. A number of newspapers adopted them for their headline schedules. Then their use declined again, and most newspapers dropped them by the early 1970s in favor of the more modern Romans such as Bodoni and the Sans Serif faces such as Helvetica.

In the mid-1980s Square Serif faces enjoyed a revival. Since these types are structurally rugged and have but little, if any, variation in stroke widths, they are good types for reverses, surprints (type over illustrations), and photomechanical composition and platemaking techniques. They also print well in offset and gravure.

Scripts and Cursives

The fifth race includes letter styles that resemble handwriting. The typographer calls them Scripts or Cursives. Since Scripts and Cursives are so similar, we can group them together in the Scripts and Cursives race.

Fig. 3-20 This is Bernhard Cursive Bold. Lucian Bernhard, a German designer, produced many types. The American Type Founders use his designs to produce a number of his typefaces in the United States.

But, why Scripts and Cursives? Why not just call them one or the other?

A little explanation is needed here. Some typographers classify *italic* types (Roman types with slanted letters) as a separate race. The slanted letters that we refer to as italics were actually a style all their own when they first appeared in the 1500s.

A printer and letter designer, Aldus Manutius, who established his famous Aldine press in Venice in 1490, wanted to publish small books of significant works. They might be called the first "everyone's library" books. Manutius sought a letter style for his type that would permit as many words as possible on a page.

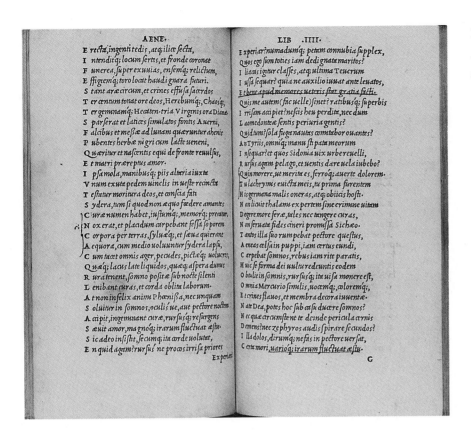

Fig. 3-21 The first italic type appeared in Virgil's *Opera*, printed by Aldus Manutius in 1501. (Courtesy The Pierpont Morgan Library, New York. PML 1664)

He devised a cursive letter style that was the first italic typeface. A *cursive* (from the Latin *currere*, which means "run") was a free construction in which there were no exact or repetitious letter forms (the letters were free flowing like handwriting). The scribes and printers in the Germanic countries called the italic style "cursive" and that is how it got its name.

When the Aldine books appeared in the original italic face, other printers quickly copied it and the design spread throughout Europe. (However, slanted letters in the Sans Serif and Square Serif races are called *obliques* and not italics.)

Freestyle Script   *Early Americana*

Fig. 3-22 Some Scripts or Cursives are designed so that the letters appear to join (right); others have noticeable gaps between the letters.

In some books on type the authors insist that Scripts are letters resembling handwriting and are connected while Cursives have a noticeable gap between the letters. On the other hand, just the opposite is cited as the distinguishing feature of the two types (that Scripts have gaps between the letters and Cursives do not) in other texts on typography and graphics. It is usually sufficient, however, to recognize and define the letter styles that resemble handwriting as either Scripts or Cursives. The safest way to approach these types is to select the style you want to use and refer to it by its family name.

Fig. 3-23 A traditional wedding invitation set in Shelley Allegro script. Note how this typeface helps create the image of a formal occasion.

Mr. and Mrs. Reginald Harper

request the honor of your company

at the marriage of their daughter

Jacqueline Marie

to

Mr. Alfred Bruce Saylor

on the evening of

Friday, the first of October

at half after seven o'clock.

Six hundred and ten Richelieu Place

Kingsland, Wyoming

Fig. 3-24 This is an example of italic type, not to be confused with a Script or Cursive. It is the italic version of Century Expanded, a member of the Roman race. (Courtesy American Type Founders)

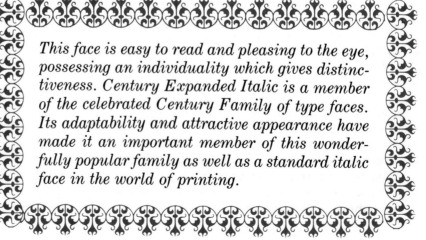

This face is easy to read and pleasing to the eye, possessing an individuality which gives distinctiveness. Century Expanded Italic is a member of the celebrated Century Family of type faces. Its adaptability and attractive appearance have made it an important member of this wonderfully popular family as well as a standard italic face in the world of printing.

Scripts and Cursives play an important role in good design. They can give a special "tone" to a printed piece. They are excellent for announcements and invitations and they can be used for titles, headings, and subheads. They can also add interest, contrast, and life to a printed page. But because they have a low readability rating, they (and italics, too, for that matter) should be used sparingly—only for a line or two or three—not for many sentences or paragraphs.

Miscellaneous or Novelty

The sixth race is not actually a race. It is a catchall category for all those unusual designs that do not have the clear-cut characteristics of the other races. These can include hand-designed types and types designed for special effects, logos, trademarks, and novelty arrangements. They are "display" types and most are unsuitable for use in blocks of reading matter. They also are called Decorative.

Selection and use of these types should be governed by the same criteria we would use in selecting and using a Script or Cursive.

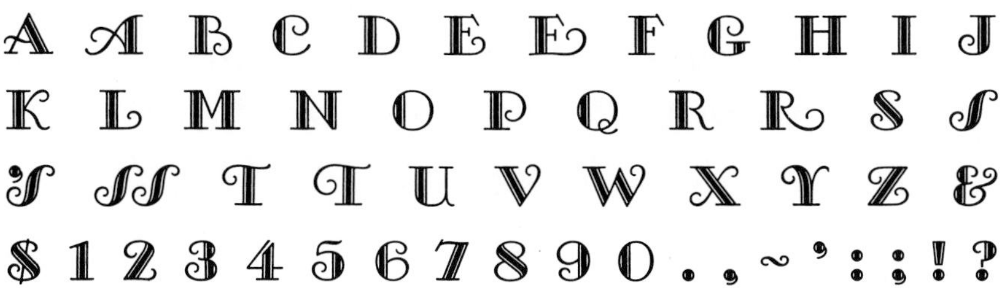

Fig. 3-25 Types of unusual or decorative design that cannot be clearly classified by race are placed in the Miscellaneous or Novelty category. This typeface is called Gallia.

Families of Type

Roman, Black Letter, Sans Serif, Square Serif, and Scripts and Cursives and Miscellaneous—are the basic races of types. But not only can each letter design be identified by its race, each has a name as well. Quite often, the name of the designer is used. There is Bodoni, named after the designer of this popular modern Roman type, Giambattista Bodoni. There is (Nicholas) Jenson, (William) Caslon, (Claude) Garamond, and (Frederic) Goudy, to mention a few faces that are in use today.

Sometimes the type name may be descriptive of its design or function. There is Cloister, a Black Letter type reminiscent of the cloisters of medieval monks; the gaudy Lilith, named for the female demon of Jewish folklore; and there is Bankers Gothic, a no-nonsense type often used for business forms and letterheads.

Or, the type names may reflect a geographical area, such as the Square Serifs called Cairo, Memphis, and Karnak.

Within these families there are further divisions that are called series. A series of type within a family consists of all the sizes of a structure, or variation, within that family. Such variations might be letters that are

Fig. 3-26   Variations of type designs within a family are called a series. These are members of the Garamond family.

Garamond Light
Garamond Book
Garamond Bold
Garamond Ultra
Garamond Light Italic
Garamond Book Italic
Garamond Bold Italic
Garamond Ultra Italic
Garamond Light Condensed
Garamond Book Condensed
Garamond Bold Condensed
Garamond Ultra Condensed
Garamond Light Condensed Italic
Garamond Book Condensed Italic
Garamond Bold Condensed Italic
Garamond Ultra Condensed Italic

Fig. 3-27   Here is a series of the Helvetica family, medium weight, from 6 to 36 point.

Point Size

6   abcdefghijklmnopqrstuvwxyzabcdefghijklmnopqrstuvwxyzabcdefghijklmnopqrstuvwxyzabcdefghijklmnopqrstuv

7   abcdefghijklmnopqrstuvwxyzabcdefghijklmnopqrstuvwxyzabcdefghijklmnopqrstuvwxyzabcdefg

8   abcdefghijklmnopqrstuvwxyzabcdefghijklmnopqrstuvwxyzabcdefghijklmnopqrstuvw

9   abcdefghijklmnopqrstuvwxyzabcdefghijklmnopqrstuvwxyzabcdefghijklmno

10   abcdefghijklmnopqrstuvwxyzabcdefghijklmnopqrstuvwxyzabcdefg

11   abcdefghijklmnopqrstuvwxyzabcdefghijklmnopqrstuvwxyzab

12   abcdefghijklmnopqrstuvwxyzabcdefghijklmnopqrstuvwx

14   abcdefghijklmnopqrstuvwxyzabcdefghijklmnop

16   abcdefghijklmnopqrstuvwxyzabcdefghijkl

18   abcdefghijklmnopqrstuvwxyzabcdefg

24   abcdefghijklmnopqrstuvwxy

30   abcdefghijklmnopqrst

36   abcdefghijklmnop

ABCDE FGHIJKLMNOPQRSTUVWXYZ
ABCDEFGHIJKLMNOPQRSTUVWXYZ
abcdefghijklmnopqrstuvwxyz *abcdefghijklmnopqrstuvwxyz*
ABCDEFGHIJKLMNOPQRSTUVWXYZ fi fl [($%&'"#@¢¹/₄ ¹/₂ . , ; : ? !
/ l)] - – — * † ‡ § £ 1234567890 *1234567890* 1234567890
ÅÃÆÇØÕŒ Ñåãæç ÂÄÊËÎÏÔÒÔÛÜ¿ ÂÄÊËÎÏÔÒÔÛÜ'
ÁÀÉÈÍÌÓÓÚÙ åãæçøõ œñ øõ ŒÑ âäêëîïôöûü¡ ÁÀÉÈÍÌÓÒÚÙ ß
áàéèíìóòúù

Fig. 3-28 A font of type. Many more than 26 characters are needed to set type; in fact, the usual bare minimum is about 150. Small caps, ligatures (fi, fl), numbers, and punctuation add considerably to the collection. Notice here the number of accented characters, which are essential for setting foreign words. This typeface is ITC Stone.

bold, letters that are condensed, or extended, and so on. Each one of these variations in all the sizes available is a series.

Finally, there is the font. A font consists of all the characters available in one size of a particular type style.

▼ Measurement of Type

Type and printing have unique units of measurement. These units are quite simple, but anyone working with type and printing should understand them thoroughly. It will be difficult to function with any sort of efficiency as an editor, designer, or communicator if you do not understand and know how printing measurements are used.

For about 300 years after Gutenberg there was no standard system of measurement for printers. Typecasters gave names to the various sizes of types they produced. Usually the types of one foundry could not be used with those of another. The problem was further compounded because the names given to types of certain sizes by one foundry might be used to designate entirely different sizes by another. One foundry might have called a type "nonpareil" while another used nonpareil for a different size of type. (Note: The term *nonpareil* became accepted later as referring to type now called 6 point.)

A system of sizing type by units of measurements called *points* was devised in the middle 1700s by a French typographer, Pierre Simon Fournier, and further refined by his countryman Francois Ambroise Didot. In 1886, under the sponsorship of the U.S. Type Founders Association, a point system was adopted as the official uniform measurement of types in the United States. England adopted it in 1898.

The point is the smallest unit of measurement in the printer's and graphic designer's world. It is commonly defined as being ¹/₇₂ of an inch. (Actually, the point is slightly less than that. Specifically, it is 0.0138-plus of an inch. But for all practical purposes communicators consider 72 points as equalling an inch.)

The thickness of spaces, leads, slugs, rules, and borders (the white space between words and lines of types and the decorations around blocks of type) and the height of type are all measured in points. If type is 36

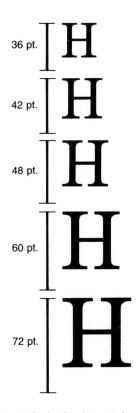

36 pt.
42 pt.
48 pt.
60 pt.
72 pt.

Fig. 3-29 The body of type is large enough to accommodate the descending lowercase letters, such as g, p, y and the ascending letters such as b, h, d. The vertical line indicates the true body size. The typeface shown is Century Expanded.

points high it is approximately ½ inch high. Type that is 72 points is approximately 1 inch high.

Some of the old type names have made their way down into the world of graphics today. Pica and agate are the most common type names still in use. A pica is a unit of measurement that is 12 points in width. Pica type is type that is 12 points in height. Agate refers to type or a unit of measurement that is 5½ points in size. There would be 14 agate lines in one inch. Actually, 14 agate lines is 5.142 points high, but printers and advertisers use the simple comparison of 14 agate lines to equal 1 inch in depth.

Pica is used to designate 12 points of space. Thus there are 12 points in a pica and 6 picas in 1 inch. The type on a pica typewriter (or pica type on a word processor) is 12-point type.

Points and picas are the heart of the printer's and the communicator's measurement system, just as inches and feet are used to measure most things in our daily lives.

Fig. 3-30

Printer's rule or "line guage" with points, picas, metric and inches.

Graphic artist's rule with inches and picas.

Sometimes the word *em* is used as a synonym for pica. This is erroneous, but it has become so prevalent that only a purist would object now. A pica is linear measure. That is, it measures height, width, and depth. An em is the measure of the square of an area.

The em is an important unit of measurement in graphic design. Formally it is the square of the type size in question. For example, a 36-point em measures 36 points, or ½ inch, on all four sides. An em that measures 12 points, or a pica, on each side is referred to as a *pica em*.

The difference between a pica and an em should be remembered. Both terms are used in marking instructions for compositors or printers, in copy fitting, and in making proper dummies and layouts. Confusion here can cause trouble later.

Another measure of area, but not a square area, is the *en*. An en is half the width of an em. An en in 36-point type is 36 points, or ½ inch high—the same height as the type. However, it is only 18 points, or half the 36-point measure, wide. Thus a 36-point en would be 36 points high and 18 points wide.

Old-timers, to keep the two terms from being confused, referred to an em as a mutton or mut, and an en as a nut.

Em's and en's are still used quite often to indicate indentation. Many video display terminal keyboards have an em and an en key. If copy for 9-point type is marked for an em indent, it will be indented 9 points when the operator hits the em key on the keyboard. If copy set in 12 points is marked for an en indention, it will be indented 6 points. The compositor does not have to stop and figure out spacing if copy is properly marked by the communicator.

Once we understand the classification of types and type measurements, we can put this information to work. However, there are a few more points to keep in mind that apply to all the styles of types.

For instance, two typefaces may have the same point-size designation, but the body of the letters in one may be considerably larger than those of the other. This is referred to as being "big on the body" or "small on the body." The *body* is the actual base size or true point size of the type.

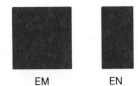

Fig. 3-31 An em is a unit of measurement that is the square of the type size being used. An en is one-half the width of an em. The square on the left illustrates the area of a 48-point em. The rectangle on the right illustrates the dimensions of a 48-point en.

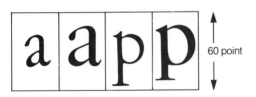

Fig. 3-32 Type is measured by the height of its body. Letters that vary in appearance can still be considered the same size. The a on the left is "small on the body" and the largest a is "large on the body."

It is possible for types that have, say, 48 points as their size designations to be quite different in appearance. The *x* height of the letter forms may be different even though the total point sizes of the types are the same.

Notice how the lowercase *h* in figure 3-33 appears to be in three different sizes though all are actually 72-point types.

Fig. 3-33 Even though these letters appear to be different sizes, they are all 72-point. (Courtesy Westvaco Corporation)

The size of letters cannot be measured accurately by placing a line gauge or graphic arts ruler on the letters themselves. To be perfectly accurate the base has to be measured. However, we can get a close reading, or workable reading, of the point size of a certain type. We can do this by measuring from the top of the ascender to the bottom of the descender of sample letters.

This is useful when we spot a type that is just the right size for the purpose in mind but the point size is not known. We can measure the letters and find the size. We can then proceed to make a layout and specify the size of type, or we can tell the printer, "Set it in 48 point" or whatever.

▼ Important Points to Remember

- Type size is measured in points. There are 72 points in 1 inch.
- Line length is measured in picas. There are 6 picas in 1 inch.
- The pica is used to express overall width or depth as well as the length of a line. There are 12 points in 1 pica.
- The most common use of the em is to indicate indentation of blocks of copy or to indicate the indentation at the start of a paragraph. The em is the square of the type size being used.
- The agate line is used to measure the depth of advertising space. There are 14 agate lines in 1 inch.

▼ Effective Design Checklist

- There is a Roman type to fit nearly every need. Old-time printers said, "When in doubt, use Caslon."
- Square Serifs print well and are good for "no-nonsense" messages.
- Black Letters emphasize tradition. They give the feeling of Gothic cathedrals, medieval castles, institutions. They are, however, rarely used in today's designs. They, and Scripts and Cursives, are also difficult, if not impossible, to read in all capital letters.
- Scripts and Cursives are like spices and seasonings. They can add interest, contrast, and life to a printed page. However, they have low readability and should be used sparingly.
- Miscellaneous or Novelty faces should be selected and used sparingly, like Scripts and Cursives.
- It is considered better to have a wide range or series of a few families rather than a wide variety of families with limited series.
- In selecting a font of type, consider the characters it contains in addition to the letters, numerals, and punctuation marks.
- Compare the x heights of letters, the lengths of ascenders and descenders of the various families, as well as the letter designs when selecting types.

▼ Graphics in Action

1. Explain the printer's measurement system. Illustrate your explanation with elements, letters, or lines of type clipped from publications, or

with simple sketches. Use rubber cement to attach your illustrations. Neatness and accuracy count.

2. Find several lines of type of varying widths in publications. Mount them on plain white paper and indicate the widths in inches and picas.

3. Collect examples of types representing each of the races. Mount the examples on plain white paper and indicate the widths of the lines in inches and picas.

4. Explain how type is classified according to race, family, and series. Illustrate your explanation with examples clipped from publications.

5. Draw a frame about 1 inch wide around a sheet of 8½ by 11 or 9 by 12 tracing paper. Color the frame any color you desire. Draw a very light line about an inch below the middle of the page, horizontally, with a soft pencil (HB or 2H work well). Trace your name or nickname (limit it to about five letters or less) using the line as a baseline for the letters. Select types from any source (such as magazines or newspapers) that are at least 72 points in size if all capitals and about 120 points if capitals and lowercase. Try to keep the spacing between the letters equal. Then go wild with colored pencils, crayons, or felt pens and color your name any way you want. This could create a design worth framing! This is also the first step toward one method of lettering for layouts.

Creative Typography

Courtesy Anchor Paper Company. Design: Yanovick & Associates

The proper selection and use of the hundreds of type designs available is the second step in creating an effective printed communication. The first step is a look at one word in that sentence: *creating*.

As a matter of fact, all the work done by the communicator is the result of creativity. We start with little more than an idea. But even before that we have to come up with the idea—originate it, consider it, examine it, revise it, try it. We have to be creative.

Among professional communicators more and more attention is being paid to effective creativity. There are some important reasons for this. We have reached the point in technological advancements where our computers, laser printers, and scanners are capable of producing highly refined work in typesetting, in creating graphics, and in arranging the elements of a graphic communication. We have the tools that have expanded our ability to execute what we create. At the same time, we are working in an overcommunicated environment. If we want our communication to penetrate the consciousness of our target audience, we have to come up with new and novel and creative ways to get our message to do its job.

But, you say, I am just not the creative type. I have to work and struggle. I have a hard time getting started. How do I begin? How did people like Einstein and Galileo and Thomas Edison come up with their ideas? I wish I could get ideas that work, too.

Fig. 4-1 Creativity in action. You are the editor or designer of a magazine and you want a layout for an article titled, "If Anything Can Go Wrong, It Will." How would you do it? Here a model railroad engine and track tell the story. Note, too, the reversed R in the title. (Courtesy *JD Journal,* Deere & Company)

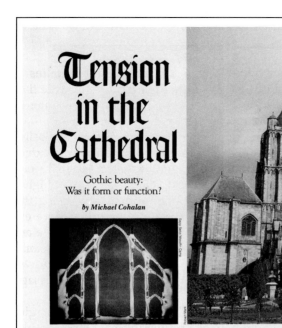

So, before we take a look at how we can select and work with type and art to create the most effective communication possible, let's listen to the experts for a minute or two.

"Human creativity uses what is already existing and available and changes it in unpredictable ways," explains noted psychiatrist Silvano Arieti.[1] As we shall see, we can take words, experiences, pieces of art, even a scrap of paper in the street, and use them to trigger creative ideas of all sorts.

Writer Arthur Koestler has an idea to get us started thinking about creativity. Koestler notes that children are instinctively creative. They all create without restraint. They sing, dance, draw, fantasize, play, all in a world of their own creation.[2]

But then, as children get older something happens. Their lives become increasingly structured in a mass of rules and regulations imposed by mother, father, and teacher. The creativity of their early days is restricted and repetition replaces creativity.

Now, to unleash our abilities and come up with ideas that are unique and different we have to break out of the restraints, if only for a while, that have ordered our lives. We have to bring the right brain into action.

Psychologists tell us we have two brains. The left brain operates our talking, walking, and all our mechanical actions. The right brain is the repository of our imagination, intuition, and creative thinking. We use

Fig. 4-2 Type can be selected to reinforce the message, as in the title and initial letter for this magazine article about Gothic cathedrals. The use of Black Letter for the title and for the initial letter aids unity. The subhead and byline in Roman are subordinated to the title and illustrate proper mixing of the races—there should be define contrast and one should dominate. (Reprinted by permission of *Sciences 84*, copyright, The America Association for the Advancement of Science. Cathedral photo by R. Mark, Princeton University)

Creative Communication

The creative communicator can be frustrated by accumulating either too little or too much information.

Effective creative solutions require adequate information. Superficial research results in superficial design.

On the other hand, too much information that is not relevant to the goal of the layout can muddy the waters and cause confusion.

The creative communicator should make an effort to gather enough information to understand the design problem thoroughly. The irrelevant information, however, should be eliminated.

Creativity is aided by concentration on information pertinent to the solution of the problem.

our left brain in much of our day-to-day living; to be creative we need to unlock the right brain—the home of our creative abilities.

We need to call up our right brain and let it work with the left to develop creative problem solving. This starts the whole brain operating—using things we have learned about the fundamentals of effective communication that are stored in our left brain with the innovative and creative potential of the right brain.

The advertisement or brochure layout or the magazine cover design start with an effort to solve a problem; for example, how to get our target audience to buy a product or idea or how to get it to select our magazine from the scores in the rack.

One approach suggested by Edward de Bono, author, physician, and psychologist, is first to list every possible solution to the problem. Come up with as many as you can. If you become stuck, de Bono says, then use the random input technique. In this technique the designer selects a word or an object completely at random and adds it to what he or she has already worked out. Then the designer works backward and often comes up with a fresh idea.

Suppose you have to create a cover idea for a promotional brochure for a computer company. You don't want a boring image of a person just sitting and looking at the computer. You're looking for a fresh way to get at the problem. What can you do?

One approach would be to start with a word picked at random from a dictionary. It has to be random. If it isn't you are choosing it and that won't work, de Bono says. Let's try it. Let's take the first word, lefthand column, on page 869 of *Webster's Ninth New Collegiate Dictionary*. The word is pen. Can you play with this word and come up with a fresh idea of how this word could be used to illustrate a promotional brochure for a new computer?

Andy Hertzfeld, who headed the team that built the Macintosh computer, says about creativity and ideas: "I get inspired by the chance to make a difference in the world. An idea has to be able to help people in some fashion to make it worth pursuing." [3]

Hertzfeld believes the best way to fight creative blocks is to work on two or three problems at once. "When you're blocked on one thing you can go work on something else. Often a nice solution to your former problem will occur when you're working on something else," he notes.

Quite often success comes from "something old, something new." This is the approach of four graphic designers in New York. Paula Scher, Louise Fili, Carin Goldberg, and Lorraine Louie say they obtain their creative ideas from such approaches as the study of European design going back to the first fifty years of the twentieth century, unorthodox typefaces, and a flagrant disregard for the rules of proper typography. [4]

Unorthodox attitudes about the rules of proper design and typography permit them to take risks and experiment.

Goldberg says the study of design history is important for anyone seeking new and creative ways to tackle graphic problems. "Without a sense of design history graphic designers are lost in space," she comments. A striking image is needed in today's overcommunicated environment, and art and design history provide her with an important resource. [5]

 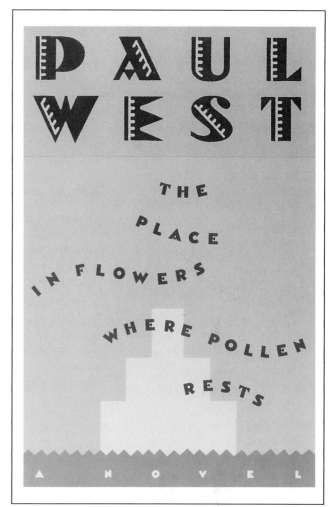

Another approach to design creativity is used by members of the Society of Environmental Graphic Designers. They like to consider themselves *wayfinders*. They define wayfinders as those who are finding their way or helping others to find their way.

The communicator might use this approach to develop a plan of action. Also, it could be applied to creating written and graphic communications that will help the target audience to find its way and thus reach the communicator's objective.

Developing a sense of creativity should be a continuing process. Browsing through magazines can trigger ideas. So can looking at objects as you stroll around your community or take a relaxing hike. Ideas are everywhere. All it takes to find them is putting the right side of your brain to work.

Many creative ideas develop by ignoring accepted procedures. In this chapter we attempt to examine some rules and accepted practices and then some examples of creative typography that might bend or break some of these rules. Perhaps we can get both sides of our brains working to help us become more effective communicators.

Fig. 4-3 Carin Goldberg's striking book jackets reflect her strong sense of design history. In the example on the right, Goldberg redesigned and improved Max Salzman's eccentric 1922 typeface, Zierdolmen.

▼ Readability Is The Foundation

The proper selection and use of the hundreds of type designs available is important in creating an effective printed communication. A production is a terrible waste of time, money, effort, and material if it is not read and understood.

Type should be selected and arranged to achieve maximum readability of the message. Readability should be the foundation on which the printed communication is built.

Frederic W. Goudy, the most prolific American type designer, put it this way:

> Letters must be of such a nature that when they are combined into lines of words the eye may run along the lines easily, quickly, and without obstruction, the reader being occupied only with the thought presented. If one is compelled to inspect the individual letters, his mind is not free to grasp the ideas conveyed by the type.[6]

Type selection and display should not only aid readability but also reinforce the goal for the communication. That is, the type should be appropriate to the message. If the communicator is shouting, the type should help. If the communicator is trying to reason with the reader, the type should reflect this goal, and it can. There are "stern" types for stern and remonstrative messages, and there are "happy" types for happy messages.

An effective message must first gain attention. That is one reason why newspapers use headlines. The type is selected and arranged to get the attention of the potential reader. Proper selection and arrangement make a publication stand out from other publications.

Proper selection and use of type can also help create identification. *Time* magazine, for instance, can be spotted quickly because of its consistent use of certain graphics and types. There is no mistaking which newspaper is the *New York Times.* Harlequin books are easily spotted among the

Fig. 4-4 Text matter set in a slightly bold Sans Serif (middle) is considered to have greater readability than the light Sans Serif (top). Roman (bottom) is the preference of most readers. (Courtesy U&lc.)

Four score and seven years ago our fathers brought forth on this continent, a new nation, conceived in Liberty, and dedicated to the proposition that all men are created equal.

Four score and seven years ago our fathers brought forth on this continent, a new nation, conceived in Liberty, and dedicated to the proposition that all men are created equal.

Four score and seven years ago our fathers brought forth on this continent, a new nation, conceived in Liberty, and dedicated to the proposition that all men are created equal.

Fig. 4-5 Type talks. Notice how the letter designs help create the image the words transmit.

What Typefaces Can Express

1. Type can be light or **heavy**

2. Type can be unassuming or graceful

3. Type can whisper or **shout**

4. Type can be monotonous or sparkle

5. Type can be **UGLY** or beautiful

6. Type can be **mechanical** or formal

7. Type can be *social* or **ecclesiastical**

8. Type can be **FAT** or THIN

9. Type can be **decorative** or plain

10. Type can be easy to read or *hard to read*

Key to the type faces listed above, left to right in order:

1. 18-point Avante Garde Extra Light and Lubalin Graph Bold
2. 18-point Baskerville and Palatino
3. 8-point Belwe Light and 18-point Gill Sans Ultra Bold
4. 18-point News Gothic and University
5. 18-point Auriol Black and Weiss
6. 18-point Lubalin Graph Medium and Cloister Open
7. 18-point Snell Roundhand Bold and Old German Bold
8. 18-point Bauhaus Bold and Century Bold Condensed
9. 18-point Nicolas Cochin Black and Caledonia
10. 18-point Century Schoolbook and 6-point Novarese Medium Italic

paperbacks because their typography says "light, romantic novel" and helps define this particular publisher's audience.

It should be kept in mind that the proper use and arrangement of type is both an art and a science. It is an art in that within limitations there is wide freedom of action. To begin with, there is only a blank page. The designer has comparative freedom in designing the elements that will occupy this page. The communicator is an artist using type, borders, illustrations, and color to create a physical environment for the written message.

Typography is a science as well. It is a science in that it uses rules that have been proven by research and testing. These rules must be understood and applied with creative freshness to produce the most effective

WEISS
TORINO
Baskerville

Fig. 4-6 Examples of Roman types that might be used for display purposes. They suggest dignity, stability, and integrity, and they rate high in legibility. Many Romans have their own personalities and individuality. They can help establish an image for an organization or product.

UNIVERS
Eurostile
Anzeigen Grot
Optima

Fig. 4-7 Although Sans Serifs are basically versions of a style introduced in the nineteenth century, they are considered by many designers the most modern of type designs. Their precise, simple lines give high legibility to titles and headings. Optima especially, with its Roman appearance, seems warm and human.

printed pieces possible. So, the communicator who desires to produce the most effective printed communication must apply a mixture of art and science.

Let us consider the science of typographic design first and then look at the art of using type.

Below are ten suggestions based on the experimentation, research, and testing that have proved to be effective in working with type:

1. Use the right type style.
2. Set the type on the proper measure.
3. Watch the spacing.
4. Remember the margins.
5. Select the proper type size.
6. Mix type styles carefully.
7. Use all-cap lines sparingly.
8. Do not be boring.
9. Avoid oddball placement.
10. Spell it out.

If we keep these ten rules in mind, they will help us produce an end product that is suitable, readable, and legible. These suggestions are quite easy to apply and can be summed up as: Keep the reader (or target audience) in mind when making layouts or giving instructions to the compositor or printer.

▼ Use the Right Type Style

The first step in selecting the right type is to divide the content of the message into words that speak out in short, crisp phrases and words that must be read in mass.

The short phrases (headings, titles, subheads) should be set in large type for emphasis. This type is called *display* as contrasted with *text* or *body* types for masses of words, or reading matter. The headline of a news story, for example, will be set in display type and the story itself in body type.

Generally, types within the size range of 8 or 9 to 12 or 14 point are considered suitable for reading matter. However, 8-point types in some families such as Goudy Old Style are quite difficult to read and the trend is toward using the larger sizes. Types larger than 14 point are display types.

Both classifications of type must be considered, initially, as separate entities. Remember, however, that for unity the two must work together and harmonize. Sometimes the most effective choice for display is a larger size of the same design used for the body type, or a larger size from another series of the same family as the body type.

The second step in selecting the most effective type design for a message is to read the copy. What sort of message is this? If it is a hard-sell advertisement, then display type in a heavy-hitting design is the obvious choice. Clean-lined, bold Sans Serifs would be a good choice. A bold Square Serif has a forceful voice.

Suppose we want to say "Here is a long-established organization that has become an institution in our city"—what type should we select? The *New York Times* and many other newspapers say this by using Black Letter types for their nameplates. An old-style Roman, such as Caslon or Goudy, would help get the idea across, too.

If the message is crisp and modern but dignified as well, a modern Roman might be a good face to choose. Bodoni or Craw Modern would do the job.

High fashion dictates a Script or Cursive or perhaps a light-faced Sans Serif. Vogue is a family that speaks to high society. It was designed especially for the audience of *Vogue* magazine.

The improper type design can sabotage the message. A newspaper would look ridiculous setting a banner head concerning an exposé at city hall in Black Letter type. Yet Black Letter would be a fine choice to say "Merry Christmas."

When it comes to body or text type for reading matter, there is controversy. Some designers maintain that Sans Serif is the modern type and those who eschew it are old-fashioned. Others stick by the Romans that have been used since printing was introduced in England.

The argument in favor of Romans is twofold. First, we are used to it because so much of what we read—books, magazines, newspapers—is set in Roman types and always has been. Second, the thick and thin strokes of Romans and the rounded shapes of the letters cause less eye fatigue and make reading more pleasant.

Studies seem to side with the Roman advocates. Although his evidence is inconclusive, Miles A. Tinker, a psychologist, has written that readers prefer Roman faces even though there seems to be no real difference in reading speed, fatigue, and so on. He also points out that readers usually prefer a type that is slightly on the bold side.[7] But a number of later studies confirm the longstanding belief that Roman faces are preferred. An American Newspaper Publishers Association study found that Roman types can be read seven to ten words a minute faster than sans serifs. Other studies show similar results, including one that reveals sans serifs are read 2.2 percent slower than Romans and another that indicates it takes 7.5 percent more time to read sans serifs.[8]

▼ Set the Type on the Proper Measure

How wide should the columns in a publication be? If we have a block of copy we want to include in an advertisement layout, what width is best? Can we just arbitrarily select a width that looks nice? Does width make a difference? It certainly does. If columns of type are too wide, reading will be slow and difficult. Readers will tire easily, and they can lose their places in the message as their eyes shift from one line to the next. They may become discouraged and stop reading.

If, on the other hand, type is set in columns that are too short, the constant eye motion from left to right and back to left again will slow reading and make it a tiring chore. In addition, type set too narrow causes many awkward between-word spacing situations. Words must also be

Fig. 4-8 Square Serifs, or Egyptians, are hard-hitting, bold, no-nonsense typefaces. Playbill creates a "turn of the century" mood.

Fig. 4-9 Scripts and Cursives try to capture handwriting. They can give an image of austere formality or casual informality, create an effect of graciousness or rugged individualism. They should be used sparingly.

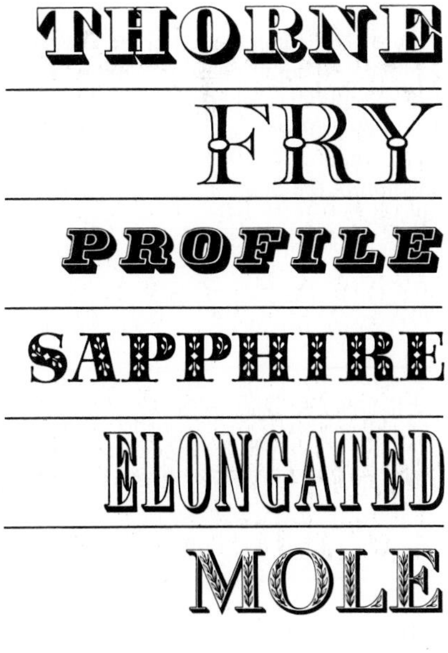

Fig. 4-10 Miscellaneous or Novelty types can be effective if used sparingly for display lines. They run the gamut from dainty to solid. They are unsuitable for more than a line or two. (Courtesy of Westvaco Corporation)

divided between lines more frequently and this can disfigure a printed piece.

Setting type on the proper measure is easy, and there are several rules of thumb that can be used. The proper measure (or line width) is, of course, determined by the type size and style selected. For small type sizes, say, 6 or 8 point, narrow columns are suitable. But larger type requires wider columns.

Some typographers have developed a theory of *optimum line length* This refers to the width of a line that is considered ideal for greatest reading ease. It has been found that a good way to determine the optimum width for lines of any typeface and size is to measure the lowercase alphabet (all the letters lined up from *a* to *z*) and add one-half to this. For instance, if the lowercase alphabet measures 18 picas, the optimum line-width for that type would be 27 picas.

Fig. 4-11 The ideal line width (1½ times the lowercase alphabet).

abcdefghijklmnopqrstuvwxyzabcdefghijklm

(Ideal line width: 1½ times the lowercase alphabet)

Typographers using this theory believe the minimum width any type should be set is the width of one lowercase alphabet, and the maximum width should never exceed the width of two lowercase alphabets.

Another rule of thumb that works out to approximately the same measure is to double in picas the point size of the type being used. For instance, 18-point type would ideally be set about 36 picas wide and 6-point type would be set no more than 12 picas wide.

Fig. 4-12 Examples of 8-point type set on three different measures. [Reprinted from Ralph W. and Edwin Polk, Practice of Printing (Peoria, Ill.: Bennett Publishing Company, © 1971.) Used with permission of the publishers. All rights reserved.]

The width of a column of matter influences its legibility. The ideal width for any piece of composition is based on the breadth of focus of the eye upon the page. For small types the focus will be narrow, and it will widen out as the type faces increase in size. If the column is set in too wide a measure, as is the case with this paragraph, the lines will be scanned with somewhat of effort, and it will be found harder to "keep one's place" as he reads. Also, it will require some effort to locate the starting point of each new line. A large amount of matter set like this would be tedious to read.

On the other hand, if the column is too narrow, fewer words may be grasped at a time, and thus, too frequent adjustments must be made for the numerous short lines of the type, seriously hindering the steady, even flow of the message. In addition, a greater proportion of words must be divided at the ends of the lines, and the spacing of the lines is necessarily uneven and awkward, also affecting the legibility.

This group is set the proper width for the comfortable, easy reading of 8 point type. The eye may easily take in a line at a time, and in this way the message may be read without any mechanical encumbrances. Larger types set to this width will present the same difficulties to the reader that are experienced in the 8 point example, set in the narrow measure, above. There is a suitable width for each size and style of type.

abcdefghijklmnopqrstuvwxyz

————— 15 picas —————

(minimum width)

abcdefghijklmnopqrstuvwxyzabcdefghijklmnopqrstuvwxyz

————— 30 picas —————

(maximum width)

The thing to remember is, the smaller the type used, the shorter the lines in which it is set. As the size of type is increased, the width of the lines should be increased but never to more than double the lowercase alphabet.

Type set on an improper measure can detract from the attractiveness of a printed communication. But, more important, it can be a definite deterrent to easy, pleasant reading.

Fig. 4-13 This 12-point Square Serif type should not be set in lines narrower than 15 picas or wider than 30 picas for ordinary compositions.

▼ Watch the Spacing

Too much spacing can make the message unattractive and difficult to read, too little can jam the words and lines together so that the message's appearance is destroyed and reading is a real task. Most people will skip over poorly spaced printed matter unless the content of the message is so overpowering that they will put up with unnecessary channel noise (in this case poor spacing) to get at the information.

White space, when applied to type masses, is the space between words and the space between lines. Words that are crowded too closely together are difficult to read. Words that are spread too far apart can create gaps

Fig. 4-14 Some examples of appropriate word spacing.

A line of 8pt. Garamond is too difficult to read with only a thin space

Its legibility is greatly improved when the words are separated

A narrow type needs much less space between the words

A wide type needs much more space

A type with a small x-height requires only a thin space

A type with a large x-height needs a thick space

of ugly white within the printed block. Lines that are too close together also cause problems. The reader must sort out where a new line starts. When lines are packed too tightly the reader may start to reread the same line. When lines are too far apart there are time-consuming gaps as the eye travels from one line to another.

Thus we need to decide how much space to use between lines of a printed piece. Printers call this *leading* (pronounced *ledding*). The term originated when thin strips of lead were used between lines of metal type as spacing material. When a printer suggests a paragraph should be leaded out, it means that more space is needed between the lines.

For reading matter, the 2-point lead (or space) to separate the lines of type has been considered standard.

Spacing between lines depends on several things. One is the size of the type being used. Another is the design of the type, whether it is big or small on the body. Type that is small on the body—that is, type that has a rather small *x* height for its point size—has more natural leading between lines than type that is big on the body, that has a large *x* height. Type styles that have large *x* heights need more spacing between lines.

Here, again, there isn't exact scientific data to prove the ideal. However, experimentation and practice have given some clues. In a general way, the longer the type line, the more spacing is needed between lines. For example, an 8-point type is selected for a 12-pica-wide line and 1 point of space is designated between lines. If the width of the line is increased to 18 picas, the message would be easier to read if the space between the lines would be increased to about 2 points. So a "subhint" for this general suggestion might be, when the line width is increased the space between lines should be increased, too.

This is even truer of display type, though we need to restate the rule as, the larger the type, the more space needed between lines. While 1 or 2 points of space between lines is fine for reading matter, when it comes to display type (type between 14 and 36 points and larger), at least 6 points between lines will be needed.

Fig. 4-15 Space for readability has two aspects: space between words and space between lines. The first copy block illustrates improper space between words. The other examples show the effects of various amounts of space between lines. (Courtesy Westvaco Corporation)

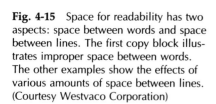

These rules illustrate space between lines.
Top to bottom: 1 point, 1½ points, 2 points, 3 points.

It may be said of all printers that their job is to reproduce on paper the exact face of the letters which they have set into pages. This face is of a definite, constant and measurable size and shape; with any one press and any one paper there is a right and exact quan-

It may be said of all printers that their job is to reproduce on paper the exact face of the letters which they have set into pages. This face is of a definite, constant and measurable size and shape; with any one press and any one paper there is a right and exact quantity of ink and pressure necessary to re-

It may be said of all printers that their job is to reproduce on paper the exact face of the letters which they have set into pages. This face is of a definite, constant and measurable size and shape; with any one press and any one paper there is a right and exact quantity of ink and pressure necessary to re-

It may be said of all printers that their job is to reproduce on paper the exact face of the letters which they have set into pages. This face is of a definite, constant and measurable size and shape; with any one press and any one paper there is a right and exact quantity of ink and pressure necessary to re-

It may be said of all printers that their job is to reproduce on paper the exact face of the letters which they have set into pages. This face is of a definite, constant and measurable size and shape; with any one press and any one paper there is a right and exact quantity of ink and pressure necessary to re-

Space between lines
is just about right
in this headline

There's too much

space between lines

in this headline

Space between lines
is just about right
in this headline

We need to be aware, however, that too much space between lines creates islands of black in a sea of white. If the lines of a head are too far apart, unity will be destroyed. The reader can become confused or waste time figuring out what the head is trying to say. The lines need to be properly spaced and also kept together so that they can be seen as a unit.

Fig. 4-16 The traditional approach of line spacing is illustrated by the first headline. The last arrangement illustrates minus or negative leading, in which lines are overlapped. This approach can create an effective design unit, but care must be taken so that ascenders and decenders do not collide and destroy readability.

▼ Remember the Margins

The *margins* (the white space) that surround a printed product are important elements for creating an attractive and effective communication. Margins act as frames for the page. If the margins are too small, the framing effect can be destroyed. The page will look cramped and crowded. Reading will be more difficult. If the margins are large enough to be strong frames for the page, they help unify the page, hold the layout together, and make reading easier and more pleasant.

Try this. Take a column of type and trim the margins as close to the type as you can. Read it. Now read a column with the margins intact. See how much easier it is to read? Although the margins are nothing but white space they hold the column together.

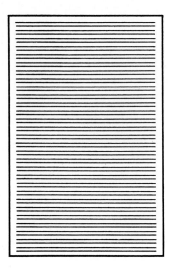

Fig. 4-17 Margins are too small on the left and the page looks crowded and unattractive. The larger margins on the right frame the type and give it light, at the same time unifying the page.

Margins can help set the mood of a printed communication. The traditional book margin is known as the *progressive margin*. This means that the smallest margin on the page is the *gutter* margin (the area between two adjacent pages). The top-of-the-page margin is larger. The outside margin is larger than the top, and, finally, the bottom-of-the-page margin is the largest of all. In other words, the margins are progressively larger around the page.

Progressive margins help give a look of careful consideration that denotes quality to the reader. Some "class" magazines use progressive margins for this quality look. Progressive margins are also suitable for brochures

Fig. 4-18 Progressive margins. The margins increase in size counter-clockwise on the left-hand page and clockwise on the right-hand page. The margin in the middle between the two pages is known as the gutter.

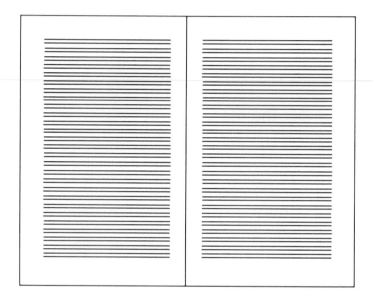

Fig. 4-19 Proper use of margins with a border. The outside rectangle indicates the page, the inner is a border around the type. On the left, the margins between the type and border and the border and edge of page are the same. They should be unequal. On the right, the border is closer to the type for greater unity of type and border—it is better to have smaller margins inside than outside.

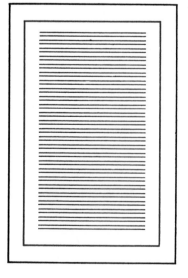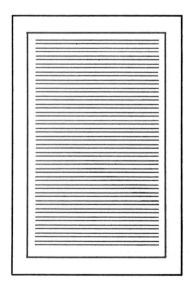

describing museums, art festivals, cultural events, or menus of exclusive restaurants.

How much space on the page should be devoted to margins? A good rule of thumb to keep in mind is: About 50 percent of the printed page should be devoted to margins. This sounds excessive—half the page for margins! But consider this. Suppose we are designing a layout for a 9 by 12 page. That means there are 108 square inches of space on the page. If the 50 percent rule is followed, 54 square inches will be devoted to margins. On a 9 by 12 page that works out to a margin of less than 1½ inches around the printed part of the page—not excessive at all for a 9 by 12 page.

▼ Select the Proper Type Size

So far, we have picked out a type style for our message, determined that the columns or blocks of type will be one and a half times the lowercase alphabet in width, designated 2 points of space between the lines of body copy, and have designed the printed area to be about 50 percent of the page. Now, what size type should we select?

If the type is too small, it will be difficult to read. If the body type is too large, we cannot get all the copy on the page. If the sizes selected for headings or display lines are too large, they will dominate the page and it will look like a sale bill. If they are too small and cramped, they will not give the emphasis desired and they will not attract the attention of the people we want to read the message.

As with all decisions regarding the most effective way to produce the best possible piece of printing, the decision as to what size of type to use depends on other factors. These include such decisions as the type design, the line width, the amount of leading, and the type of audience. The decision about type size cannot be made alone but must be coordinated with other legibility factors.

However, there are some general rules that are worth considering.

Remember being cautioned to "read the fine print" in contracts and legal documents? It is not just to save paper that some printed pieces contain much small print. It is a fact that the smaller the type the less likely it is to be read carefully. This is not to imply that small print means something devious is afoot. It is just to point out a truism of printing—the smaller the type, the more difficult it is to read. Rather than tackle the formidable task of reading small print, many people will skip it.

At the other end of the scale, reading matter set in too large type takes longer to read. It occupies a larger area and simply requires more time to cover.

As always, there are no fixed rules. However, in a general way it has been found that the most legible size for type in newspapers, magazines, and books is between 9 and 12 points. Reading matter set in smaller or larger type is more difficult to read. A number of tests made by researchers through the years, some of them back in the late 1800s, have indicated that 11-point type is the most legible size for constant reading.

There are, however, situations in which a larger type should be selected. Reverses (white letters on a black background) require larger sizes as well as clear letter designs (such as sans serifs or clean-line Romans) for easier reading. Children and older people find larger type sizes more pleasant to read. Thus the audience should be kept in mind.

Type style should also be considered when specifying type size. Most decorative and novelty faces and Scripts and Cursives should be set in larger sizes. The unfamiliar letter forms make reading and comprehension more difficult. Larger sizes help overcome these problems.

Condensed types, except in short messages, should be set in larger sizes than the easier-to-read regular widths.

Sometimes it is a temptation to cram as much material as possible into the space available by using smaller type sizes. Text type set in 10 point will take up about 25 percent less space than text set in 12 point. It is tempting to take advantage of this obvious economy. But remember that the objective is to have the message read and understood. If 12-point type will make the message easier to read and understand, it should be considered. If economy is important, the copy can be edited a little more tightly to preserve readability.

Fig. 4-20 Legibility of reading matter increases as the size of the type is increased, but 12 point is the maximum size for legibility. [Reprinted from Ralph W. and Edwin Polk, *Practice of Printing* (Peoria, Ill.: Bennett Publishing Company, © 1971). Used with the permission of the publisher. All rights reserved.]

The small sizes of body type are not as legible as larger ones. This paragraph is set in 6 point type, and it is far too small for ordinary reading matter, for it causes undue eye strain to make out the letter forms. Many of the ads now appearing in newspapers and periodicals contain types no larger than this and some are even harder to read than this paragraph is. Whenever possible, the use of such type should be avoided.

This paragraph appears in 8 point, and as you read it you are impressed with the greater ease with which it may be read. One may read it faster than the 6 point and with much less eye strain. However, it is still somewhat small for perfect ease in reading, as will be seen by a comparison with succeeding sizes.

When we get into 10 point type, we begin to get a sense of more comfortable, easy reading. The letters now are large enough that the eye can take them in at a rapid glance, without any strain or tension. This is a size of type found in many books and other reading matter, and we are quite familiar with it. Consequently, we read it with ease and speed. Ten point, leaded, is used for the body of this book.

We come to the highest degree of legibility when we consider the 12 point size of body type. A number of best authorities designate this size of type, leaded adequately, as offering the maximum of legibility in the mass, and being the most inviting to the eye of the reader. At any rate, the range of sizes for most satisfactory reading is around 10 and 12 point type.

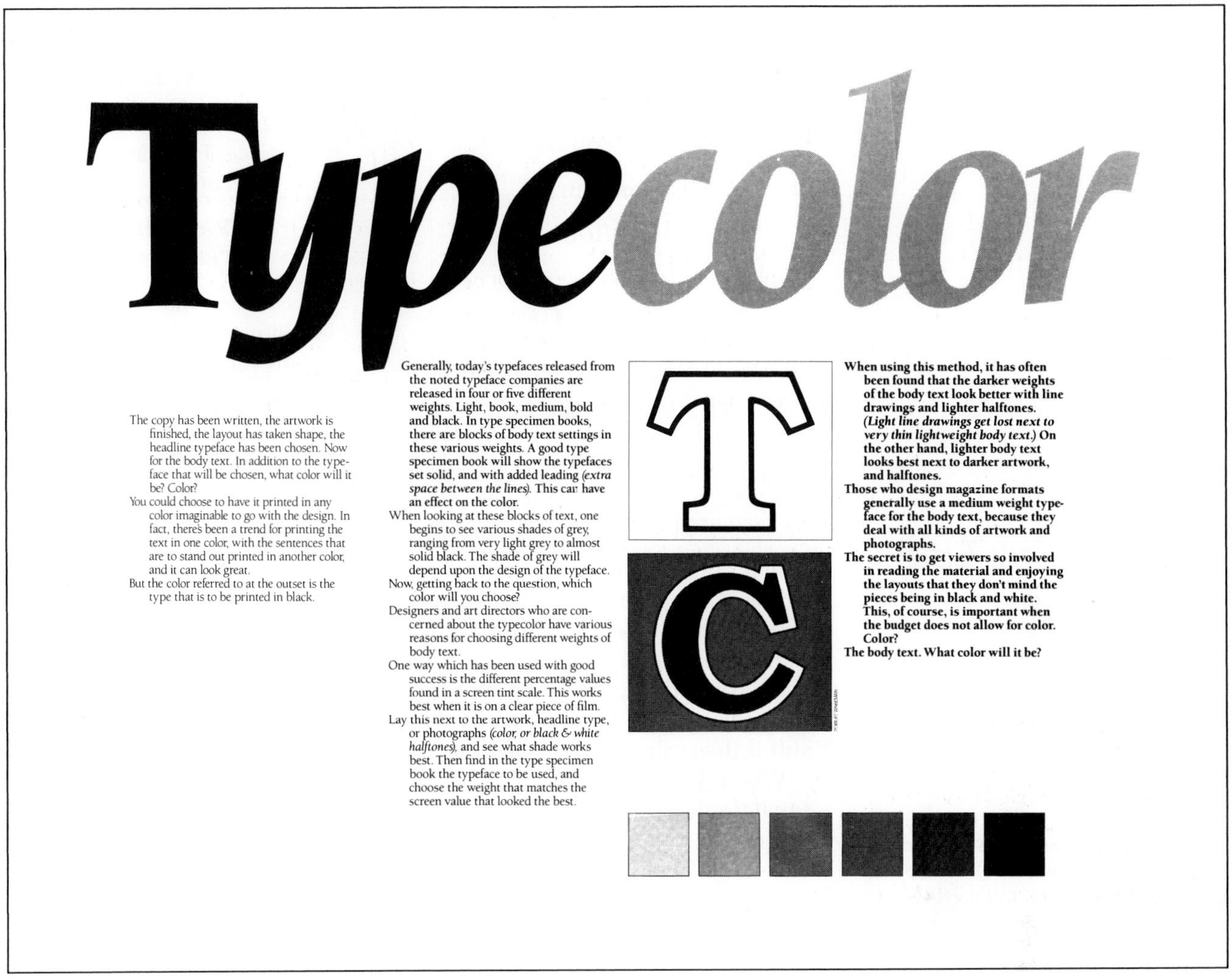

Typecolor

The copy has been written, the artwork is finished, the layout has taken shape, the headline typeface has been chosen. Now for the body text. In addition to the typeface that will be chosen, what color will it be? Color?

You could choose to have it printed in any color imaginable to go with the design. In fact, there's been a trend for printing the text in one color, with the sentences that are to stand out printed in another color, and it can look great.

But the color referred to at the outset is the type that is to be printed in black.

Generally, today's typefaces released from the noted typeface companies are released in four or five different weights. Light, book, medium, bold and black. In type specimen books, there are blocks of body text settings in these various weights. A good type specimen book will show the typefaces set solid, and with added leading (*extra space between the lines*). This can have an effect on the color.

When looking at these blocks of text, one begins to see various shades of grey, ranging from very light grey to almost solid black. The shade of grey will depend upon the design of the typeface.

Now, getting back to the question, which color will you choose?

Designers and art directors who are concerned about the typecolor have various reasons for choosing different weights of body text.

One way which has been used with good success is the different percentage values found in a screen tint scale. This works best when it is on a clear piece of film.

Lay this next to the artwork, headline type, or photographs (*color, or black & white halftones*), and see what shade works best. Then find in the type specimen book the typeface to be used, and choose the weight that matches the screen value that looked the best.

When using this method, it has often been found that the darker weights of the body text look better with line drawings and lighter halftones. (*Light line drawings get lost next to very thin lightweight body text.*) On the other hand, lighter body text looks best next to darker artwork, and halftones.

Those who design magazine formats generally use a medium weight typeface for the body text, because they deal with all kinds of artwork and photographs.

The secret is to get viewers so involved in reading the material and enjoying the layouts that they don't mind the pieces being in black and white. This, of course, is important when the budget does not allow for color. Color?

The body text. What color will it be?

▼ Mix Type Styles Carefully

Old-time printers knew what they were doing. They said, "If you start a job with Bookman, finish it with Bookman." They knew that the most attractive printed product was the one with the least amount of mixing of typefaces. Modern typographers call this the rule of *monotypographic harmony.*

Monotypographic harmony means that a harmonious layout will result if one family of type, or one design, is used throughout. Contrast and emphasis can be obtained by using type from different series of the same family, which will not sacrifice harmony. An attractive printed product will thus be produced with a minimum of typographic noise.

But there are times when faces must be mixed. Perhaps a certain emphasis or contrast is desired and a boldface or italic is not available in the family selected. Or we may decide a change of face is needed for contrast.

Fig. 4-21 Type comes in various weights—bold, medium, light, and so on—the weight you specify for the body text creates a "color" or shade of gray when the type is printed in black ink. Which weight or typecolor is best? In selecting a typeface for body text, designers suggest laying a tint scale—see illustration above—next to the artwork, headline type or photographs to see what shade works best. Then, select the typeface with the weight that harmonizes with the other elements in the layout.

Fig. 4-22 Many type styles in one layout, (left) cause disharmony and confusion. The example on the right, set in one family, is neat, attractive, and unified.

Display Lines

Differ from other elements in a

Piece of Printing

in that they attract

Attention

of the reader who should not be

Confused

by directing his attention to

TOO MANY DISPLAY LINES

Display Lines

Differ from other elements in
a piece of printing in
that they

ATTRACT THE
ATTENTION

of the reader who should not
be confused by directing his
attention to too many

Display Lines

𝔗𝔥𝔦𝔰 𝔦𝔰 𝔅𝔩𝔞𝔠𝔨 𝔏𝔢𝔱𝔱𝔢𝔯

This is Sans Serif

𝔗𝔥𝔦𝔰 𝔦𝔰 𝔅𝔩𝔞𝔠𝔨 𝔏𝔢𝔱𝔱𝔢𝔯

This is Sans Serif

Fig. 4-23 The Sans Serif at top is subordinated so that the two races are harmonious. When the Sans Serif is increased in size as in the bottom example, the two races begin to compete and harmony breaks down.

When families are mixed, care should be taken to avoid mixing families of the same race unless they are of a strongly contrasting design. For instance, the basic racial structures of Caslon and Garamond letters are the same: limited variation in the widths of letter strokes, slight diagonal slant at the thin point in bowls, and generally rounded serifs. However, each treats these basic characteristics differently. Confusion can thus result if these similar but different types are used together.

When faces are mixed, some means of contrast should be used to separate the two. One can be used for headlines and the other for body copy, for example. Or, one can be used in boldface or italic and the other in its regular form.

When two inharmonious types are used, one should be small. If a line of 24-point Black Letter is mixed with several lines of Sans Serif on the title page of a booklet, the Sans Serif lines should not be more than 10 or 12 point. The Black Letter, then, will dominate, and dissonance will be reduced to a minimum. This technique can often result in a very pleasing arrangement.

▼ Use All-Cap Lines Sparingly

There was a time when printers and typographers believed that the most legible letters were the old Roman capitals. It has since been found, without question, that lowercase letters are more legible than capitals.

Use capital letters, but use them sparingly. The reasons for this can be summarized by two points: Lowercase letters have more character and they speed reading.

"Lowercase letters have more character" means that the letter forms of each letter in the lowercase alphabet are distinctive and not as easily confused as capitals, which have many similarities. Contrast, for example, the lowercase *c* and *g* with the capital *C* and *G*. There is less likelihood

cg   CG

Fig. 4-24 Lowercase letters have more character and are less likely to be confused than capital letters.

ALIGN-   ALIGN-
MENT   MENT
OF TYPE   OF TYPE
SHOULD   SHOULD
ALWAYS   ALWAYS
BE DONE   BE DONE
OPTICALLY,   OPTICALLY,
RATHER   RATHER
THAN   THAN
MECHANI-   MECHANI-
CALLY   CALLY

MECHANICALLY ALIGNED   OPTICALLY ALIGNED

Fig. 4-25 Quite often the alignment of letters and lines should be done optically and not left to the typesetting equipment to do it mechanically. Notice how the lines MENT, BE DONE, RATHER, and MECHANI- have been moved slightly to the right to present better visual alignment. (From *Form & Function*, Adobe Systems Inc.)

of the reader confusing the letter forms in the lowercase line than in the all-capital line.

Research has shown that lowercase letters increase reading speed and are more pleasing to the reader. They do not tire the eye as easily as all-capital letters. However, words or lines set in all caps can be important design elements in headlines, headings, title pages. They can also be used to emphasize points in a layout.

In the instances where all caps would enhance a layout, designers suggest that they should be limited to two or three lines at a time and the lines should be short. The lines should also have ample leading.

Lines set in all caps must always be used with care. You will find that, unlike lowercase letters, capitals will naturally have inconsistent letterspacing when they are set into words. Often you will need to make very subtle changes in the letterspacing so that the space between the letters appears consistent and equal. It may help to imagine that you are filling the spaces between letters with grains of sand. The goal of proper letterspacing is to place an equal amount of sand between each letter. This may require reducing space in some cases and increasing—always slightly— in other cases.

Fig. 4-26 Lines of Black Letter or ornamental types should never be set in all-capital letters: They can be virtually unreadable.

CREATIVE TYPOGRAPHY

Sometimes designers will use capitals with extreme letterspacing (see, for example, the chapter numbers on the opening page of each chapter of this textbook). If you want to try this, remember to adjust letterspacing optically first, then mechanically add equal space between the letters. You will also need to increase word spacing to keep the type legible. Legibility is always the goal.

CHAPTER TWELVE

C H A P T E R   T W E L V E

NEGATIVE SPACE

Look at the space in, and around, and in between the letters above. This is called the negative space. But good negative space does not come about simply by spacing out each letter equally, or by doing so mechanically. This creates irregular space between letters. The negative space must be given some uniformity of visual weight around the letter, in order to be balanced and pleasing to the eye.

Good designers and typographers gain, over the years, experience in knowing how much negative space to put between each two letters. This means more than just putting letters next to each other. Capital letters are more difficult to do than the lowercase letters. The same holds true for the thinner weights. The larger the letter, the more important the negative space. The negative spaces between the serif typefaces are more graceful and easier to accomplish, design-wise, than the sans-serif typefaces. The latter almost require the skill at geometry of an architect.

Certain fundamental principles should be taken into consideration when dealing with negative spaces. Letter combinations that have almost no space at all between them. Vertical stroke against vertical stroke. This includes such combinations as **AW, AV, IM, IN, II,** etc. These are often put too close together, especially when setting sans-serif typefaces. The trained eye will allow enough space in such combinations to complement the other letters in the word.

Again, there are certain letter combinations with unavoidably large negative spaces. For example, **LA, TT, IT, TY, LL, LE,** and the like. They can be overlapped slightly, or altered, to cut down on the negative space between letters, and the rest of the word can be spaced out visually to please the eye. This does not mean, however, that all other letters in that word are to have the same amount of space between them as the examples above. But one has to take into consideration the space inside and around each letter to get the proper negative space. (*The combinations would be verticals next to curves; curves next to curves; and open curves next to verticals.*)

Today's great type designers are masters at what they do. They spend hours creating masterpieces out of letters of the alphabet. Each letter must have its uniqueness, but not be too expressive, so as to overpower the other letters in the alphabet. There should be a good relationship between the upper and lowercase characters as well. The type designer is also concerned with the balance of the black positive areas of each letter, and how to capture harmonious whites (*negative space*) inside the letters, as well as between them.

One well-known designer and art director in the New York City area always looks at the negative space first, before looking at the positive areas.

So, when taking all of the above into consideration, the creative person can enhance the beauty of the typeface design by the proper use of the negative space. The result? Typographical masterpieces!

Fig. 4-27 The creative person can enhance the beauty of type set in all caps by the proper use of the negative space—or the space in, around, and between letters. Lowercase letters are carefully designed to fit together properly in any combination and almost never need any letterspacing. Capital letters, however, will naturally have inconsistent spaces between them when they are set into words. Proper letterspacing requires study and a little practice.

▼ Do Not Be Boring

It is getting harder and harder to grab and hold the audience's attention. People today are bombarded from morning to night by efforts to make them listen or read. And there are scores of activities to fill leisure hours. In addition, people are becoming more and more conditioned to receiving information in short bits and takes. The average television news item is from 90 to 110 seconds long—barely 100 to 150 words. A half-hour television drama packs a complete story line from introduction to climax into about 22 minutes.

When modern readers are confronted with columns of reading matter, most will rebel. The reading matter looks forbidding and boring—there's nothing to catch the eye but line after line of type similar in size and tone. There is a temptation to skip it.

Long body matter can be made more attractive. There is no better way than to use that old standby, white space. It is so easy and yet so many editors and designers seem reluctant to take advantage of this natural way to brighten a page.

If fairly narrow columns are used on a page (say, ones that do not exceed about 12 picas), three columns of type can be used in a four-column space, increasing the white space between columns. Five columns of type can be set in a six-column space.

Occasional indented paragraphs can also help. Or, if type is set all flush left with no paragraph indentation (the usual indentation is 1 em for lines up to 18 picas wide), an extra lead or two of space between paragraphs can be used. Shorter paragraphs can also help, especially when narrow columns are being used. Long paragraphs of reading matter slow reading and look formidable.

Subheads with a contrasting face or boldface or a larger size of type than the body type can help break up the gray flow of type. Occasional italics where appropriate can also make the page more inviting. They will give emphasis where it is wanted, as well.

Sometimes variety can be obtained if copy is set *ragged right*. This means setting the left lines even but allowing the lines to end at natural breaks instead of justifying them. Studies have shown that ragged right does not slow reading. But ragged left should be used with caution. It is fine for limited use such as in the cutlines that identify illustrations, but it does slow reading, and most readers will skip long messages set in ragged left.

Sometimes, long body copy can be broken up with typographic devices—bullets, stars, dashes, and other dingbats. But they should be used cautiously. When overdone they disfigure the page and make the layout look amateurish.

Today, the concentrations are not quite as spectacular, but this same part of Utah, now known as the Bear River Migratory Bird Refuge, is still a powerful magnet for nomadic creatures. In autumn, visitors can see a half-million ducks and geese in a single day at the 65,000-acre federal preserve.

Located at the eastern edge of the Pacific Flyway—one of four major north-south migration routes in the U.S.—the refuge also serves as the crossroads for birds flying into the central states. Some 200 species.

Today, the concentrations are not quite as spectacular, but this same part of Utah, now known as the Bear River Migratory Bird Refuge, is still a powerful magnet for nomadic creatures. In autumn, visitors can see a half-million ducks and geese in a single day at the 65,000-acre federal preserve. Located at the eastern edge of the Pacific Flyway—one of four major north-south migration routes in the U.S.—the refuge also serves as the crossroads for birds flying into the central states. Some 200 species.

Fig. 4-28 Type set ragged left (left) should be avoided except in short sections because it slows reading. Type set ragged right (right) has the same reading ease as justified (even margins both left and right sides) type.

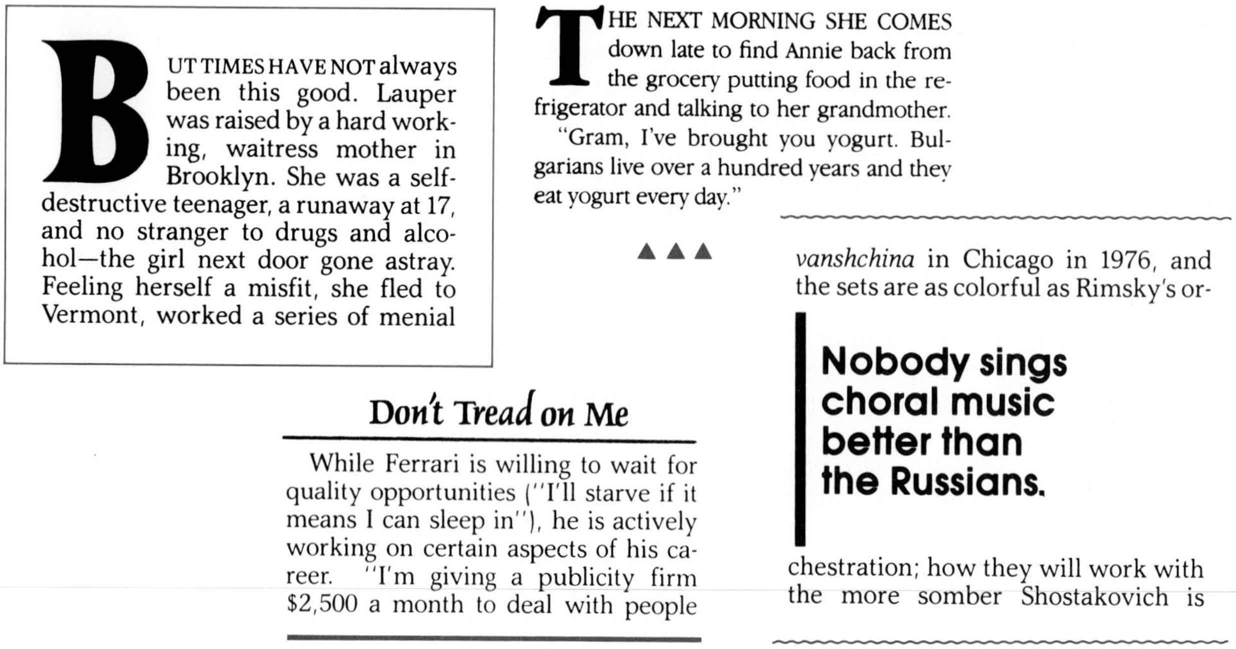

BUT TIMES HAVE NOT always been this good. Lauper was raised by a hard working, waitress mother in Brooklyn. She was a self-destructive teenager, a runaway at 17, and no stranger to drugs and alcohol—the girl next door gone astray. Feeling herself a misfit, she fled to Vermont, worked a series of menial

THE NEXT MORNING SHE COMES down late to find Annie back from the grocery putting food in the refrigerator and talking to her grandmother.

"Gram, I've brought you yogurt. Bulgarians live over a hundred years and they eat yogurt every day."

▲ ▲ ▲

Don't Tread on Me

While Ferrari is willing to wait for quality opportunities ("I'll starve if it means I can sleep in"), he is actively working on certain aspects of his career. "I'm giving a publicity firm $2,500 a month to deal with people

vanshchina in Chicago in 1976, and the sets are as colorful as Rimsky's or-

Nobody sings choral music better than the Russians.

chestration; how they will work with the more somber Shostakovich is

Fig. 4-29 Some examples of ways to break the flow of type to make a page look more inviting.

Initial letters can add spice to a page. These large letters at the start of paragraphs are a valuable typographic device. They can brighten up the page and they can help guide the reader.

Designers have a few suggestions for using initial letters. Initials should never be used at the top of a column except at the beginning of an article. When an initial is used at the top of a column it is a signal to the reader that this is the point where reading should begin. If an initial is used at the top of a column that does not begin the article, the reader might still think it is the beginning point and be confused.

Also, care should be used in placing initials so they do not appear side by side in adjacent columns when the copy is set in type. Initials should be scattered so they create unity and balance and their weights serve a design purpose.

▼ Avoid Oddball Placement

This is the easiest rule of all to follow. All this recommendation means is that types were meant to be read from left to right in straight lines. Curved lines, horizontal lines, and lines on an angle should be avoided. Of course, there are times when it can be effective to set type in a curve or on an angle to create a special effect or in certain situations such as designing logos and seals.

Oddball placement should not be done just to be different. And it should be done with caution. Lines of type in curves or on an angle should be kept to a minimum—no more than two or three words, if possible.

Communicators should avoid typographic affectations. Of course, Edward Estlin Cummings (e. e. cummings) used eccentricities of language, typography, and punctuation as a trademark for his works of poetry, and he became famous. But for straightforward professional printed communications, unusual gimmicks should be avoided unless such devices will really improve communication. If they are used, they should be limited and considered, as one designer points out, like jewelry for the typographic dress. They can be easily overdone.

▼ Typographic Color

Decorative elements that can add "color" to layouts include several miscellaneous typographic devices. These are borders, both straight line and decorative, ornaments, and decorative letters. The straight line borders, whether they are a single line or several lines together, are called *rules*. Rules with three or more lines of the same or varying thicknesses are known as *multiple rules*. Rules that are composed of a series of short vertical lines are called *coin-edge rules.*

There was a time when flowery ornaments and borders were considered a must in printing. The monks who illuminated hand-lettered manuscripts with colorful initial letters and curlicues produced beautiful pieces of art. Their skills were carried over into the early days of printing when letterpress-printed pages were often embellished with hand-drawn colorful ornamentation.

Fancy ornaments and borders in printing went along with ornamentation in other art forms. During the age of gingerbread turrets and scroll work in buildings, fancy letterings and ornamentation were popular in printing. But today the trend is toward simplicity, boldness, and functionalism. We are told to keep it simple, but give it punch. If it doesn't serve a function, leave the rule or ornament out.

Ornaments should not be discarded out of hand, however. Proper use, which is to say restraint, in adding borders and ornaments can add decorative relief to a printed page that might otherwise be mechanical and deadly severe.

The fancy curlicues and shaded borders of another era have been replaced with a simpler decorativeness that better fits today's design techniques and typography. A little added ornamentation can be much like a dash of spice in a cooking recipe—it can make the typographic production sparkle.

Borders should extend the basic design of the art and type selected. An *Oxford rule* (parallel thick and thin straight lines), for instance, goes very well with the thick and thin lines of Bodoni or other modern Roman types with good contrast between thick and thin strokes and straight-line serifs. Single-line solid rules go well with the monotonal and simple Sans Serifs. A simple line drawing requires a simple straight-line rule. A decorative illustration requires a decorative border. The design and size of ornaments and borders should be in harmony with the design and size of the types and illustrations in the layout.

Fig. 4-30 Creative integration of art and type can cause unity and attention in a layout. (Courtesy Metro Associated Services, Inc.)

Fig. 4-31 Borders and rules should be selected to harmonize with the art and type styles used in the layout. (Courtesy Westvaco Corporation)

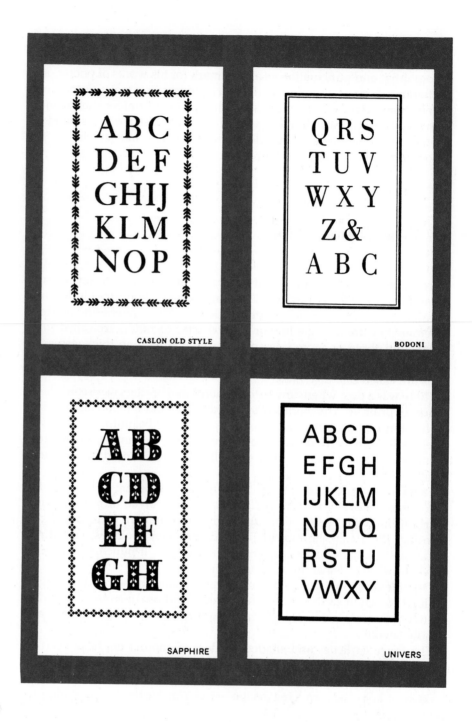

When borders and decorations are printed in color, the color value should be in tune with all the other elements in the layout. Decoration that may appear harmonious printed in black ink will appear weak and ineffective if printed in pastel or light colors. Also, if solid or screened backgrounds in color are used, care should be taken to be sure the colors are light. Deep color for tint blocks or screened backgrounds will overshadow the type and make it difficult to read.

Generally, borders and decorations to be printed in color should be specified in a little heavier weight than if they are to be printed in black. This will give them more body and they will harmonize better with the black type.

When borders are placed around type masses, the borders should be closer to the type than to the margins of the page. Borders and type should be handled as one unified typographic element.

Swash letters are usually "stretched-out" versions of regular letter forms that end in an ornamental flourish. Swash letters are used to add a touch of distinction to a logo or a company name or to create a graceful image for an invitation, an announcement, or a column heading.

Other exaggerated letter forms, such as abnormal extensions of ascenders and descenders or decorative initials, are also available.

But all of these ornamental letter forms should be used sparingly. They should be used only for the first or last letter of a word and not for letters within words. Letters with extended ascenders or descenders are an exception to this rule. However, they, too, must be placed very carefully in the layout so that they add to the unity of the whole arrangement.

The opportunity to be creative with type is almost limitless. For example, modern typesetting equipment makes it possible to reduce spacing between words and lines to the point where the letters overlap and leading doesn't exist between the lines. When spacing is reduced to the point where lines overlap beyond the normal baseline of the letters, the procedure is called *minus leading*. When spacing is reduced between the words

ABCDELMN OPQSTUVW Notice We Lad

Fig. 4-32 The swash letters (above) are members of the Caslon family, in italic posture. Note how these decorative letters can be integrated with the regular Caslon italic letters.

Fig. 4-33 Considered one of the greatest typographic designers of all time, the late Herb Lubalin won many awards for his designs. These are some examples of his creativity with type.

or letters to the point where they butt against each other or overlap, the technique is called *minus spacing*.

In addition, type can be elongated, obliqued, shaded, expanded, placed in circles, and shaped in an almost limitless number of variations with electronic typesetting equipment.

These techniques allow communicators to be creative by adding unity, punch, boldness, or attention value to the type.

In selecting borders and ornaments, in adding typographic color to a layout, we must always remember that the best typography and design is that which is unobtrusive. Borders, ornaments, and type should blend with art, margins, white space, paper, and ink in a way that does not call attention to any one element. No graphic device should have such strong display value that it interferes with the purpose of the message.

Fig. 4-34 Computers and advances in phototypesetting equipment have opened opportunities for communicators to adopt new spacing techniques. Reverse leading, in which characters are aligned at their tops, is shown at the top. The word generics is composed with minus (or negative) letterspacing. The bottom example is composed with minus (or negative) leading. Care should be taken, though, not to destroy legibility when employing these techniques.

Dance Music From 30s
To 80s Contemporary Rock Disco
Jewish Israeli Folk Dance Hamish Show
Tunes Big Bands Pop Kleizmer Chassidic Latin
Soft Rock Middle Eastern Spanish Italian Yiddish
Schmaltz French Irish Greek Russian Jazz Classical
Oriental Muzak Continental Dance Music From 30s To 80s
Contemporary Rock Disco Jewish Israeli Folk Dance Hamish
Show Tunes Big Bands Pop Kleizmer Chassidic Latin Soft Rock
Middle Eastern Spanish Italian Yiddish Schmaltz French Irish
Greek Russian Jazz Classical Oriental Muzak Continental Dance
Music From 30s To 80s Contemporary Rock Disco Jewish Israeli Folk
Dance Hamish Show Tunes Big Bands Pop Kleizmer Chassidic Latin
Soft Rock Middle Eastern Spanish Italian Yiddish Schmaltz French
Irish Greek Russian Jazz Classical Oriental Muzak Continental Dance
Music From 30s To 80s Contemporary Rock Disco Jewish Israeli Folk
Dance Hamish Show Tunes Big Bands Pop Kleizmer Chassidic Latin Soft
Rock Middle Eastern Spanish Italian Yiddish Schmaltz French Irish Greek
Russian Jazz Classical Oriental Muzak Continental Dance Music From 30s
To 80s Contemporary Rock Disco Jewish Israeli Folk Dance Hamish Show
Tunes Big Bands Pop Kleizmer Chassidic Latin Soft Rock Middle Eastern
Spanish Italian Yiddish Schmaltz French Irish Greek Russian Jazz Classical
Oriental Muzak Continental Dance Music From 30s To 80s Contemporary Rock
Disco Jewish Israeli Folk Dance Hamish Show Tunes Big Bands Pop Kleizmer
Chassidic Latin Soft Rock Middle Eastern Spanish Italian Yiddish Schmaltz
French Irish Greek Russian Jazz Classical Oriental Muzak Continental Dance Mu-
sic From 30s To 80s Contemporary Rock Disco Jewish Israeli Folk Dance Hamish
Show Tunes Big Bands Pop Kleizmer Chassidic Latin Soft Rock Middle Eastern Span-
ish Italian Yiddish Schmaltz French Irish Greek Russian Jazz Classical Oriental Muzak
Continental Dance Music From 30s To 80s Contemporary Rock Disco Jewish Israeli Folk
Dance Hamish Show Tunes Big Bands Pop Kleizmer Chassidic Latin Soft Rock Middle Eastern
Spanish Italian Yiddish Schmaltz French Irish Greek Russian Jazz Classical Oriental Muzak Con-
tinental Dance Music From 30s To 80s Contemporary Rock Disco Jewish Israeli Folk Dance
Show Tunes Big Bands Pop Kleizmer Chassidic Latin Soft Rock Middle Eastern Spanish Italian Yiddish
Schmaltz French Irish Greek Russian Jazz Classical Oriental Muzak Continental Dance Music From 30s To 80s
Contemporary Rock Disco Jewish Israeli Folk Dance Hamish Show Tunes Big Bands Pop Kleizmer Chassidic
Latin Soft Rock Middle Eastern Spanish Italian Yiddish Schmaltz French Irish Greek Russian Jazz Classical Ori-
ental Muzak Continental Dance Music From 30s To 80s Contemporary Rock Disco Jewish Israeli Folk Dance
Hamish Show Tunes Big Bands Pop Kleizmer Chassidic Latin Soft Rock Middle Eastern Spanish Italian Yiddish
Schmaltz French Irish Greek Russian Jazz Classical Oriental Muzak Continental Dance Music From 30s To 80s
Contemporary Rock Disco Jewish Israeli Folk Dance Hamish Show Tunes Big Bands Pop Kleizmer Chassidic Latin
Soft Rock Middle Eastern Spanish Italian Yiddish Schmaltz French Irish Greek Russian Jazz Classical Oriental
Muzak Continental Dance Music From 30s To 80s Contemporary Rock Disco Jewish Israeli Folk Dance Hamish
Show Tunes Big Bands Pop Kleizmer Chassidic Latin Soft Rock Middle Eastern Spanish Italian Yiddish Schmaltz
French Irish Greek Russian Jazz Classical Oriental Muzak Continental Dance Music From 30s To 80s Contem-
porary Rock Disco Jewish Israeli Folk Dance Hamish Show Tunes Big Bands Pop Kleizmer Chassidic Latin Soft
Rock Middle Eastern Spanish Italian Yiddish Schmaltz French Irish Greek Russian Jazz Classical Oriental Muzak
Continental Dance Music From 30s To 80s Contemporary Rock Disco Jewish Israeli Folk Dance Hamish Show
Tunes Big Bands Pop Kleizmer Chassidic Latin Soft Rock Middle Eastern Spanish Italian Yiddish Schmaltz French
Irish Greek Russian Jazz Classical Oriental Muzak Continental Dance Music From 30s To 80s Contemporary Rock
Disco Jewish Israeli Folk Dance Hamish Show Tunes Big Bands Pop Kleizmer Chassidic Latin Soft Rock Middle
Eastern Spanish Italian Yiddish Schmaltz French Irish Greek Russian Jazz Classical Oriental Muzak Continental
Dance Music From 30s To 80s Contemporary Rock Disco Jewish Israeli Folk Dance Hamish Show Tunes Big Bands
Pop Kleizmer Chassidic Latin Soft Rock Middle Eastern Spanish Italian Yiddish Schmaltz French Irish Greek Rus-
sian Jazz Classical Oriental Muzak Continental Dance Music From 30s To 80s Contemporary Rock Disco Jewish
Israeli Folk Dance Hamish Show Tunes Big Bands Pop Kleizmer Chassidic Latin Soft Rock Middle Eastern Span-
ish Italian Yiddish Schmaltz French Irish Greek Russian Jazz Classical Oriental Muzak Continental Dance Music

Fig. 4-35 The designer may use creativity in type arrangement to transmit an idea by form as well as by meaningful words. (Courtesy Granite Graphics)

▼ Effective Design Checklist

- The design of a letter can "talk" to the reader. Select a type style that will reinforce the message.
- In general, Romans are preferred for body type.
- One and a half times the width of the lowercase alphabet is considered the proper line width for easy reading.
- Words crowded too closely together or spaced too far apart are difficult to read.
- One or two points of leading or space between lines is about right for reading matter (body type).

- As the length of the line is increased, the space between the lines should be increased.
- Generally, the larger sizes of body types are the easiest to read. Stick to between (and including) 9 and 12 point for body type.
- Margins that together occupy about 50 percent of the area are considered about right.
- Use a larger-size type for reverses. Reverses are more effective if the types are simple in design. Fine serifs should be avoided.
- Most Miscellaneous, Novelty, decorative, and condensed types should be set in larger sizes for better legibility.
- Type styles should be mixed with care. When two very different styles are used, one should dominate.
- In most cases, families of the same race of types should not be mixed.
- All-capital lines are difficult to read and their use should be limited.
- Long amounts of body copy should be broken up for easier reading. Subheads, indented paragraphs, and other devices can help.
- Ragged right does not seem to reduce readability, but ragged left does and should be used sparingly.

▼   Graphics in Action

1. Find examples in publications of improper use of type and explain what is wrong with them. Mount the examples on plain white paper with your analysis included.
2. Explain the points to keep in mind for selecting and arranging type. Illustrate the points with examples clipped from publications.
3. Find examples of type styles that appear to be poor choices for the messages used. Then find examples that appear to be excellent selections. Analyze the examples and support your decisions regarding them.
4. Select typefaces from those in the appendix (or other sources of type specimens) that seem best suited to help say the following (or similar) phrases:

    Country Barn Dance
    Paris in the Spring
    Trans-Siberian Express
    Notre Dame Cathedral
    Final Clearance Sale

5. Tear out an assortment (fifteen to twenty or so) of small pieces about 1 by 2 or 2 by 3 inches each of body matter type from publications (nothing larger than 12 point). Paste them (use rubber cement) in a free-flowing arrangement to make a collage on plain white paper. Compare the types and select the ones that appeal to you most. Explain the reasons for your selections (type design, size, space between lines, big on body, and so on). Identify the types discussed by race (and family, if you can).
6. Build an idea file of creative uses of typographic color from examples found in publications. Include a section for poor examples or techniques to be avoided.

7. Draw a rectangle approximately 5 by 7 inches on an 8½ by 11 or 9 by 12 sheet of white drawing paper. Sketch your name in a letter style that will help identify you as possessing the following characteristics. Embellish the lettering to better communicate the message. Use a separate sheet of paper for each example.
    a. I weigh 289 pounds.
    b. I weigh 98 pounds.
    c. I am a carpenter.
    d. I am a (use your profession or major).
    e. My hobby is (use actual hobby).

Notes

[1] Silvano Arieti, *Creativity, The Magic Synthesis* (New York: Basic Books, 1976), p. 4.

[2] "The 15-Second Course in Creativity," *Step-by-Step Graphics*, November/December, 1987, p. 108.

[3] "How Designers Create," *Design, the Magazine of the Society of Newspaper Design*, May 1988, p. 6.

[4] "The Women Who Saved New York," *Print*, January/February, 1989, p. 61.

[5] Ibid, p. 71.

[6] Frederic W. Goudy, *Typologia* (Berkeley: University of California Press, 1940), p. 80.

[7] Miles A. Tinker, *Legibility of Print* (Ames: Iowa State University Press, 1963), pp. 44–66.

[8] Daryl R. Moen, *Newspaper Layout and Design* (Ames: Iowa State University Press, 1984), p. 176.

Art and Illustrations

Photograph: Mark Jenkinson

Of course, it takes more than type printed on paper to make a successful communication. There are a number of other elements, or tools, that can be used to make a message more attractive and effective. These elements include art, ornaments and borders, rules, paper, color, and even the ink used in printing. Each can add a dimension to the final product.

Conversely, these elements can ruin the effectiveness of a printed piece if applied haphazardly or used improperly. The communicator who works with type, much like an artist, interior decorator, architect, or any other creative person, must know the characteristics of all the elements and what each can contribute to the finished product. And the communicator must know how to use them.

It is not necessary to know the technology of making these various elements, however. For instance, it is not necessary to know the properties of the chemicals used in engraving, or the parts of the plate-making device for making a printing plate. But it is necessary to know what can be done to a photograph to make it printable. The communicator must know how to prepare material for production.

Quite often art, illustrations, and photographs can be obtained and used effectively at a relatively low cost. We consider a number of sources in this chapter such as suppliers of stock photographs, clip art services,

Fig. 5-1 The key to effective use of art, as in all design, is creativity. "That classic charm" is a poster illustrating creative use of clip art. Designed by McCool & Company, Minneapolis. "Vision" combines photography with type (note the use of a reverse). Communicators will find the philosophical message in "Vision" worthwhile to contemplate. This is one in a series of self-promotion posters produced by Bertoldi Design, Dean Bertoldi designer/art director.

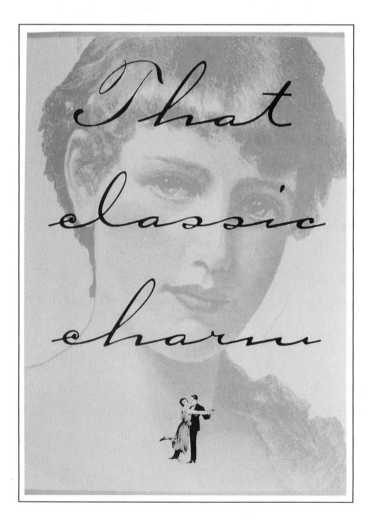

and computer illustration programs. Proper use of art requires knowledge more than money, in many cases. A photograph that reinforces and amplifies the message costs no more than those used simply as decorations.

Fig. 5-2 Enlighten, clarify, adorn—these are words Webster uses to define illustration. This effective illustration for a book cover using only type and rules was created by Paula Scher.

▼ This Is A Graphics Age

We are living in an age of graphics. Greater attention is being paid to the role of art in printed communications. Artists are moving into the editorial rooms of newspapers. Wire services are supplying more graphics to their clients.

"Graphics more and more are becoming part of a newspaper's face. They are a different visual dimension at a time when newspapers are increasingly more than just type with a picture here and there to break up the gray," a lead article in the *AP Log*, the in-house publication of the Associated Press, reports.[1]

"We live in a more visual world; television does great graphics; people are more cognizant of color photography; newsmagazines have established trends in informational graphics," Richard Curtis told members of the American Society of Newspaper Editors.[2] (Curtis is the managing editor for graphics for the newspaper *USA Today*.)

Informational graphics, Curtis pointed out, are graphics that both illustrate and inform. He noted that editors who learn about graphics can produce publications that will be better vehicles for communicating information. Some information is communicated more effectively through graphics than words.

Graphics can improve readability; they can clarify, attract attention, add realism to writing. They can explain complicated information and provide data in an easy-to-understand way. They can give the reader a sense of place. They can explain the size of items quickly and easily.

Graphics can not only amplify information, but also set a mood, and they can help create a pleasing design. Of course, graphics can entertain. How often have you leafed through a magazine and read the cartoons before settling down to the serious articles?

From Idea to Printed Page

To obtain the best results, we must understand how art gets from an idea to the printed page. We need to learn ways to make art and photographs more effective. In addition, we need to know how to crop and size art, the various ways cutlines can be handled, and how to pass along instructions to the printers so that we obtain the results we want. Finally, communicators need to know how to instruct a computer to produce finished layouts.

If this all sounds formidable, don't despair. It is a rare professional illustrator who can handle *all* the art forms and techniques available. Some designers may be superior photographers. Others may be great at retouching and finishing.

Fig. 5-3 Graphics can translate complicated information quickly and create high reader interest. This graphic was produced by the staff of *USA Today* to demonstrate effective visual communication for an American Society of Newspaper Editors seminar.

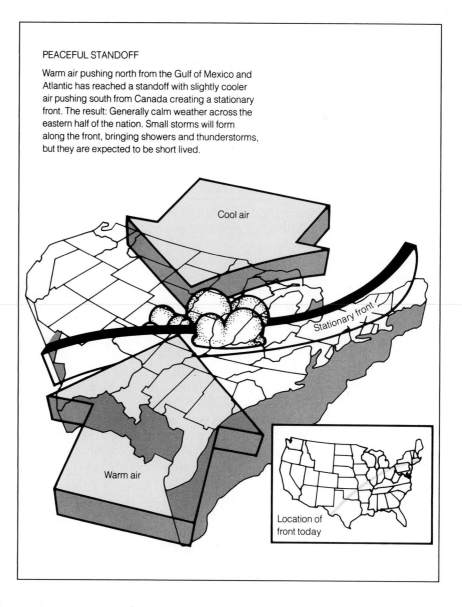

PEACEFUL STANDOFF

Warm air pushing north from the Gulf of Mexico and Atlantic has reached a standoff with slightly cooler air pushing south from Canada creating a stationary front. The result: Generally calm weather across the eastern half of the nation. Small storms will form along the front, bringing showers and thunderstorms, but they are expected to be short lived.

Cool air

Stationary front

Warm air

Location of front today

Fig. 5-4 This illustration of an owl (from a clip art service) was used in a television graphic to give added impact to the late show announcement.

It is unlikely that the editor of a small publication, or the public relations professional, or the newspaper editor will also be a highly skilled designer who can handle successfully all the various techniques involved in originating and preparing art for printing. However, an effective communicator must have a knowledge of what can be done graphically. Basically this means knowing how to obtain, process, and place art in a layout so that it will be affordable and effective.

Where to start?

A communication must attract the attention of its target audience. Judicious employment of an illustration can help.

Do not forget that this is a visual age. Children are propped in front of a continuous electronic visual parade before they can read or write. More people watch television than read books. This can work to our advantage. Since people are oriented toward the visual, we can use this orientation to make them want to read. Our use of art that amplifies words can help us achieve this goal.

Printed communications must be powerful indeed to hold readers without the use of art. People will usually avoid words alone, unless those words really tell them what they want to hear. For instance, studies by the Gallup Applied Science Partnership, using a device called Eye-Trac, indicate readers use photos and graphic images as entry points into a printed page.[3]

But if art is used, it must be of high quality. It must be simple, well done, and get to the point quickly. Superfluous art can be worse than no art at all. It can distract the reader and dominate the layout. It can communicate the message "Here's a sensational picture, look at it and don't bother to read the words on this page."

Keeping Graphics Simple

Keeping graphics simple leads readers to a quicker and more comprehensive understanding of the information, the graphics staff of *USA Today* told editors. They pointed out in their publication *ASNE Today* (May 1983) that "if the graphic is easy to produce and easy to read, the information within will be easily understood."

Art can help the communicator do a number of things. It can help set the mood for the message. For instance, it can promote a feeling of peace and contentment. It can say "Sit up and pay attention to this," or "Sit back, relax, and enjoy."

Want peace and serenity? How better to portray it than with a scene of a couple observing a quiet lake at sunrise or a calm sea at sunset?

Want to show discontent? A picture can create this mood more effectively than perhaps the legendary 1,000 words. But the picture must be selected carefully and properly cropped and processed and unified with the other elements in the layout. An improperly selected and presented piece of art can be as ineffective as 1,000 poorly written words. Also remember that these mood pieces of art must tie in with the message being printed. They must not just be thrown in because they are attractive and the communicator believes readers will like them.

Art can also be effective in showing a situation or explaining a situation in sequence. And it can be used to establish identity. Most people are familiar with a number of trademark characters that have been used effectively by corporations. This identity value can be extended to printed communications, too. For example, the *Minneapolis Tribune*, when it was redesigned, adopted a simple symbol of a web-fed press and used it in nameplate, masthead, section pages, and column headings to create continuity and identity.

The first step in working with art, then, is to evaluate the message we want to tell. Can graphics do it better? Can art help the audience understand what we are trying to say? Will art attract attention? Will it make the layout more interesting, more entertaining? Will art help guide the reader through the message in proper sequence? Will it help create identity? Will the layout be improved with art?

If the answer is yes to any of these questions, the next step is deciding on the right piece of art and obtaining it.

Fig. 5-5 Simplicity is the key to effective symbol design. The *Minneapolis Tribune* used this simple symbol of a roll of paper going through the press to give immediate recognition to its publication. (Courtesy *Minneapolis Tribune*)

▼ Essentials of An Effective Graphic

Graphics serve two purposes: to restate the main point of a story visually or to serve as a stand-alone visual to accompany an article and elaborate on a point related to but not included in the article.

The essentials of an effective graphic are listed by Bill Dunn, graphics editor of the *Orange County* (California) *Register*, who has traveled the country making presentations to graphic seminars, include:

- *Selection*. What is the proper type of graphic? Should it be a chart, a diagram, or a map, for instance?
- *Headline*. An easy-to-read explanation, the shorter the better.
- *Explainer*. All graphics must contain a short paragraph that explains not only what the graphic is about but also why the information is important to the reader.
- *Body*. The "meat" of a graphic. This element transforms data into visuals that explain information clearly.
- *Source*. All graphics must carry a source line that identifies the primary origin of the information.
- *Credit*. All graphics must have credit lines. They should be included as part of the finished artwork by the artist.

▼ Approaches to Using Art

On large publications the editor doesn't have to worry about obtaining the right art; that's the art director's job. But even though editors may not have to be involved with the technical side of this part of the layout process, they do need to understand art and artists. Both people need to work together, and this is not always easy.

Too often editors are oriented toward words whereas artists think "design" first and communication of the message last. It is not unusual for editors and art directors to have differing opinions as they view the finished product from their individual perspectives.

This situation is changing as each develops a greater understanding of the value of teamwork in producing effective printed communications. Editors are becoming more knowledgeable about art and design, and designers are developing a greater understanding of how words are used and arranged for readability. This spirit of teamwork will go far in providing effective and attractive publications.

There are about 9,000 business and trade publications in the United States. Many of these are company papers and magazines aimed at employees or other special audiences. In addition, scores of newsletters are issued by various organizations and interest groups. Add to this small daily newspapers and the more than 8,000 weekly newspapers and you have a vast number of publications.

Many entry jobs in communications are in these small publications, and many successful and satisfying lifetime careers can be had in editorial and production positions with these publications.

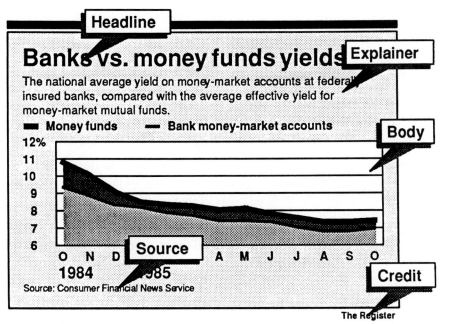

Fig. 5-6 An effective graphic includes five critical parts. The **headline** is an easy-to-read label (use as few words as possible); the **explainer** is a short paragraph that explains what the graphic is about and why the information is important; the **body** is the meat of a graphic; a **source** line must be included to give attribution of the information; the **credit** tells who produced the graphic. (Courtesy *Orange County Register*)

Fig. 5-7 (Below) Three basic types of graphics are: charts, diagrams, and maps. A bar chart is used to compare different items on a common scale. A pie chart is a visual presentation of separate portions that fit together to make 100 percent.

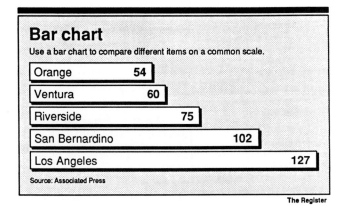

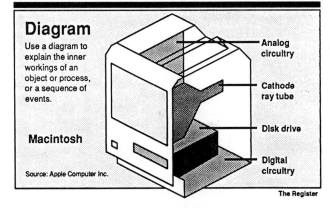

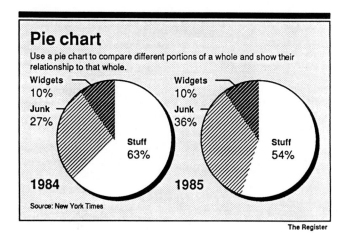

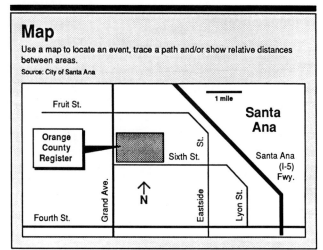

There are giant publications with their large editorial and design staffs and small publications with a single person performing all the editorial and design functions. There are firms and publications that can afford to hire free-lance artists and designers to assist them on a limited basis. And in all these situations the editor must know something about selecting and processing art.

Which Art to Use?

Once the decision is made to use art, the next step is to decide whether a photograph or line art will be most effective. *Line art* is a term for straight black-and-white images as opposed to photographs with continuous tones from black to white, including all the grays in between.

In selecting art, the key word, as in all things typographical, is *reader*. The audience must be kept in mind at all times. Consideration should be given to the sort of art a particular audience will relate to and find appealing. (This can be determined by research carried out during the planning stages for the communication.)

In addition to the content questions regarding the most suitable art for a layout, there are some technical qualities to check. Will the art produce well on the type of paper selected or the printing process to be used for the job? Photographs usually do not produce well on antique or coarse finished paper. They need paper with a smooth, even high gloss finish to bring out all of their definition and contrast. (Paper is discussed in Chapter 7.)

Fig. 5-8 Symbolic art can create identity and unity. The Southern California Edison Company uses symbols to indicate types of energy in its annual report. The symbols are grouped to communicate the complete energy message and then used individually to identify sections of the report dealing with each type of energy.

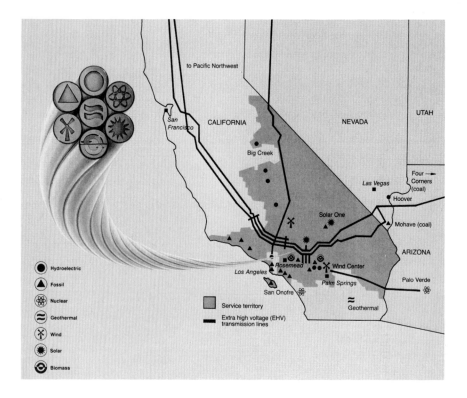

One approach to selecting the proper art is to consider the content of the message. Some art directors like to use photographs for nonfiction to reinforce the sense of realism they wish to achieve. Photographs transmit a sense of authenticity. Some art directors like to use drawings for fiction. Schematic drawings are often helpful. Charts and graphs help explain complicated economic information. Maps illuminate national weather conditions and foreign affairs.

Nigel Holmes, chart designer for *Time* magazine, has the job of reducing complex news and events involving statistics to attractive and easy-to-understand art forms, mainly charts. The challenge, Holmes says, is "to present statistics as a visual idea rather than a tedious parade of numbers. Without being frivolous, I want to entertain the reader as well as inform him."

Sources of Art

Today about the only restricting factor is the limit of the imagination of the communicator. Those who do not have an art director to do the job or the funds to hire a professional artist, photographer, or photo researcher should learn to locate art themselves.

A number of firms supply *stock art*. This is art that is prepared and sold in a variety of sizes, shapes, and subjects. The firms usually offer *clip art* books or catalogs. These are available in scores of subjects from seasonal topics to special events. Each book contains art of various sizes that can be cut out and placed in a layout. They are camera ready and simple to use.

Then there are numerous clip art programs for computers, and communicators using desktop publishing programs can add a graphics program that will provide scores of illustrations.

Fig. 5-9 Clip art is available in proof books for actual clipping and placing in layouts or in computer programs for integrating in a layout created on the screen. This is an example of computer clip art from ImageBase, an electronic art program by Metro Creative Services.

Fig. 5-10 A wide range of historical clip art is available from Dover Publications at relatively little cost. These examples are from: (top) *Old English Cuts and Illustrations,* by Bowles and Carver, and (bottom) *Baroque Ornament and Design,* by Jacques Stella.

Many designers consider clip art an inferior substitute for art produced for a specific layout situation. They note that often it is difficult to find just the right illustration for the creative idea in a clip art collection and that the quality of work in some collections is inferior.

However, the facts of life for the communicator often include a tight budget. Art is becoming increasingly costly and the ideal of using the services of a professional designer or artist may not be the reality. There are alternatives.

Good sources of illustrations, often at no cost at all, include public relations firms, chambers of commerce, state departments of tourism and economic development, and museums and historical societies. Quite often art can be borrowed from other publications; usually they ask for only a modest fee or that a credit line be given.

▼ The Two Basic Types of Art

Art comes in two basic forms—*illustrations* and *photographs. Line art,* such as pen-and-ink drawings, consists of definable lines of black and white space. There are a number of techniques that the artist can employ to give a shaded effect to these drawings, but basically they consist of just black lines or dots and white space.

A photograph consists of black and white and all the tones between—thus the term *continuous tone* art. The difference between continuous tone art and line art is important to understand because each requires different treatment.

No matter what printing method is used, continuous tone art must be converted to line art, or art that gives the illusion of continuous tone. This is done by photographing the image through a finely ruled screen. This produces a negative that consists of tones reduced to a dot formation. Since it is believed that half of the original image is lost in this process and half of the full tone remains, the resulting printing plate is called a *halftone*.

In the world of desktop publishing advances are being made in scanners and digital printing that can simplify continuous tone art production. These are discussed in the chapter 10.

Fig. 5-11 One of the first art decisions to be made is whether to use continuous tone art, as in the photograph above (left), or line art. Line art (right) can give the illusion of continuous tone through shading but it is only blacks and whites. (Line art courtesy Dover Publications, Photograph: Richard Anderson)

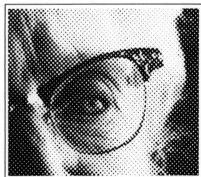

Fig. 5-12 When this photograph is processed for printing it will be rephotographed through a screen and the resulting negative used to produce a printing plate. The enlarged area shows the configuration of dots that gives the illusion of continuous tone. (Photography: Richard Anderson)

Fig. 5-13 (Right) Look closely at this photograph and you will see that it consists of continuous tones from black to white (or light to dark). When it is screened, the resulting dot structure permits printing to a very close approximation of the tones of the original. (Photography: Richard Anderson)

Fig. 5-14 The prints below illustrate the effects of various halftone screens. The 85-line screen is used for printing on rough paper such as newsprint, and the very fine 200-line screen is used for very smooth-finished or high-gloss papers. 150-line screens were used to print the halftones in this text. (Photography: Richard Anderson)

85-line screen

100-line screen

150-line screen

200-line screen

The screen used for making halftones comes in a variety of numbers of lines per inch, ranging from about 65 to 200. The greater the number of lines per inch, the finer the detail of the printing plate. For high-quality reproduction, a fine screen and a high-gloss, smooth-finish paper is used. A plate made with a screen of about 135 lines per inch will produce an excellent result.

If a fine-screen halftone is printed on coarse paper, such as newsprint, the resulting print will look muddy. Most newspaper halftones are made with about an 85 screen for best results.

In making plates from line art no screen is used and the resulting plate is an exact duplicate of the art.

Photography is playing an increasingly important role in printed communications. It is a major tool in converting art to printing plates, and it can be used to produce a vast array of special effects for presenting this art on the printed page.

One example is called *line conversion*. In this process, continuous tone art is converted to line art for printing. Examples include reproducing a photograph so it looks like a drawing; making various finishes that look like antique matting; exaggerating dot structures to make the art look as if it is constructed of all sorts of circles, lines, squares, and shaded effects.

All this is done by placing a screen or film between the lens and the unexposed film and taking a picture to produce a negative for plate making, just as you would to make a halftone.

Although such special effects have surprise value or help set a special mood, they should be used sparingly. An example of how these special effects can be useful is the conversion of a photograph to line art for printing on antique or very coarse paper that might not print the photograph well. Also, it is usually a good design principle to avoid mixing

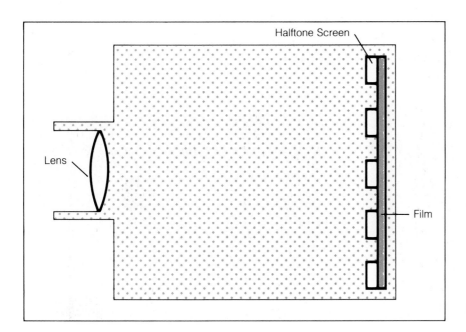

Halftone Screen

Lens

Film

Fig. 5-15 To make a halftone (or a line conversion), a screen is placed in front of the film in the camera and then the continuous tone art is photographed to produce a screened negative.

85-line halftone

Cross-hatch

Straight line

Circle

Tone Conversion

Mezzotint

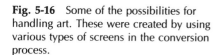

Fig. 5-16 Some of the possibilities for handling art. These were created by using various types of screens in the conversion process.

Fig. 5-17 Duotones are halftones printed using two screens—often using 2 colors. They work best with photographs that have a wide range of grays. This example is a double-black duotone. See also Figs. 6-17, 6-18 on page 136. (Photography: Richard Anderson)

line and halftone art in the same layout. If, for instance, "mug shots" (head-and-shoulder photographs of people) are to be used in a brochure along with some line drawings, it might be much more attractive to do a line conversion of the photos.

Communicators should be familiar with the capabilities of the photographer, production department, or printer to produce these special effects and take advantage of them when they will result in more effective layouts.

▼ A New Breed of Journalist

A new breed of journalist is invading the newsroom. These "graphic journalists" are playing an increasingly important role in the world of communication. An understanding of how they work can benefit anyone involved in the profession.

Let's take a look at a day in the life of the graphic journalist. Her first step is to do some creative brainstorming about a project or idea. The planned project is discussed with editors and others who might be involved. Then serious research begins. The graphic journalist does on-site observing and collects visual resources. Sometimes she works with an artist outside the office to get a better understanding of a complex project.

Once information has been collected, the graphic journalist goes over it, evaluates it, and lists questions that might need further information gathering or clarification.

The graphic journalist writes the words to be used in the graphic. It is her job to see to it that all words are clear and conform to the style of the publication. Often the graphic will be gone over by the editor or whoever is responsible for the production. After this, the graphic is completed and a final rendition is made for production.

Michele Fecht was the first person to hold the position of graphic reporter/researcher at the *Detroit News.* She explains her role: "Primarily, I work in the same way that a news reporter works. I gather information and bring it back and work it out with an artist and a graphics editor. Research is a big part of it—making phone calls from the office to find visual resources."

Pam Reasner, graphics coordinator for the *San Francisco Chronicle* says, "My primary responsibility is to work with editors who want graphics to make sure that what they're asking for is 'doable.' I hone the idea and work with an artist to get it done on time, make sure that it's good. I'm really a liaison between the artists and the editors." [4]

▼ Obtaining Photographs

There are two things to consider in obtaining effective photographs for publication: the creativity of the photographer and the quality of the image the photographer captures.

Creative Communication

Try this: On a blank piece of paper, using only your imagination, sketch a trademark that says: Made in (use your state or city).

Now try this: Select a concrete item such as a desk or a tree and see how many ideas you can come up with for its use in a layout.

Or this: Select an abstract word such as *love, kindness,* or *wealth* and see how many ideas you can come up with for illustrations to "say" this word.

Or this: Select a saying such as "A penny saved is a penny earned" and see how many possibilities you can come up with for illustrations that might accompany that saying in a layout and help communicate the message.

You've been creative!

Try to analyze how you arrived at your solutions to these problems.

Perhaps a good way to consider creativity in photography is to go back to the Greek origin of the word *photography*. *Photos* means "light" and *graphos* means "writing." Thus, the photographer "writes" with light while the writer records images with words. Both are the techniques of creative communication.

Philip C. Geraci, writing in *Photojournalism: Making Pictures for Publication*, has this to say about the role of photographs in communication:

Editorially, photographs help to tell the publication's story, not better than words, for both words and pictures are essential to give the reader full understanding. But pictures generally say it first. That's the purpose of a photograph: to sum up in a rectangular frame the mood and essence of the story.[5]

Detailed advance planning is an important part of obtaining a quality photograph. If you are selecting a photographer for an assignment, write out exactly what you desire, look over his portfolio to see if it includes the types of illustrations you have in mind. If permissions are needed, secure them in advance as well. Often, you will need permission to restrict the place to be photographed, such as roping off an area. If all this is taken care of in advance, things will go smoothly and you will save money.

Here are a few hints on how to obtain good photographs for printing:

- Even lighting is best without extreme highlights or extreme shadows.
- Dark shadows do not print well on coarse paper, such as newsprint.
- A light source on or near the camera, if it is the only light source, can make the subject appear flat and shapeless.
- Generally speaking, the outdoor photograph that will reproduce best is one taken in the evenly lit hours of the day—especially for printing on newsprint.
- For reproduction on smooth and glossy paper outdoor photographs are often planned for early morning or late afternoon hours specifically to get shadows.

An ideal photograph for printing is one where details are clearly visible in both highlight and shadow areas, there is good contrast within the mid-tone range, and shadow is held to an absolute minimum. Prints that are very contrasty or very flat and light should be avoided.

For black-and-white printing, photographs taken with black and white film are best. Prints made from color transparencies tend to have too much contrast and often they are not as sharp as original black-and-white prints.

Professional graphic designers who work with photographs learn to tell at a glance whether a print will reproduce well. Contrast is a quality they judge. The density range is also considered. Density is the lightness or darkness of a photograph.

A densitometer is a photoelectric device that measures density by recording the density range, which is the difference between the density of the lightest highlight in a photo and the darkest shadow. For optimum printed reproduction on newsprint, for instance, the recommended densitometer reading is 1.4 to 1.8 for a black-and-white print and 2.5 to 2.8 for a color transparency.

The size of the print is a factor to consider in ordering a photograph. The 35mm is the most popular all-purpose camera. You might choose a

2¼ by 2¼ inch or larger print if the photograph will be used in a large area on a layout. In general, the larger the film format, the sharper the final photograph will be. Ideally, the size of the photographic print should be as close as possible to the size of the image to be printed. For instance, an 8 by 10 inch photograph should be ordered for a full page that is larger than 8½ by 11 inches. If the original is too small the photograph may become grainy and blurred when enlarged. Also, a large print is easier to retouch than a small one.

Here are some hints for the most satisfactory photograph:

- Light- and medium-colored subjects are best when shot against a dark background for black-and-white photography.
- Dark subjects are best photographed against a background that is just moderately light.
- If a background is too light it may merge into the page it is printed on.
- If there are dark and light subjects in the same photograph, select an intermediate-tone background that will give both fairly good contrast.
- More contrast is obtained in color photography even though the subject and background are of similar tones.
- Try to obtain several originals of different exposures to select the one best suited to your purposes and the paper and printing method to be used.

▼ Preparing Art for Printing

Regardless of whether a photograph or a drawing is selected, there are certain steps that must be taken to prepare the art for printing. These include cropping, sizing (or scaling), and retouching.

Cropping Art

Cropping is the process of removing unwanted material or content or changing the size or direction of the art. It isn't just haphazard chopping away at a photograph or drawing. It is judicious editing with an eye toward enhancing the effectiveness and design characteristics of the art.

Art is usually cropped:

- To emphasize the center of interest.
- To eliminate an unwanted portion.
- To compensate for technical errors.
- To adjust the shape to fit a given layout.

Skillful cropping can be used to alter the proportions of the background and foreground. For instance, the horizon line can be raised or lowered to change the emphasis. The center of interest can be moved to a better location. If the art has motion in a certain direction, the center of interest should be given "elbow room" to move in that direction. For example, imagine a photograph of a sailboat moving across the sea. If the boat is

Fig. 5-18 Cropping Ls are an important tool for editors or designers working with art. They can be used to emphasize elements, to isolate parts of an illustration, and to determine where crop marks should be made. (Ann Standorf)

Fig. 5-19 The square and the wedge graphically illustrate the simulation of various movements by placing the same art in different positions. First, the wedge appears to be moving into the square, then it appears to be dropping into the square, then falling out, and finally it seems to be moving out of the square. (Courtesy Metro Associated Services, Inc.)

Fig. 5-20 Direction of art is important. In the layout on the left, directional art is used correctly. The motion, and therefore the reader's eye movement, will be into the copy. On the right, the direction of the art moves the reader's eye out of the layout. (Courtesy Metro Associated Services, Inc.)

moved out of the exact center of the photograph, the feeling of movement is increased and the illustration becomes more dynamic.

It isn't often that a photograph or drawing can be used in the form in which it is received. By careful and thoughtful cropping it is possible, for example, to convert a picture containing several people into one showing a single character, two characters, or whatever number is desired. Quite often a very dramatic head-and-shoulder shot of an individual can be obtained by cropping the person out of a group picture. This is especially true if the photograph is an informal action shot rather than a stiff group picture.

One illustration can serve many purposes if creative thought is given to cropping. For instance, in an illustration of a family group, a child can be isolated if a picture of a single child is needed. A person's hand pointing might be isolated from the rest of the body if a hand pointing is needed. Legs and feet can be separated from the bodies if art just showing people's legs and feet is needed. The creative possibilities are limitless.

The easiest way to examine cropping possibilities is to make two right angles out of heavy paper or light cardboard. These are called *cropping Ls*. The Ls should be about 1½ inches wide and each leg should be 8 or 10 inches long. Good Ls can be cut from a manila file folder. They are used to frame the various parts of an illustration.

Once you determine the desired cropping it must be recorded so that whoever processes the art for printing will know exactly how you want it to be changed. The two most often used methods for doing this are with a marker or an overlay. Grease pencils or felt markers can be used to make *crop marks*, arrows or lines, in the margins. Markings should not be made on the face of the art. Grease pencils, also called china markers, can be purchased at most stationery or art supply stores. An overlay is a sheet of tracing paper taped to the illustration on which cropping and sizing can be indicated.

Most desktop publishing programs have cropping Ls in the computer "tool box" and cropping and sizing (or scaling) can be done on the computer screen.

Scaling or Sizing Art

Once art has been cropped it must be scaled (or sized), or reduced or enlarged, to fit the desired spot in a layout. For instance, most photographs used in publication work are either 5 by 7 or 8 by 10 inches. But

Fig. 5-21 Cropping can eliminate distracting influences and strengthen the art's impact. The blue lines show a few of many ways this photograph might be cropped. (Photo: Stock Boston)

it is quite unusual for those sizes to be exactly right for the spot where the photograph is to be used in the layout. Therefore art must be reduced or enlarged. Also, sometimes the dimensions of the art piece must be altered. A rectangle that is 5 by 7 might be cropped so it can be used in a space 6 by 4.

There is a general mathematical process that can be used to determine how much must be cropped from the width or depth to change the proportions so the art can be sized to fit the available space. This involves the principle of *proportion*. That is, when a four-sided area is reduced or enlarged, the sizes of the sides remain in direct relationship with each other. If we enlarge a 3 by 3 area to 5 inches wide, for instance, it will be 5 by 5 or 5 inches deep as well. If an 8 by 10 photo is reduced to half its original size, the width—top and bottom—will be 4 inches and the depth will be 5 inches. This principle can be stated in an equation that is simple to calculate:

$$\frac{\text{New width}}{\text{Old width}} = \frac{\text{New depth}}{\text{Old depth}}$$

The unknown, or the dimension being sought, is indicated by x.

Suppose the designer has a photograph that is 8 by 10 and a space 6 inches wide (36 picas) has been allocated for it in the layout. The depth of the new, or "sized," art is not known, so it is x in the equation. The formula will look like this:

$$\frac{6 \text{ (new width)}}{8 \text{ (old width)}} = \frac{x \text{ (new depth)}}{10 \text{ (old depth)}}$$

To solve the equation, the algebra instructor tells us, we use cross-multiplication. That is, 6 times 10 equals 60 and 8 times x equals $8x$.

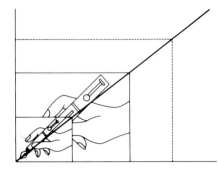

Fig. 5-22 Art can be sized by using a simple proportion drawing or placing a straight edge to create the diagonal and the new width.

The equation now becomes $8x = 60$. We continue by dividing 60 by 8 to get the new depth of 7½ inches.*

There is another and simpler, but more cumbersome, way to work proportions.

First, draw a diagonal line from the upper left to the lower right or from the upper right to the lower left corner of the original art. (Don't draw on the art, of course.) Any straight edge can be placed in this position. A rule is best as it will give the depth in inches or picas (see Fig. 5-23).

Next, measure the width of the new art across the top of the original art. Indicate the width on the bottom of the original art as well. Now line up another ruler at these two points. The point where the two rulers intersect is the depth of the new art. If the art is to be enlarged rather than reduced, both proportion lines should be extended beyond the art until they intersect.

A little more challenging situation is when we have a rectangular piece of art that has a depth greater than the width and we want to use this art in an area that has a width greater than its depth. How can this art be cropped to change the shape? Should the cropping be done from the width or the depth?

First of all, since the depth must be changed to be shorter than the width, we have to crop the original depth. But how much?

We can use the proportional equation again. Let us assume that an 8 by 10 photograph must be used in an area 6 by 4 inches. Portions must be cropped from the original depth, which we will designate as x (the unknown) in the proportion formula:

$$\frac{6 \text{ (new width)}}{8 \text{ (old width)}} = \frac{4 \text{ (new depth)}}{x \text{ (old depth)}}$$

If we solve this formula, $6x$ will equal 32, and x will equal 5⅓. However, and here is the tricky part, since the original depth was 10 inches and the new depth must be 5⅓ inches, 5 ⅓ must be subtracted from 10. The result is an illustration that is now 8 by 5⅓. To obtain this, 4⅔ inches must be cropped from the original depth of the illustration. This cropping can be done from either the top or bottom, or parts of each, depending on the composition of the illustration.

These sizing results can be checked by working the equation using the solution (in this case, 8 by 5⅓) for the new "original" art by substituting x for either the new width or the new depth. If, in our example, the new width works out to 6 or the new depth to 4, we know the problem was solved correctly.

Quite often instructions for sizing art are given in percentages of enlargement or reduction. A communicator might crop the art and then write "reduce to 80 percent" on the instructions. It is easy to find the percentage of reduction or enlargement. Just take the original width of the art or photograph and divide it into the desired width for the layout. That is, if a photo is 5 by 7 and it is to be placed in an area that is 3 inches

---

*Note: Some designers state the proportion formula slightly differently. However, the results are exactly the same. They use:

$$\frac{\text{Old width}}{\text{Old depth}} = \frac{\text{New width}}{\text{New depth}}$$

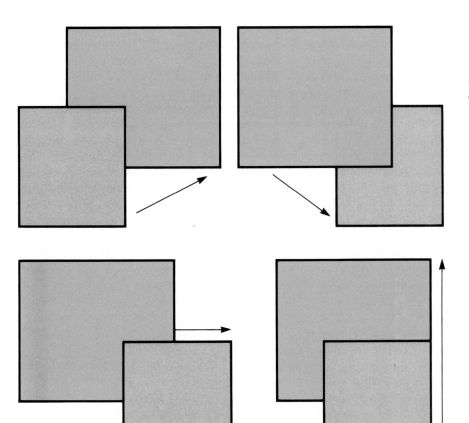

Fig. 5-23 The correct way to cut one illustration into another is shown on the left. The smaller illustration should be placed over the larger in a spot where it will not interfere with the center of interest.

Fig. 5-24 When cutting one illustration into another, a common edge, as on the right, should be avoided as the reader may not notice that two illustrations are involved. (Courtesy Metro Associated Services, Inc.)

wide, we simply divide 5 into 3 to find the percentage of reduction. In this case we would write "60% of original" on our instructions. If, on the other hand, this 5-inch-wide photo is to be used in an area that is 8 inches wide, we would divide 8 by 5 and write "160% of original."

Some communicators find it simpler to size art using picas rather than inches for the dimensions.

There are tools that help with these sizing problems. They are simple to use and can be mastered in a minute or two. The most common is a *proportion wheel*, which can be purchased at a reasonable cost from most art and graphic supply firms. There are *proportion slide rules*, too. One or the other should be in every editor's or designer's tool kit.

Always keep in mind that it is better to reduce than enlarge. Reduction makes flaws and imperfections less discernible and helps to make the definition of tones sharper and more contrasty. A good rule of thumb is to try to reduce most art at least 50 percent.

Fig. 5-25 A proportion wheel can be mastered in minutes. It can be used to size art and calculate the percentage of enlargement or reduction.

Retouching Art

Retouching is the process of eliminating unwanted material or flaws in artwork. It can be used to emphasize details or repair defects. However, it takes a skilled artist to do effective retouching.

Usually an airbrush is used for retouching. This is an atomizer that uses compressed air to spray watercolor paint on art. The retoucher masks the areas to be retained and sprays the areas to be eliminated or makes the background a uniform shade or tone. Bleaches are often used to remove unwanted portions of photographs.

Sometimes it is more economical, and possibly more effective, to correct photographs in the darkroom rather than to use the services of a retouch artist.

A photographer can alter the image by using different types of photographic paper to change the contrast. Techniques such as *dodging* and *burning* can also be used. *Dodging* is a method of lightening areas of a photograph by reducing the light reaching certain areas of the print. *Burning* is used to darken areas by allowing more light to reach areas of the print.

And now the computer has moved into the darkroom to take over many of the retouching and enhancement tasks. With the use of digital transformation, technicians can do amazing things to photographs.

Through electronic retouching the *National Geographic* slightly moved one of the great pyramids at Giza to fit the art into the shape of its cover. The *Asbury Park Press* removed a man from the middle of a news photo and filled in the space by copying part of an adjoining wall.

The editors of a book, *A Day in the Life of California,* used an electronic enhancement device called a Scitex to alter the cover of a jacket by electronically moving a surfboard and making the smoggy sky a more appealing blue. This ability to make changes in photographs that cannot be detected has created a twenty-first century ethical problem for the communicator.[6]

▼ Editing Photographs

A communicator's skills should include editing photographs as well as editing the written word. Here is a typical photo-editing problem. The editor of a newsletter has received three mug shots. They are all flawed, some more than others. The editor wants to correct the flaws and also make the people's heads all the same size.

Here are the photographs:

A   B   C

The first step is to examine the photographs. Photograph A is a good likeness of the individual. The head size is appropriate and there is good contrast. There are adequate tonal values—good blacks and bright whites with ample detail in the middle tones. However, the dark background should be lightened.

The only solution to distracting backgrounds as in this photograph as well as Photograph C is to retouch the original, either with an airbrush or by carefully outlining the photograph with a neutral shade of opaque watercolor. In this case an airbrush was used.

One way to correct the photograph is to coat a piece of tracing paper with rubber cement and put it over the entire surface of the photograph. The masking sheet is cut from around the head with a razor blade or sharp knife and the tracing paper is removed from everything but the head. Then the photograph is airbrushed. The tracing paper is removed from the head and all the rubber cement is rubbed off.

Photograph B presents two problems. It is overexposed and too contrasty. The highlights are washed out, there are very little middle tones but there are solid blacks. The high contrast can be softened and middle tones can be introduced.

How can this be done? First, the editor must be familiar with the printer's requirements for a normal halftone. This would include the type of paper, the press, and the screen that the printer regularly uses. To make the negative for Photograph B agree more with the negative for Photograph A, the printer should decrease the main exposure slightly and slightly increase the flash to enlarge the dots in the shadow end of the photo.

The other problem with Photograph B is the T-shirt. This is distracting and inconsistent with the shirt and tie attire of the other subjects. Most likely the best solution would be to crop all three as tightly as possible.

Photograph C also presents two problems—a distracting background and the fact that he is holding a coat slung over his back. Both problems can be solved fairly easily. The background can be removed as was done for Photograph A and the problem of the pose can be solved by close cropping.

Photographs B and C should be enlarged after cropping so the heads are about the same size as the head in Photograph A.

Here are the results:

A   **B**   **C**

Sometimes photographs are received that are damaged in some way. They may have creases or other defects. Careful retouching is necessary.

This calls for a very fine artist's brush, some India ink, and white tempera color. White spots in deep shadows can be retouched by using the black India ink, imperfections in the white area with the white opaque tempera and in the middle tones with a mixture of the two materials. After some practice, even those without artistic talent can do a good job of retouching photographs.

▼ Captions and Cutlines for Art

Are they captions or cutlines? In newspaper design they are cutlines in the United States. In magazine design and in newspaper design in England they are captions. Regardless, though, of what they are called, there are a number of ways to handle cutlines or captions when planning printed communications. Some are more suitable for newspapers, and others work better for magazines and brochures.

The basic design criterion is to select the arrangement believed best suited for the job and to stick to it. Also, cutline and caption styles should not be mixed within one publication.

Newspapers consider cutlines (in newspaper parlance a caption is a head *above* an illustration) a must. Some insist that a cutline accompany every illustration. There are several ways these cutlines are handled:

- The cutline is boldface with the first two or three words in all caps.
- The cutline is in the same face as the body matter but in a different point size.
- The cutline is in an entirely different style of type, but one that harmonizes with the headline type.
- The cutline has a sideline head. This is a line or two of display type placed with the cutline but to its left.
- The cutline has a catchline. A catchline is a display line placed between the illustration and the cutline. The catchline can be set flush left to line up with the cutline or it can be centered. Most newspaper designers seem to prefer the centered catchline.

Whatever arrangement seems to harmonize best with the whole graphic design of the newspaper should be used consistently. Cutline styles should not be changed from illustration to illustration, though a single centered line might be used for mug shots and another arrangement used for all the other illustrations. The point is to be consistent.

Many designers in the magazine world prefer not to use captions. They see them as an annoyance that clutters up a layout. Many believe that a good illustration needs no caption, that it tells the story by its composition and content.

When captions are used, they appear in two basic design formats. One is adjacent to the illustration and the other is a combination caption that serves several adjacent illustrations on a page or in a layout.

The arrangement of the caption and its placement should be based on two criteria. First, the caption should not just be "thrown in." It should be part of the overall design and treated as an important element. That

Camera catches bank holdup

A bank robber holds a sawed-off rifle over a woman customer as he and two accomplices hold up a branch of the Deposit Guaranty National Bank in Jackson, Miss. The robbery was recorded by an automatic bank camera. The amount taken was not known.

WINDING WAY—The Coast Guard Cutter Bollard weaves a path through Connecticut River ice Monday near Middletown. The Coast Guard keeps a channel open all winter from Old Saybrook to

MINI MUSICIANS: Young violinists from the Suzuki Music Academy of Chicago wait to play during a recital Thursday at the Daley Center. While Michael Hsu seems to have a case of

Meters to go

Park Ridge Police Dept. community service officer Robert Sundberg marks a tire in a two-hour parking area to enforce the parking time limit. Soon that procedure will be used in uptown Park Ridge when the two-hour meters are removed, probably within the next few weeks, Community Development Director Richard Schaub, said.

Fig. 5-26 Captions for art. A caption style should be chosen that is compatible with the design of the publication or brochure. Consistency in caption style is important, and caption lines should not exceed the limits for readability for the style and size chosen.

is, the caption should add something to the design and should be unified with all the other elements in the layout. Second, the caption should not be written or located in a way that confuses the reader. Numbers, arrows, and other devices should not be used in captions. They clutter and disfigure a layout.

Art that is carefully selected, prepared, and presented can be as important an element in a layout as the words. In this visual age it is a vital tool of the communicator.

▼ Effective Design Checklist

- Have a specific function in mind—do not use art just for decoration.
- Remember the reader; try to select art from the reader's perspective.
- Avoid art clichés. These are things like people shaking hands, speakers at the rostrum. They are dull, dull, dull!
- Crop carefully and with a purpose. Don't crop unnecessarily and ruin a well-composed illustration.

- Avoid unusual shapes in art. Circles, stars, and other decorations should be avoided unless there is a strong design reason for using them. Straightforward rectangles and squares are best.
- Select art for its content. Do not select it for its shape. The shape can be altered but poor content is hard to improve.
- Use mortises only after careful consideration. (Mortises are cut-out areas in art in which type or graphic elements are inserted.) Usually all mortises do is disfigure art.
- Use silhouettes for a change of pace. In the silhouette the figure or center of interest is in outline form.
- Use tricky treatment with caution. Mortises, surprints (type over art) and combination plates (line and halftone art together) should be used with great caution. The trend today is straightforward, simple, close-cropped art.

▼ Graphics in Action

1. Select three articles from magazines or newspapers that have no illustrations. Decide if the articles could have been presented with more impact, more reader interest, or clearer understanding with art. If so, decide the content and method of presentation that would be best for this art. Find a source for the art to be used.
2. Select an illustration. Make and use cropping Ls and see how many uses you can find for variations of this one illustration. List these layout possibilities. If possible, make duplicates of the art so each possibility can be marked for cropping.
3. Find in publications as many different art treatments, such as line conversions and different screening techniques, as possible. Evaluate these treatments and discuss if the art would have been as effective if used straightforwardly.
4. Obtain a selection of photos and crop them for effectiveness. Often the local daily newspaper is glad to supply wire photos for this purpose as it receives far more than it can use. Discuss in class why you cropped the photos the way you did.
5. Once photos have been cropped, size them to fit areas that are 4 inches (or 24 picas), 6 inches (or 36 picas), and 8 inches (or 48 picas). Or, crop for content and size to change the dimensions. For instance, crop an 8 by 10 photo to fit a 7 by 5 area in a layout. When sizing is complete, calculate the new dimensions in both inches or picas and in percentages of reduction or enlargement.

Notes

1   *AP Log*, March 29, 1976, pp. 1–4.
2   Richard Curtis, *ASNE Today* (Washington, D.C.: USA Today, 1973), p. 1.
3   *The Louisville Chronicles*, Society of Newspaper Design, October 15, 1988, p. 3.
4   *Design*, May 1988, p. 34.
5   Philip C. Geraci, *Photojournalism: Making Pictures for Publication* (Dubuque, Iowa: Kendall/Hunt Publishing Company, 1978), pp. 1–2.
6   *Sacramento Bee*, October 9, 1988, Forum p. 6.

CHAPTER SIX

Color: A Powerful Communication Tool

Photograph: Terry Vine

The trend in printed communications is toward bold and colorful graphics. Communicators must use every tool available to make the message stand out from its competition. The proper use of color, the judicious use of borders and ornaments, and the proper marriage of ink with paper can all add dimensions to effective printing.

The first step in becoming adept at selecting and using these elements is to develop a sense of how they can help. This graphic sense, or awareness, can be sharpened by studying the work of others. Keep a constant critical eye out as you browse through newspapers, magazines, and even the direct mail communications that fill the mailbox. Try to second-guess the person who designed them. Why was a subhead put here, a border there, an ornament at that position? Why was spot color used there? What did it add to the printed piece to invest in full color rather than spot color or simply printing it in black on white?

When you pass the magazine rack in the supermarket, look at the graphic array. Which magazines seem to be especially attractive? Why? Which magazines seem to communicate at a glance what they are all about? How do they do this?

The head buyer of a major grocery store chain once said he looked at the label on the can before he examined the contents. If the label wouldn't sell, he would reject the product, no matter how fine it might be. The same can be said about printed messages. They must have eye appeal.

Evaluation for eye appeal includes an examination of the printed communication to see if the basic criteria for good selection and use of type have been applied. The communication should be checked for the application of the principles of design. Does it have balance, unity, contrast, proportion, harmony? If the balance is formal, is that appropriate for the type of message?

Then, the piece should be examined to see if these principles have been applied in an unusual or different way to make it stand out from the rest. Are the principles of good typography and graphics used in an unusually arresting way?

Soon you will be able to automatically "score" printed communications as you view them. For instance, an ornament that throws a page out of balance will be spotted in an instant. A boxed item placed in such a way that easy reading of the text is upset will become a source of irritation. You will immediately notice an ornament that clashes with the other elements on the page and destroys unity. And you will be aware of when color dominates the page and diminishes effective communication.

▼ A Powerful Communication Tool

The blue of the sky, the red of the sunset, neon lights, paints, wallpaper, color television, advertisements—we live in a colorful world.

It wasn't always like this. It has only been within the past hundred years or so that we have been able to take advantage of color as a tool in graphic communications. Only a limited number of dyes and pigments were known before the nineteenth century. Now there are thousands of

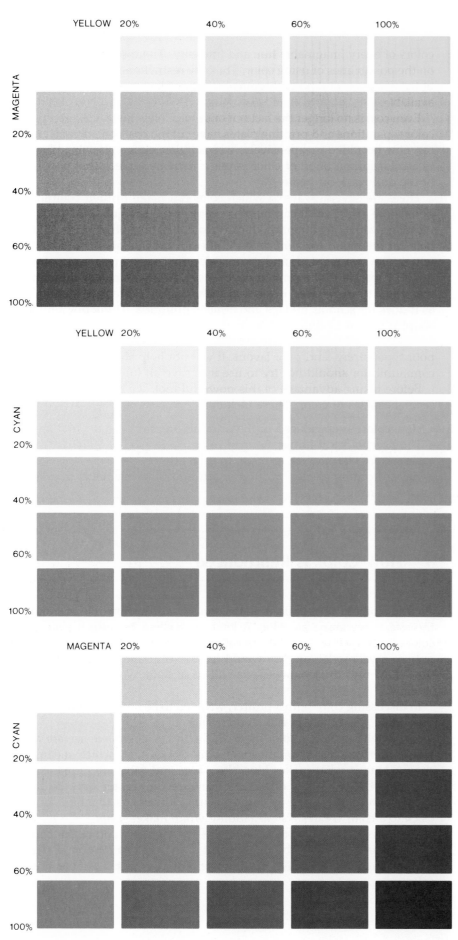

Fig. 6-1 These charts show some of the colors that can be created by combining screen tints of two process colors. Because color printing is not a precise science, examples such as these can be used as a guide only; identical screen combinations printed on different paper or with different inks can vary somewhat in appearance.

colors of every imaginable hue and intensity. There seems to be no limit on the possibilities of using color. The only restrictions are cost, the ability of the designer, and possibly some limitations imposed by the equipment available.

Even cost is no longer the factor it once was. New methods of preparing color separations and printing plates have cut the cost considerably. Color can now be added inexpensively, with a little creative planning. Printing in a color ink on another color paper can create a piece that breaks out of the dull black on white.

There are several ways in which color can make printed communications more effective. It can help accomplish the first job of any communication—attracting attention. It can create the atmosphere desired. It can help set the mood for a message. Color can provide accent and contrast where they are wanted, and it can help emphasize important points. It can add sparkle to the printed page. It can direct the reader through the message. It can be used in printed materials to help create identity just as it does for schools in flags and athletic uniforms. Think of Campbell's soup! What colors do you visualize?

However, color is not a cure-all. It will not compensate for poor writing, poor typography, and poor layout. It will not help shoddy printing. The communicator should not try to use it to save an inferior design.

Before taking advantage of this powerful tool, we need to understand several aspects of color. These include:

- How color is reproduced in the printing process
- The psychological implications of color
- How colors harmonize or relate to each other
- How to combine color with type, art, and other elements in a layout for best results

▼ How Color Is Reproduced

All color comes from sunlight. Reflection and absorption of light produces the effects we know as color. A lemon is yellow because it absorbs all colors except yellow and reflects yellow. In an unlighted room we would not see a yellow lemon. We would not see it at all. Under a dim light, the yellow rays the lemon reflects will be so weak we will see the lemon as gray.

In discussing color with a printer, we need to be familiar with six terms: hue, tone, value, shade, tint, and chroma.

Hue is what makes a color a color. That is, all colors we see are hues. Hue is derived from the ancient Gothic word *hiwi*, which means "to show." Hue is what makes blue blue. *Tone* and *value* are terms used to designate the variations of a hue. They are the lighter tints or darker shades of a color created by adding white or black ink to a hue. Adding black to a color creates a *shade*; adding white makes a *tint*.

Chroma is a term used to indicate the intensity of a color. The chroma of a color is determined by the amount of pigment saturation in the ink that produces the color. Increasing the chroma creates a more intense color.

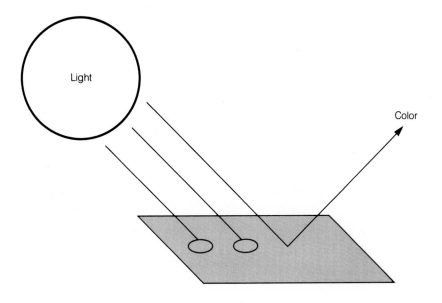

Fig. 6-2 Light rays reflected from a surface create color. A yellow sweater, for example, will absorb all light rays except yellow. It reflects the yellow and appears yellow.

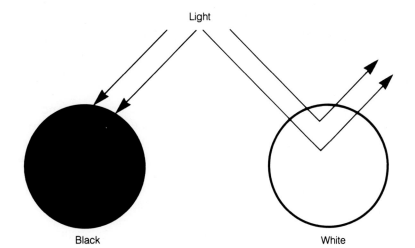

Fig. 6-3 Black is the absence of color. A black object absorbs all light rays and thus appears black. White is the presence of all colors. A white object reflects all light rays and thus appears white.

Fig. 6-4 The artist mixes colors on a palette; the printer mixes colors on the printing press.

Fig. 6-5 This enlargement of an image on a color television screen demonstrates additive color formation. At close range you can see red, green, and blue primaries. (Photo: François P. Robert)

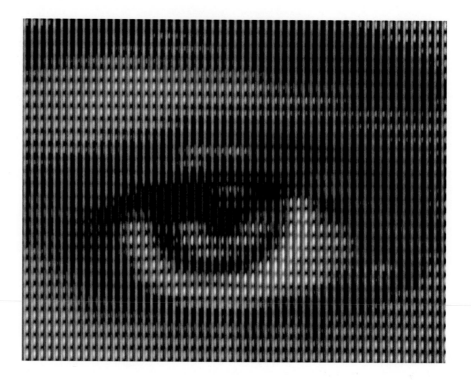

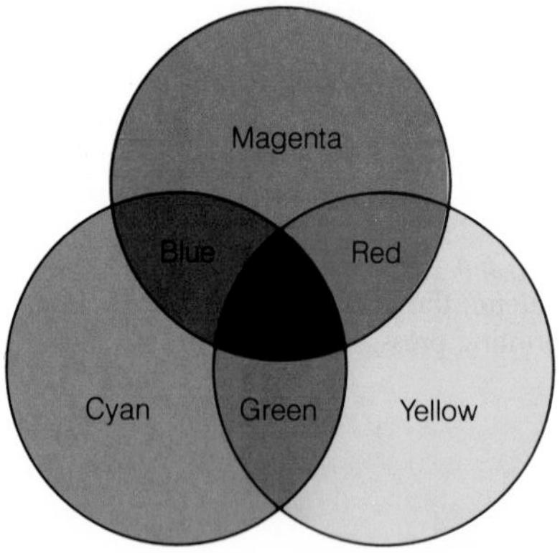

Fig. 6-6 Ink color is called subtractive. The primary ink colors are magenta, cyan, and yellow. Secondary colors are mixed from the primary colors. Magenta and cyan make blue. Green is made from cyan and yellow. Red is made from magenta and yellow. These are the six basic inks needed to make most colors in printing.

When an artist wants to create different colors or shades and tints of colors, she or he mixes paints on a palette. The printer does the same thing when a spot of color is needed for printing. The inks are mixed on an ink plate and then placed on the press, unless premixed inks are purchased from the manufacturer.

We have been taught that yellow, red, and blue are the three primary colors. All other colors can be created by mixing these primaries. There are, however, *additive* and *subtractive* primary colors.

The *additive primaries* are blue, green, and red. They are called additive because they produce white light when added together. Additive primaries are the dominant colors of the rainbow. Another good example is a color television image. Color televisions use a red, green, and blue color projection system to create the sensation of color.

The process is different when a printer needs to produce the full range of colors in a full-color illustration. Three colors slightly different from blue, green, and red are used. The printer uses cyan, yellow, and magenta for full-color work. Cyan is a blue green, and magenta is a red violet. These darker colors are called *subtractive primaries*. They are called subtractive because they absorb light. And they are primaries because a full range of colors can be produced by mixing the inks together in various proportions.

The printer makes four plates through a process called *color separation*. Each of these plates will print a color of ink in the density required so that when it is combined with an impression from another plate it will create the tone or shade desired. In effect, the printer uses the printing press for a palette. If the people using this process are skilled, the reproduction will be difficult to distinguish from the original.

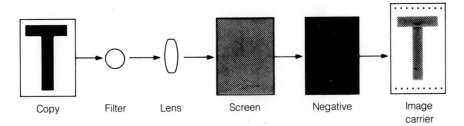

Copy   Filter   Lens   Screen   Negative   Image carrier

Fig. 6-7 Color plates are made by photographing the full-color art through filters to separate the primary colors. If the original art is continuous tone, such as a color photograph, a screen is used for each exposure as in the creation of half-tones from black-and-white photographs.

The printer uses inks called *process inks* in producing printing in full color. Process inks are special transparent inks that are available in magenta, cyan, and yellow. When these ink colors are superimposed, they produce three other colors in the spectrum. Yellow and magenta combined produce red. Yellow and cyan produce green, and magenta and cyan produce blue. Thousands of tints and shades can be reproduced through the combinations of these inks. Black is added to give depth to the dark areas and shadows.

The distinctive sensation we see and identify as a certain color is actually the reflection of light waves that weren't absorbed by the object. For instance, if white light strikes a surface and the surface absorbs green light waves, we see the red and blue light waves that were not absorbed. We see, then, the color magenta, which results from the combination of red and blue.

If the surface absorbs blue light waves, we see the combination of red and green lightwaves, or yellow. And, if the surface absorbs red light, we see cyan—a combination of the reflected green and blue light.

So, in making plates for full-color reproduction, one plate is made with a green filter and used in printing with magenta ink. A red filter is used

Fig. 6-8 In printing colors four inks are used: yellow, magenta, cyan, and black. The mixing of the inks is done on paper as it goes through the press.

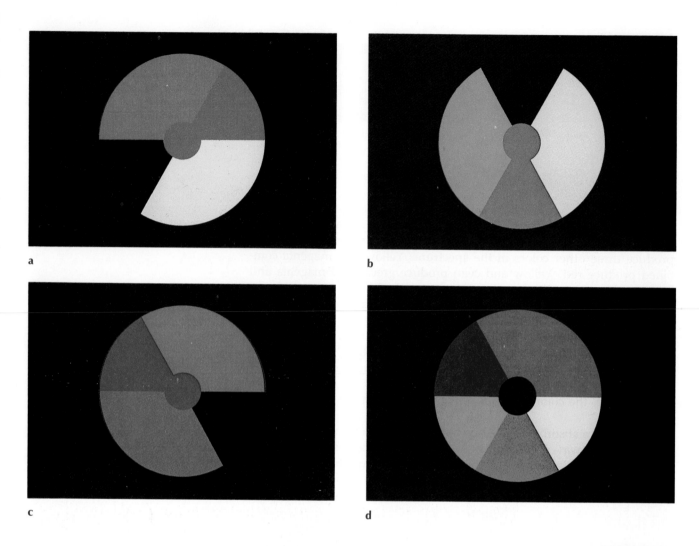

a

b

c

d

Fig. 6-9 Superimposing transparent pairs of yellow, magenta, and cyan will produce three other colors in the spectrum. Yellow and magenta produce red (a), yellow and cyan produce green (b), magenta and cyan produce blue (c). The combination of all three colors produces a near-black neutral (d). Thousands of tints and shades result from combinations of these inks in dots of differing sizes (e). (Courtesy Eastman Kodak)

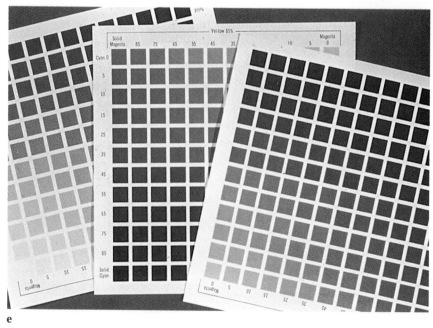

e

a

Fig. 6-10 Color artwork is "read" by a scanner, which creates negatives for cyan, magenta, yellow, and black printing plates. (a) Shown here is a scanner from the Hell Chromacom System. (b) Another part of the Chromacom system is the Combiskop. It can make precise color corrections and adjustments to scanned artwork, and it can create a wide range of special effects.

(Photographs: Jeffrey Grosscup)

b

YELLOW

YELLOW AND MAGENTA

YELLOW, MAGENTA, AND CYAN

YELLOW, MAGENTA, CYAN, AND BLACK

Fig. 6-11 The four-color process plates printed in sequence, or progression. Proofs of this sequence are called progressive proofs or "progs."

for the cyan printer, and a blue filter for the yellow printer. The fourth plate, the black ink printer, is made to add density. Without the black ink impression, the reproduction would appear weak and dark brown rather than black in the shadows.

The process is much more complicated than this brief description, but the point to remember is that it takes four plates and four separate impressions or runs through the press to produce a full-color reproduction. It is expensive, but the price is coming down as new equipment for making color separations is developed.

Fig. 6-13 Color bars are included in four-color process proofs. These bars are used to check such things as the density of the color, the amount of ink used, and the quality of color reproduction in screened areas.

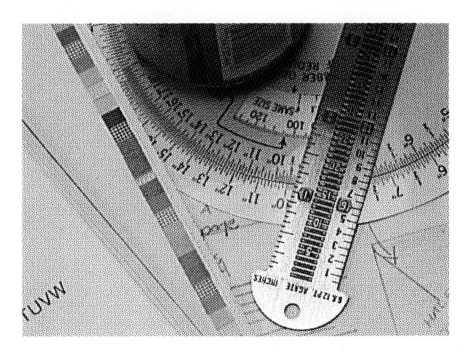

Fig. 6-14 This enlargement of the upper left corner of Fig. 6-11 illustrates the halftone dot structure of four-color process printing.

Creative Communication

Here is a plan for creative problem solving. (It is from a booklet produced by the Royal Institute of British Architects, described by Bryan Larson in his book *How Designers Think*.)

The plan includes four steps: assimilation, general study, development, and communication.

The first step is the accumulation and ordering of general information and information specifically related to the problem. This is followed by study of the problem, its nature and possible solutions.

Next, the best solutions are isolated and developed and refined.

Finally, the solutions are discussed with those involved in the project or those who can offer advice and informed evaluation.

▼ The Psychological Implications of Color

The increase in use of color has led communicators to look to psychologists for help in making the most effective use of this powerful tool. Just as sugar manufacturers have learned, for example, that their product will not sell in a green package, and manufacturers of beauty preparations know that brown jars will remain on the shelf long after others are gone, so communicators must know which colors to use in which situations.

In the case of sugar, tests by marketers have shown that it sells best in a blue container or at least in one where blue is predominant. Blue is the color of "sweetness," whereas green is seen as astringent, like a lime.

Airlines know that proper color schemes in airplane cabins can help relax nervous passengers.

There are warm and cool colors. Fire and sunshine make red, yellow, and orange—warm colors. The shadows of deep forests and the coolness of water make blue, violet, and dark green cool. Night brings on inactivity while day brings brightness and hope. Thus dark blue is the color of quiet while bright yellow is the color of hope and activity.

Experiments have shown that people exposed to pure red are stimulated. Depending on the length of exposure, blood pressure increases and respiration and heartbeat speed up. Red is exciting.

On the other hand, exposure to pure blue has the reverse effect. Blood pressure falls and heartbeat and breathing slow down in a blue environment. Blues are calming.

Advertisements for air conditioners use cool colors. Those for furnaces are more effective if warm colors are used.

Unnatural use of color can cause adverse effects. Printing a luscious grilled steak in green not only fails to add to communication in a favorable way but it can also detract by creating a strong sense of repulsion on the part of the reader.

"I have worked with designers who ran halftones of people in blue, green, pink, or some other terribly unnatural color," remarked one designer who reviewed this manuscript. "The results were uniformly terrible."

Blue is the favorite color of the majority of people. It can be used with no fear of adverse psychological effects (unless, as stated, it's used unnaturally). Yellow generates the buoyant happiness of a sunny day. Orange is a happy color, too. And brown is one of the most versatile colors for printing. Men associate it with wood and leather. Many women associate it with leather goods and furs. Like blue, it has no inherent weaknesses and can be used for a wide variety of purposes. Green is also a universally popular color. And purple suggests robes of royalty, the dignity of church vestments, and the pomp and splendor of high ritual.

There are two important points to remember when selecting color for communications. First, warm colors advance and cool colors recede when printed. Reds tend to dominate and can overpower other elements on the page if the designer is not careful. Second, colors in printing should be used as much as possible in their natural associations—green forests, blue sky, sunny mornings. (There may be times when the unnatural use of color can create the most effective communication. But remember the risk of repulsion when color is used unnaturally.)

▼ Selecting Colors for Harmony

Which colors go well together when printed? Many people can tell instinctively if colors look compatible when printed together, but others need help in choosing color schemes for communications. Luckily help is available.

In 1899 a Boston teacher, Albert H. Munsell, began research that resulted in a system for distinguishing color. He charted color values on a numerical scale of nine steps ranging from black to white. Munsell's system was adopted by the National Bureau of Standards and slowly added to until it contained 267 different color names.

Out of this came the *color wheel.* Around the wheel are the colors comprising the primary triad of red, yellow, and blue in an equilateral triangle. Halfway between the primaries are the secondary colors. In all, the wheel divides the color spectrum into twelve hues. Five basic color combinations have been devised and the communicator can use these combinations in deciding which colors to use in creating harmonious layouts.

The combinations are:

- *Monochromatic:* This is the simplest color harmony and is made of different values of the same color. These values may be obtained in printing by screening artwork at different percentages. Monochromatic harmony works well in printing. Care should be taken not to screen type so that legibility is lost.

Fig. 6-15 Color combinations used for creating harmonious layouts.

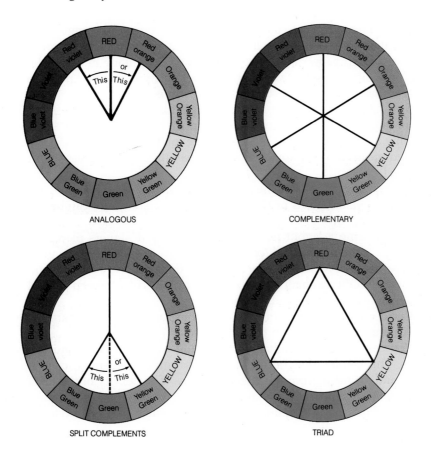

ANALOGOUS

COMPLEMENTARY

SPLIT COMPLEMENTS

TRIAD

- *Analogous:* These are two colors that are adjacent on the color wheel, such as blue and blue green, or red and red orange.
- *Complementary:* These are colors that are directly across from each other on the wheel, such as red and green. A complementary selection adds drama because of the contrast of warm and cool colors.
- *Split complements:* These are colors that are selected by choosing a color on the wheel and finding its complement but using a color adjacent to the complement. A split complementary harmony for red would be blue green or yellow green.
- *Triad:* This is a combination of three colors, each of which is at the point of a visualized equilateral triangle placed on the wheel. As the triangle is turned to any position on the wheel, its points will designate the three colors of a triad.

Another tool for color selection is a *matching system.* This is composed of samples, or chips, showing the various colors as they will appear in print. It is similar to the collection of swatches found in most paint stores. One widely used such guide is the Pantone Matching System, referred to as PMS. This guide shows the colors as they will appear on coated or uncoated papers.

▼ Some Ways to Use Color

There are many ways to take advantage of the powerful communications possibilities of color at very little added cost. The cheapest, of course, is to use colored paper. We are so conditioned to thinking in terms of black ink on white paper that we often overlook this possibility. And we often forget that we could use colored ink on white paper.

Spot color is the process of adding individual colors in printing. One color added to the basic color can do much to make printing stand out at only about a 35 percent increase in the total cost of the job. Each color added means an additional run of the sheet through the press, but it can be well worth it.

Spot color can increase the impact of a printed piece. It can be used to emphasize illustrations or type. More than one spot color can be used in the same layout.

In preparing a layout with spot color, many designers place a tissue overlay on the mechanical (completed layout) to specify where the color will be used, reverses, the percentage of the screen to be used, and the color itself. Often a small swatch or sample of the color desired is attached. A separate overlay is used for each color.

(Incidentally, if process inks are used for spot color and they are printed on colored paper, that can change the color.)

You can also use simultaneous printing of a solid color and various tints. This can be done on the same press run by screening, or printing type and art made from screened negatives. The density of the resulting tones will depend on the density of the screens. This can be any percentage from solid (100 percent) through half-solid (50 percent) to almost white (10 percent).

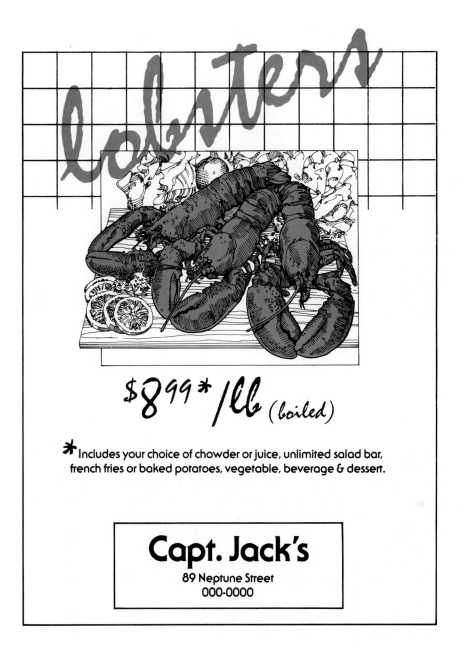

Another easy way to obtain effective color is by creating a duotone. A *duotone* is a two-color halftone print made from a screened photograph. Two plates are needed. One is printed in the desired color and the other in black. The result can be a highly dramatic added element to the layout.

The duotone can enable the communicator to add color, warmth, and depth to photographs at a much lower cost than full color—to achieve a unique, artistic interpretation of photographs that is different from the image transmitted with either black-and-white or four-color reproductions.

There is nothing complicated about making a duotone. First you select a continuous-tone photograph. Then two separate printing plates are made from two screened negatives of the photograph. The screened

Fig. 6-17 A duotone. Compare it to Fig. 5-18 on page 108.

Fig. 6-18 A "fake duotone" created by printing a black-and-white halftone over a 10% screen.

negatives are made by shooting with the screens angled 30 degrees apart, which eliminates the chances of creating an unwanted pattern (called a *moiré* effect) when the plates are printed. By printing the two halftone images in exact register a duotone is created.

A so-called fake duotone is obtained simply by printing a black-and-white halftone on a colored area. This area can be a screen, a solid, or colored paper.

Not all duotones are made by printing one plate in a color and the other in black-and-white. It is possible to print both plates in different

TEE TOTALER

*Whatever the weather, you'll score big
on comfort and performance in this
all-weather jacket...just 'fore' you!*

Notes

1 *Magazine Design & Production,* March 1989, pp. 32–34.

Paper and Ink

Art: Culver Pictures

No printed communication should be planned without some thought being given to the paper to be used. The proper paper can help the communication do the job. Paper plays an important part in achieving readability and mood as well as durability of the message.

If an "old time" or "antique" theme is planned, consider soft, textured paper. Old style types printed on antique-finish papers will enhance the "feel" of yesteryear. Old style types blend with old style paper. The angular, uneven strokes of the letters harmonize with the roughness and unevenness of soft-finish paper.

A modern look can be achieved better with smooth papers. The detail in a piece of art or the contrast in a photograph will show up much better on a hard-finish paper.

The formal, precise style of many modern and thin typefaces appears at its best on paper that is uniformly smooth and even, with a hard finish. Types with very light letter forms show up best on coated and enameled book papers. But hard-finish papers can accentuate the crude details in the construction of old style letters and make them appear ragged and awkward.

Highly glossed finishes on papers lessen legibility. These papers can even cause eye fatigue and strain. They create a glare that makes reading tiring and difficult. A dull-finish stock is best for long reports and magazines with lots of reading matter.

Anyone concerned with printed communications should know how paper is made, some of its characteristics, and the various types of papers available. Some understanding of paper weights and standard sizes is helpful also.

▼ How Paper Is Made

Before the early 1700s practically all paper was made by hand. The raw materials—rags—were placed in tubs or vats. They were mixed and beaten into a pulp. The wet pulp was dipped from the tubs by hand and placed in molds made of fine wires stretched across wooden frames.

The milky pulp settled on the frames and drained and became sheets of paper. The damp sheets were then placed in a press, which flattened them and squeezed most of the remaining water out. Finally, the sheets were hung on wires to dry and stiffen.

Although sophisticated machinery has been developed to manufacture paper today, the basic process hasn't changed much. Wood is the most widely used raw material, though some of the high-quality papers are made partially or entirely of cotton and linen fibers. A number of treatments, chemicals, and fillers can be added in the manufacturing process to produce papers with a variety of finishes and other characteristics.

For instance, rosin size is added to create water-repelling qualities so the paper can be used for pen and ink writing, offset printing, or resistance to weather. Fillers, such as clay, are used to improve smoothness, to prevent printing from showing through the sheet (known as the *opacity* of the paper), and to help the ink adhere to the sheet. Dyes and pigments are added to produce papers of various colors.

Fig. 7-1 Historians record the long history of the written word as beginning some time between 1085 and 950 B.C. with the manufacture of papyrus and the use of pictography by the ancient Egyptians. The Egyptians used marsh plants to create a primitive form of paper. (Courtesy Rochester Institute of Technology)

Fig. 7-2 Papermaking in Europe in the seventeenth century. The vatman, coucher, and layman are performing their respective duties. The "pistolet," or heating device, can be seen at the extreme left of the illustration. The wooden hammers of the pulp beater can be seen in the background, upper right.

Fig. 7-3 The steps in converting wood into paper are shown in the upper drawing at the right (provided by the Wisconsin Paper Council). The logs enter the pulping machine on the left and leave as a mushy pulp on the right. The process of converting pulp to paper in the machine is shown on the lower right. The pulp enters the machine on the left and emerges as a continuous roll of paper on the right.

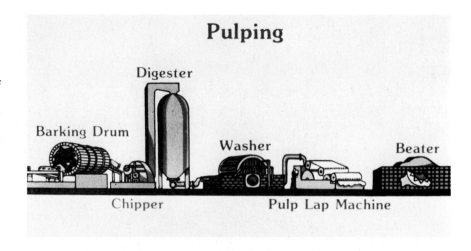

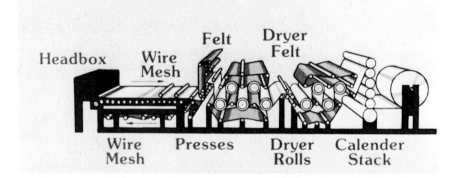

Fig. 7-4 A paper manufacturing machine at the Mead Corporation. "Gingerbelle" contains the machinery illustrated in Fig. 7-3, lower diagram.

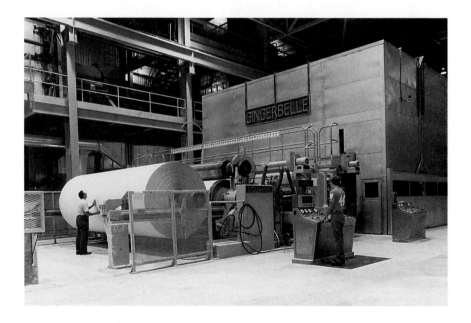

Paper can be *calendered* (smoothed) by passing it through a series of rollers, and it can be "supercalendered" for an even finer finish. Some papers are produced with a high-gloss, or coated, surface to provide excellent reproduction of photographs and art.

A variety of special finishes is available. Some papers resemble leather, others linen or tweed. Still others are available with all sorts of pebbly and special-effect finishes. These are produced by running the paper through an embosser. Other papers are produced with a watermark or faint design or emblem made by impressing it on the paper with one of the rollers during the process. This roller, called a *dandy roll*, is a wire cylinder for making *wove* (a paper with a uniform, unlined surface and a soft, smooth finish) or *laid* (a paper with a pattern of parallel lines giving it a ribbed appearance) effects.

Fig. 7-5 Finished paper emerges on the roll in the lower center of this photograph of a huge paper-making machine.

Fig. 7-6 Grain is the direction fibers run in paper, much like in a piece of wood. Grain should be parallel with the binding edge or fold in books, pamphlets, publications, or programs. Pages fold, turn, and lay flat much more easily when printing is with the grain.

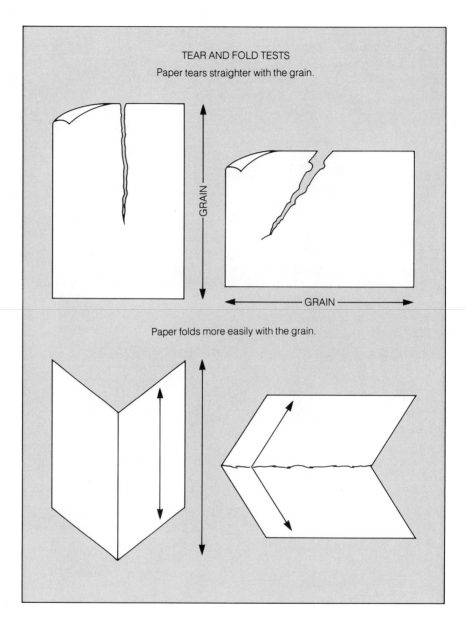

TEAR AND FOLD TESTS
Paper tears straighter with the grain.

GRAIN

GRAIN

Paper folds more easily with the grain.

▼ Characteristics of Paper

Look closely at a sheet of paper with a magnifying glass and you will discover that paper has two sides. Each side has certain characteristics. The side that was on the wire mesh as the paper traveled through the manufacturing process is called the *wire side*. The other side is the *top*, or *felt*, side. The felt side usually has a smoother finish. The front of a printed piece, especially a letterhead, announcement, or business form, should be printed on the felt side of the sheet.

Paper has a grain. The *grain* is the direction in which the fibers lie in the sheet. Grain affects paper in several ways. For instance, paper folds smoothly in the grain direction. This can be demonstrated by folding a piece of construction paper or cardboard in both directions and noting which direction produces the smoothest fold. For books, brochures, catalogs, and magazines, the grain direction should be parallel with the binding edge of the pages. The pages will turn easier and lay down better.

▼ Paper Sizes and Weights

Papers are manufactured in standard sizes and weights. They are identified and selected for printing by these measures. As an example, consider the standard paper used for letterheads known as *bond*. This type of paper is manufactured with a hard surface that will receive printing and writing inks without a blotter effect. The letters won't soak into the paper and become distorted.

Bond papers are made in a basic size of 17 by 22 inches. They will cut into four 8½ by 11 standard letter-size sheets per full sheet. Bond papers come in various weights ranging from 13 to 40 pounds. The most common available weights are 16, 20, 24, and 28. The basic weight is 20 pounds. This means that 500 sheets (a ream) of 17 by 22 bond paper will weight 20 pounds. You can cut costs—and have a flimsier looking letterhead—by ordering 16-pound bond paper. Or, you can spend a little more and have a heavier, more impressive looking letterhead by ordering 28-pound bond.

The basic size is different for different kinds of papers. It is thus important to be aware of how papers are classified and sized. The vast number of papers available may make this appear forbidding, but the classification system is really quite simple since nearly all papers can be classified into basic groups. The number of groups varies somewhat but most manufacturers recognize ten. These basic paper groups and the standard sizes for each are listed below. A discussion of the importance of knowing about standard paper sizes is included in Chapter 11, on designing printed communications.

- *Bond* (17 by 22): These are the standard papers for business forms and letterheads. They have a hard surface that works well with pen and ink, typewriter, or word processor. They are available in a large variety of colors, mostly pastels, and most manufacturers provide matching envelopes in standard sizes as well.
- *Coated* (25 by 38): These papers have a smooth, glossy surface. They are used for high-quality printing. The surface can be dull-coated as well as glossy, and there are a variety of other surfaces, such as coated on one side, for labels.
- *Text* (25 by 38): Text papers are available in a variety of colors and finishes. They are used for booklets, brochures, announcements, and many quality printing jobs.
- *Book* (25 by 38): Book papers are not as expensive as text papers. They are used mainly for books, of course, and pamphlets, company magazines, and so on. Book papers are available in a wide range of weights and finishes from antique to smooth.

- *Offset* (25 by 38): These are papers made especially for offset lithographic printing where dampening is a factor. They have sizing added to help the paper go through the offset printing process. They are similar to book papers.
- *Cover* (20 by 26): Cover papers come in an endless variety. Some are of the same texture and color as book papers but of a heavier weight. All sorts of finishes are also available. As the name implies, these are designed for use as covers on pamphlets, magazines, and so on.
- *Index* (22½ by 35 and 25½ by 30½): These papers are made to be stiff and to handle writing ink well. Index cards are examples of this type of paper. Index papers are available in a variety of colors.
- *Newsprint* (24 by 36): This is a cheap paper for printing handbills, circulars, and newspapers.
- *Cardboards* (24 by 36): These are sometimes called *tag board*. They are the heavy stuff used for posters. Lighter weights are used for tickets, cards, and tags. Cardboards come in all sorts of colors and can be colored on one or both sides.

▼ New Papers For a New Age

The new technology requires new papers, and now paper mills are producing papers especially for electronic printing methods such as laser and ion deposition. For instance, one manufacturer produces a 25 percent and a 100 percent cotton paper called ls-Tech. It is available in sizes from 8½ by 11 inches to rolls that are 51 inches wide.

These aren't old papers with new names, but new types of papers created especially for electronic printing. They are designed to cut down the possibilities of trouble as the sheet travels through the press. The cotton content of these papers gives them strength for permanence and brightness for contrast, and just enough stiffness to pass easily through electronic printing machines.

Another feature of the new papers is moisture control, which prevents static buildup that can result from improper moisture levels. Also, the sheets have uniform *porosity*. The amount of air that can pass through the sheet can affect print quality and the ability of the toner to fuse to the paper. Uniform porosity allows an even distribution of air to pass through the sheet.

▼ Specifying the Right Paper For the Job

It is important to list all the information the printer needs to obtain the right paper for your printed communication, along with the other specifications for the job. But before selecting a paper, check to see if it is available. If you select a paper the printer does not have in stock or cannot get quickly from the paper warehouse, you could lose several weeks.

Paper should be specified by listing the basis weight, color, brand name, finish or texture, and grade in that order. The basis weight is the weight in pounds of one ream, or 500 sheets of the paper, cut in its standard size. Just about all papers are listed by weight except some cover papers which might be listed by thickness. In listing the color be sure to use the exact term used by the mill. Paper mills, like paint manufacturers, may have their own names for colors and they may have several shades of one color. For instance, a tan paper from one mill comes in English Oak, Chatham Tan, and Monterey Sand. Be sure to check a current swatch book for the color, as mills change colors from time to time.

Next, list the full brand name and the finish or texture. Finally, record the grade. This may seem obvious, but it is important. The grade might be bond, offset, or coated book, for example. If the grade is not listed, you may find yourself making wasted trips or extra telephone calls to the printer or paper supplier.

As an example of paper specification, you might specify the paper for a folder in this manner: 80 pound, soft coral, Gilbert Oxford, antique, text.

Paper is one of the major cost factors in printed communications. It is thus always important to talk with the printer about papers. It is also a good idea to acquire a collection of "swatches" or samples of various kinds of printing papers. These can be obtained from paper wholesalers. A file of paper samples should be a part of every communicator's kit.

▼ A Word about Ink

Although the choice of the proper ink to use for best results is the job of the printer, communicators should be aware of the role that ink plays in the printing process.

Ink has been a part of our civilization for more than 2,000 years. The Chinese used various combinations of ingredients for writing and drawing more than 1,600 years before Gutenberg developed movable type. Early Egyptian literature mentions ink. References to ink are found in both the Old and New Testaments of the Bible. For centuries ink makers produced their product mainly by adding soot, or lampblack, to a varnish made by boiling linseed oil.

In the 1850s the discovery of coal-tar dyes, pigments, and new solvents ushered in the age of modern printing ink. Today ink making is a major industry and science. The National Association of Printing Ink Manufacturers reports that there are approximately 200 ink companies in the United States producing inks in about 400 plants throughout the country. Companies range in size from those with fewer than ten employees to those employing thousands. Industry sales of printing inks are more than $600 million now and are growing at an average rate of about 5 percent a year.

Every printing ink contains two basic parts. These are called the pigment and a vehicle. The pigment is a dry powder that gives the ink its color. Linseed is often used as a vehicle or liquid that carries the pigment and adheres it to the paper.

UNCOATED
Ink pigment and vehicle are absorbed into the paper

COATED
Ink pigment is retained on the surface and the vehicle is absorbed into the coating

Fig. 7-7 Uncoated papers allow the ink to penetrate the paper while coated papers hold the ink on the surface. Art and photographs are reproduced at their best when coated papers are used.

Fig. 7-8 Printing is the ultimate test for printing papers. Both letterpress and offset presses are used in testing at this laboratory print shop at the Mead Corporation paper mill.

The ink manufacturing plant produces two basic types of ink. One is opaque and the other is transparent. The opaque ink will cover a color underneath, either the paper or other ink, completely, so it cannot be used for full-color production. Transparent ink will let a color underneath show through; thus it performs the mixing process on the printed paper to create various colors, tints, and shades, as was discussed in Chapter 6.

Inks are manufactured for compatibility with every type of paper. There are quick-drying inks for high-speed production on rotary presses, inks that harden to resist rubbing, metallic inks that simulate gold and silver, and fluorescent inks that store sunlight and glow in the dark. There are even perfumed inks that provide a subtle aroma for special effects. And there are the inks used with our morning newspaper, which will ruin a white shirt. However, work is progressing toward producing a water-base ink that may solve this problem.

For centuries black was the only color in which ink was available. Through the development of synthetic pigments and a greater knowledge of color technology, a whole rainbow of colors is now available to the ink maker. Color matching to a specific tint, shade, or hue has become an exacting procedure.

We should never overlook the possibilities certain inks can provide for that special effect. Inks, along with all the other supporting elements, combined with type, play a part in the complete printed communication.

▼ Effective Design Checklist

- When planning printing take into account the standard dimensions of the paper to be used as well as the capacity of the press, to keep waste and cost to a minimum.

- The finish of the paper can affect the mood of the completed job. Select a paper that prints well and harmonizes with the type and tone of the message.
- Avoid selecting type designs with delicate lines for printing on rough-finish papers, especially for letterpress and gravure. This is not as critical for offset printing.
- Remember that color paper can add another dimension to the printed piece, but select an appropriate color.
- Consider grain direction when figuring paper stock. If the finished piece is to be folded, the grain should be parallel with the fold.
- Process or full colors produce most accurately on neutral white paper.
- Runability, or the efficiency with which the paper can be printed, and print quality are important factors in selecting papers. Discuss them with the printer.
- If color ink is to be used on color paper, check the compatibility of the colors of both ink and paper.
- Type is most easily read when printed on a soft, white paper.
- Consider using specialty inks such as metallic or fluorescent inks.

▼ Graphics in Action

1. Start a collection of paper swatches by visiting the nearest paper wholesaler and seeing if samples of the basic types of papers are available.
2. Obtain samples of papers of various weights and textures. Examine them under a magnifying glass and see if you can determine which is the felt and which the wire side of the sheet. Fold them with and against the grain. Note the differences, if any.
3. Research paper making and make your own paper. There are a number of books available that give instructions.
4. Select a paper that would be most suitable for the following. Consider weight, texture, and color. Explain the reasons for your choices.
   a. A brochure announcing an exhibit of famous paintings.
   b. An announcement of an open house at a computer center. The announcement will include high-contrast art.
   c. A magazine devoted to home woodcraft. It will contain much line art.
   d. A newsletter that requires the most economical sheet with the highest opacity.

Putting It Together

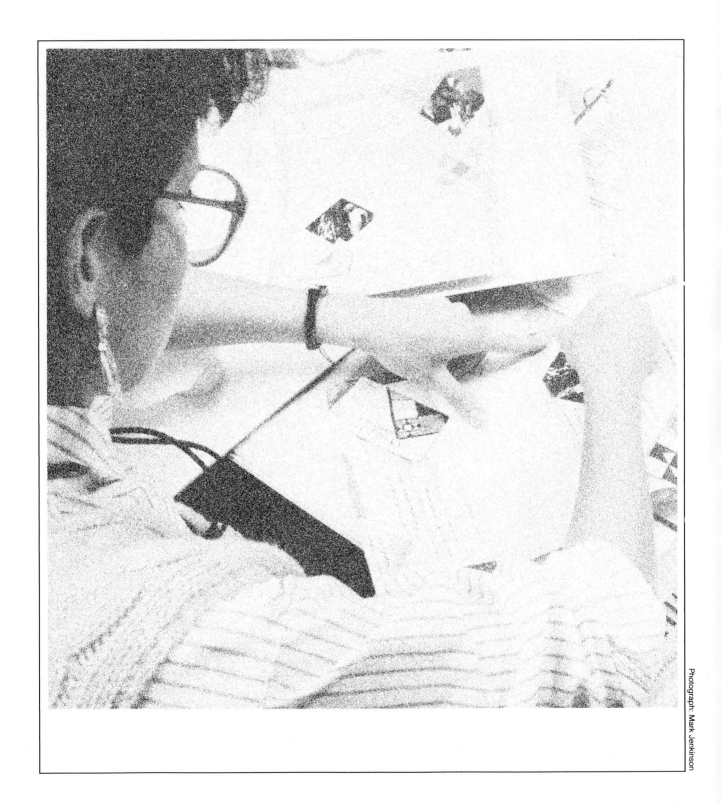

Photograph: Mark Jenkinson

Now that various elements that go into making an effective communication have been examined, we can consider the job of actually putting them together in a complete *layout* (a working drawing similar to an architect's blueprint that the editor or designer creates as a guide for constructing a printed communication). In approaching layout, keep in mind that there are three general categories of materials that people read.

First, there are printed communications that people want to read. These include newspapers, periodicals, and books. Since people are willing to pay good money to obtain these objects, we should strive to make them as pleasant and interesting as possible.

Second, there are printed materials that people must read. These include timetables, government reports, research papers, and income tax forms. Since people must read these materials whether they really want to or not, it is important to make them clear and understandable.

Third, there is material that people must be coaxed to read. This includes propaganda, public relations releases, and advertisements. A greater effort is needed here to lure people into the message and hold them until its end.

In all of these situations, the secret of using typography to its best advantage includes an understanding of the principles of design and how they can be employed in printed communications. These basic design principles include the principles of proportion, balance, contrast, harmony, rhythm, and unity. Application of these design principles will help answer the question, How do we put it all together?

We should note that some of the most effective works produced by creative people have been accomplished by deliberately breaking the rules of good design. Before attempting this sort of experimentation, however, we need to understand the rules we are breaking.

▼ The Principle of Proportion

In planning a layout where do we begin? First of all, we must settle on a shape.

Look around you. What are the most pleasing shapes you see? What are the most frequent proportions you encounter? Rectangles, right?

The ancient Greeks recognized the rectangle for all its pleasing qualities, and they built some of civilization's most attractive structures in the shapes of rectangles. In fact, this shape, as illustrated by the Parthenon, became known as the "golden rectangle." Its proportions are about three to five, and it has endured and pleased people down through the ages. Today this rectangle is encountered in doors, windows, table tops, pictures, and even the basic shape of human beings.

A square shape soon becomes monotonous. The rectangle, though uniform and precise, is not tiring because it offers variety in form.

So, let's settle on a rectangle shape for our layout for a start. Certainly, we can produce effective and interesting layouts by going to squares and breaking out of the rectangular confines. A variation in shape can be the most effective approach for certain jobs when it seems important to be

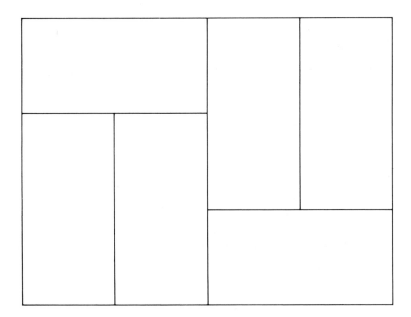

Fig. 8-1 The Japanese designed the rectangular tatami mat several centuries ago. It is still cited by graphic designers as an example of pleasing rectangular proportions.

different. But the rectangle will prove most satisfactory for most printed material. Books will fit on shelves properly, brochures will fit standard envelopes, and paper will be used most efficiently. Costs and waste can be kept to a minimum since standard paper sizes are rectangular.

In addition to the dimensions of the printed page, proportion should be considered when planning other elements in the layout. These include the margins and the relationship of the type and art with each other and the whole.

For instance, margins inside and outside a border should be unequal. Equal white space creates a monotonous pattern. Unequal margins break this monotony and present a more interesting layout. Also, the margin outside the border should be larger than that within the border. The border is part of the printed portion of the layout and it should present a feeling of unity with the other printed elements.

In general, we should select shapes of type styles and art that have a proportional relationship to the dimensions of the whole layout. Long, thin types and art go well in long, thin layouts. Short, wide type styles and art carry out the proportions of short, wide layouts. Here again, however, this is a generalization and would not hold true in every layout situation. Each should be considered individually and each should result in a layout that does the job best regardless of what has evolved as the "accepted" practice.

Throughout history the golden rectangle has been the dominant design dimension. It is derived from what mathematicians call the golden section, a formula that was a product of the revival of learning during the Renaissance in fourteenth- and fifteenth-century Europe. The true golden section is a ratio of .616 to 1.0 so that an area 5 inches wide would be 3.09 inches deep.

We have settled on standard sizes of 3 by 5, 4 by 6, 5 by 8, and so on for practical purposes such as standard paper sizes and the capabilities

Fig. 8-2 Equal margins (*left*) are monotonous. Unequal margins (*right*) are interesting and create better unity.

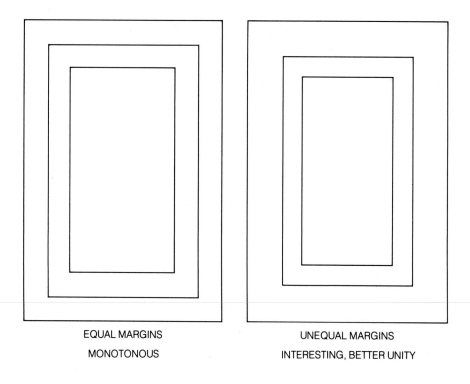

EQUAL MARGINS

MONOTONOUS

UNEQUAL MARGINS

INTERESTING, BETTER UNITY

Fig. 8-3 Most books and pamphlets are proportioned to about a 1:1.4 ratio. The height is 1.4 times the width. A square book has a 1:1 ratio. A book with a height of two times its width has a 1:2 ratio. A pleasing inequality in proportions is usually best in graphic design.

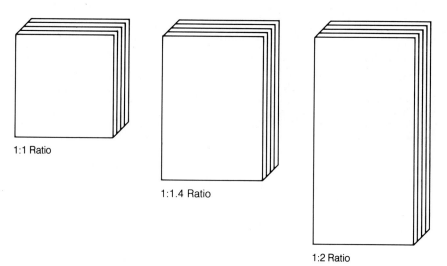

1:1 Ratio

1:1.4 Ratio

1:2 Ratio

of printing presses. Generally the most pleasing page size is considered to be one in which the length is one and a half times the width.

The dimensions of the area of a layout is an important design decision. The first thing the eye notices is the shape of a layout. A layout is said to be well proportioned if its shape is interesting to the eye and its parts are related in shape, but not monotonous in size, and the complete arrangement is attractive and effective.

Once the question of the general proportions of the layout has been resolved, we can consider where to place the elements. We begin by examining the optical center and its relationship to balance.

▼ The Secret of the Optical Center

The finished layout will be worthless if it does not stop the reader and arouse the reader's interest. We must make critical initial contact. We can do this and arrange the elements in the layout in the most pleasing and effective way possible if we put the "secret of the optical center" into action.

The *optical center* is the spot the eye hits first when it encounters a printed page. If we take any area and look at it, we will find that our eyes land on a point slightly above the mathematical, or exact center, and slightly to the left. Try it. Open a newspaper and consciously note what you see right off. It will be somewhat above and slightly to the left of the fold. It will take a mighty compelling element to pull your eyes away from this spot.

The optical center is determined by dividing a page so that the upper panel bears the same relationship to the lower panel that the lower panel does to the entire page. That is easier done than said. If an area is divided into eight equal parts, the point located three units from the top and five from the bottom is the approximate optical center.

Luckily we will not have to take a ruler and figure things that closely every time we make a layout. With a little practice in making rough layouts or planning pages, we will get the habit of orienting layouts with the optical center in mind.

We should always take advantage of this natural aid and utilize it as the focal point or the orientation center from which to construct a layout that will make the all-important initial contact. If an element is chosen for this spot that will stop the reader, such as a striking piece of art or a dramatic headline, we are on the way to achieving effective communication.

The optical center, then, is the focal point or fulcrum for placing elements on the layout, and it goes hand in hand with the second principle of layout, the principle of balance.

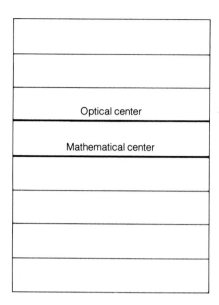

Optical center

Mathematical center

Fig. 8-4

Fig. 8-5 A line on the mathematical center actually looks awkward and below the center. Lines placed on the optical center give a pleasing sense of balance.

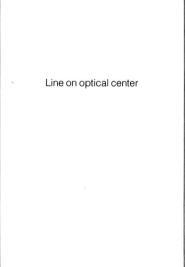

Line on mathematical center

Line on optical center

Three lines placed
in proper position
on the page

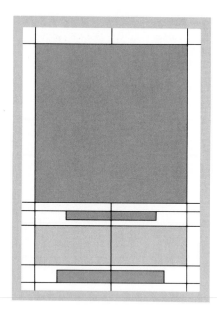

Fig. 8-6 In formally balanced layouts, type lines, illustrations, and other elements are placed left and right of center and above and below center in equal weights for a precise, orderly arrangement.

▼ The Principle of Balance

We need balance in our environment. When things are unbalanced they make us uneasy. Balanced objects look proper and secure.

Balance in printed communications is a must. We must place elements on the page in a way that will make them look secure and natural—not top-heavy, not bottom-heavy. We can do this in two ways. We can balance them formally or informally. (Some typographers call these symmetrical or asymmetrical balance.)

There are times when formal balance is just what is needed. Formal balance places all elements in precise relationship to one another. Formal balance gives us the feeling of formality, exactness, carefulness, and stiffness. Formal balance may be used for luxury car advertisements, wedding announcements, and invitations to white-tie-and-tail events. The *New York Times* and the *Los Angeles Times* occasionally have formally balanced front pages. They help support the image of a no-nonsense, precise publication.

If our communication is formal, dignified, and reserved, a formally balanced layout will help transmit this message. If our target audience has the same qualities, then formal balance may appeal to it.

A formally balanced layout has elements of equal weight above and below the optical center. To the left and to the right everything is the same. If we have a strong display type 6 picas from the top of our layout, we will need a line of the same size 6 picas from the bottom. If we have a piece of art left of center and slightly above, we need a similar element to the right and in the same position. Left and right, up and down, the formally balanced layout has elements of equal size and weight.

Fig. 8-7 Symmetrical or formal balanced layout.

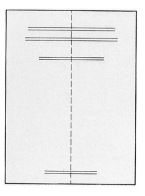

Art Auction Brunch

Formal balance has its place in layout work, but it is too stiff and uninteresting for many situations. In most layout work, balance is achieved informally. Elements of similar weight, but not necessarily precisely the same, are placed in relationship to one another so that there is weight at the bottom of the layout as well as the top, and to the left and right to balance the whole. Stability is achieved but the balance is dynamic rather than static.

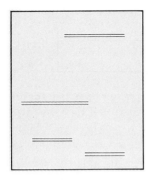

Fig. 8-8 Asymmetrical or informal balance layout. (Courtesy S. D. Warren Company, a subsidiary of Scott Paper Company)

Balance in layout is achieved through the control of size, tone, and position of the elements. It is more a question of developing a sense of balance by constant study and awareness than of following fixed rules devised by someone else.

However, there is a way to go about developing this sense of balance. The starting place, as we mentioned, is the optical center.

If we imagine the optical center as a fulcrum and place elements on the page so that this fulcrum is the orientation point, we will begin to see them fall into balanced positions. It is much like children achieving balance on a teeter-totter (or seesaw, if you prefer). If a child who weighs 80 pounds sits on one end of the teeter-totter and one weighing 40 pounds climbs on the other, what happens if they are the same distance from the fulcrum? The same thing happens with unbalanced layouts—they appear to topple over.

Applying this idea to layout work, if a single line of type is placed on a page it should be at the center of balance, the optical center, for best appearance. The same is true with a single copy block.

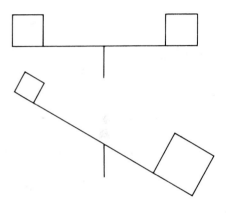

Fig. 8-9

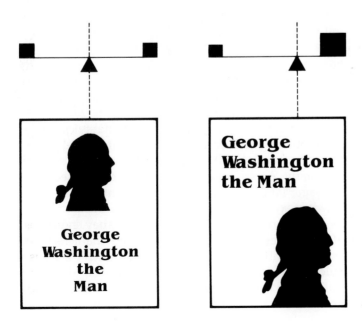

Fig. 8-10 In these layouts, formal or symmetrical balance is illustrated on the left and informal or asymmetrical on the right. A teeter-totter and fulcrum diagram illustrates the principle of balance.

Fig. 8-11 Two groups of equal size placed on a page in balance from the optical center (*left*) and one group half the size of the other placed twice the distance from the optical center for proper balance (*right*).

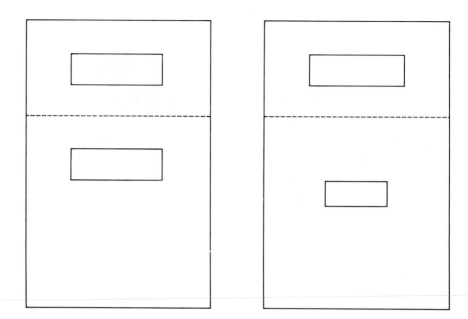

If two or more groups of design elements are placed on a page they should be balanced visually with the optical center serving as a fulcrum.

▼ The Principle of Harmony

All the elements in the layout must work together if a communication is going to do its job. They cannot deliver the message to the target audience if they are fighting among themselves. The layout and everything in it must be in harmony.

In general, effective layouts achieve harmony in three ways—through harmony of shapes, harmony of types, and harmony of tones.

Shape harmony is the first cousin to proportion and it means that the general structure of the elements is the same. The shape of the type should be the same as the shape of the page or printed area and the shape of the art and copy should follow this pattern.

Type harmony means that the letter designs of the styles selected should cooperate and blend together and not set up visual discord. Type harmony can be achieved by staying with types of one family and selecting similar series of that family. Avoid, for instance, mixing condensed and extended types of the same family. However, some judgment is needed here, as we will see when we discuss the principle of contrast.

If different types are used to provide contrast, they should be radically different. Two families of the same race that have characteristics basically similar may not be different enough to provide much contrast. They will clash, and destroy harmony. And they can just plain look bad together.

Two families from different races can provide harmony and contrast, but one should clearly dominate. If they are the same, or nearly the same,

Fig. 8-12 Shape harmony in which the general structure of the art and copy blocks are the same is a pleasing arrangement.

size, there's a chance that the effect will be jarring and destroy the harmony.

Avoid mixing lowercase and all-cap words. Capitals are formal and dignified and do not harmonize with the irregular shapes of lowercase letters. Stick to one or the other whenever possible.

If boldface type is selected for more than an occasional emphasis, other elements in the layout should be strong as well.

Borders should have shapes in common with the type letter style for type and border harmony. A Black Letter type looks best with a decorative border, and a modern Roman, with its thick and thin letter strokes, goes well with an Oxford rule, which has parallel thick and thin lines.

Tone harmony refers to the weights and designs of elements. Bold illustrations and bold types harmonize. Ornamental borders and ornamental types go well together. A straight-line rule will harmonize with a straight-line Sans Serif type.

FUTURA LIGHT

FUTURA DEMIBOLD

FUTURA BOLD

▼ The Principle of Contrast

What a monotonous world this would be if everything was the same. The changing seasons, the mountains, lakes, and oceans of our earth, the various groups of people who make up the human race, all add interest and contrast to life. Sameness is boring in life, and it can be boring in printed communications as well.

Skilled speakers use voice modulation, pauses, and gestures to make their points and hold the audience's interest. Skilled communicators also use various devices to emphasize important elements in their printed material. In addition, they add variety and interest by applying the principle of contrast.

Fig. 8-13 Elements in graphic design should harmonize. Light typefaces should be used with light art and borders, bold types with bold art.

Fig. 8-14 Contrast can be achieved and harmony maintained by varying the type sizes and styles but keeping the same family. (Courtesy Westvaco Corporation)

Contrast gives life, sparkle, and emphasis to a communication. Contrast shows the reader the important elements. And contrast helps readers remember those elements.

Contrast can be achieved by a number of typographic devices. The most obvious of these is the occasional use of italic or boldface types. However, these should be used sparingly, like seasoning in a stew. Too much contrast creates an indigestible typographical mess.

Other ways of achieving contrast include varying the widths of copy blocks, breaking up long copy with subheads, varying shapes of elements, balancing a strong display against a lighter-toned text mass, surprinting (printing type lines on top of halftones or other art), and enlarging one in a group of photos.

A good starting point in achieving contrast in layouts, and a good way to approach handling copy, is to break down the copy into "thought phrases." Thought grouping means taking the copy and positioning it on the page as it might be spoken. Headlines and lines of copy should be written in natural phrases that the reader can take in at a glance.

Suppose we are writing a title for an article and we come up with this wording:

The Miracle of Joe's Discovery

Now, suppose we want to run this head in two lines. How should we divide it? The easy way would be to divide it so both lines would be about the same length. We would end up with something like this:

The Miracle of
Joe's Discovery

But that division makes an unnatural pause in the way the reader would decipher the head, since reading is accomplished by grasping the meaning of words in groups of coherent phrases or sentences. So a better way to set the head would be to divide the wording into natural thought groups, something like this:

The Miracle
of Joe's Discovery

This head can be taken in at a glance, and the uneven lines add interest and contrast to the arrangement. (Note: The arrangement of the type lines might be dictated by the "head schedule" used by the publication. We consider this in our discussions of magazine and newspaper layout.)

A little more complex example, but one that better illustrates the principle of contrast, is the way the following simple advertisement is handled. Suppose we have this copy to arrange:

Spend your summer vacation high in the Sierra at Wayside Inn

Let's set this in display type:

Spend Your Summer
Vacation High
in the Sierra
at Wayside Inn

What is wrong? The message is almost incoherent because the lines of type cause awkward and unnatural pauses when the copy is read. Let's use thought grouping to make this easy to read and comprehend. Is this better?

Spend Your
Summer Vacation
High in the Sierra
at Wayside Inn

The next step is to bring the principle of contrast into play. We can add emphasis where it belongs to create a more interesting and memorable message:

Spend Your
Summer Vacation

High in the Sierra

at

Wayside Inn

Contrast thus relieves monotony, adds emphasis where it belongs, makes layouts more interesting, and helps effective communication.

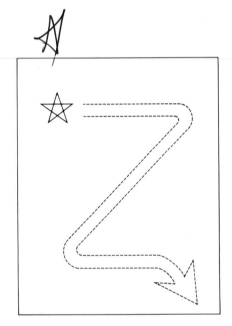

Fig. 8-15 Studies have shown that when a person scans a printed page, the usual eye pattern follows a rough Z through the page. Initial contact is made in the upper left near the optical center, then the eyes move to the right and then diagonally down the page to the lower left and then to the lower right.

▼ The Principle of Rhythm

Layouts that communicate effectively are not static pieces that just lie there on the page. They are alive and they move. In writing, just as in music, life is added by imparting action, variety, and interest to the message. While music communicates with sound, printed communications use art and words to create a beat, or a rhythm, by measuring and balancing the movement of vision.

Three ways to get layouts moving include the natural placement of elements, using repetitious typographic devices, and arranging material in logical progression. Let's consider each of these methods.

We can place elements to take advantage of the natural path the eye follows as it travels through a printed page. As mentioned, the eye hits a point slightly above and to the left of the mathematical center of an area. What happens next? The eye travels to the right, then to the lower left, and then across to the lower right in a sort of Z pattern.

If we arrange the elements on the page in a logical order along this Z path, we can take advantage of nature to keep our layout moving.

A beat can be established by repeating typographic devices. Initial letters, boldface lead-ins, numbers or small illustrations, indented paragraphs, italic subheads, all not only provide contrast but if placed in a logical order can help direct the reader through the message and thus give it rhythm and motion.

▼ The Principle of Unity

Unity holds a design together and prevents looseness and disorder. When we see a loose printed communication, our eyes can't find a center of interest and they bounce around with no place to land.

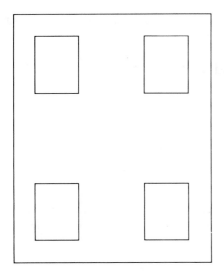

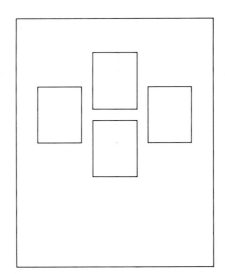

Fig. 8-16 Where should you look? Elements placed in the corners, or scattered, as on the left, create disunity as the eye bounces from one to the other. When elements are grouped together, a center of interest is developed and unity is achieved.

Simplicity is the key here. Unity and simplicity aid communication and eliminate distraction. Keeping it simple is simple. Here are a few ideas:

- Stick to a few type styles, preferably one family if possible.
- Keep the number of shapes and sizes of art and types to a minimum.
- Place art and heads where they won't interfere with the natural flow of reading matter.
- Have one element in an illustration or one illustration in a group of illustrations dominate. Create a center of interest.

Unity can be achieved in ways that are obvious and some that are not so obvious. Obvious ways include enclosing everything in a border, isolating the layout with white space, and using the same basic shape, tone, color, or mood throughout.

Using boxes in a layout can help unify it if they are all similar in thickness, design, and tone. Being consistent in the use of typography is one of the easiest and best ways to ensure unity.

Unity can be achieved by applying what some artists refer to as the *three-point layout method*. So much of life is organized around the unit of three. In religion there is the Father, Son, and Holy Spirit; in nature there's

Fig. 8-17 Unity can be achieved by applying the "rule of three." People tend to unify elements when they appear in groups of three. (Courtesy A.B. Dick)

Fig. 8-18 This advertisement illustrates important principles of design. Unity is achieved by the framing palm trees and the coupon. Attention is directed by the reverse headline near the optical center, and the obvious motion through the advertisement is toward the "action" or coupon placed in the lower right corner where the reader will exit the page. (Courtesy *Holiday* magazine, now published by Travel-Holiday Magazine, Floral Park, New York 11001)

the earth, sea, and sky; in human needs there's food, shelter, and clothing. In the academic world, the liberal arts are grouped in units of three: natural sciences, social sciences, and humanities. In our governmental structure, we have federal, state, and local governments. And our flag combines red, white, and blue.

When we see three units together we tend to unify them. In layout, headline, art, and copy are the basic three units used to create one whole. The number of units of art, the number of headlines and subheads, or the number of copy blocks may likely vary within these three basic units. In deciding the number of units for layouts, keep in mind that odd-number units are more interesting than even-numbered ones. Three and five illustrations can make a layout more interesting than two or four.

Which principle of design is most important? This question is obviously irrelevant. Proportion, balance, contrast, and rhythm must all blend together. However, if all the elements aren't combined in a unified and harmonious composition, attention value is scattered and interest declines. The completed layout must have unity and it must have balance. It must display pleasing proportions, an obvious directional pattern, and type and tone harmony.

Although principles of design can be helpful guides, the positioning of elements and the effective regulation of size, shape, and tone depend more on a sense of correctness and good taste than strict adherence to arbitrary rules. This sense is developed by experience and practice, which increase our awareness of balance, proportion, unity, contrast, rhythm, and harmony.

> ## *What Could It Be?*
>
> It could be called a malleable finite cylindraceous cell wrought of parallel axes with azimuthal terminates. It remains one of humanity's best designs because it works all the time; it is cheap, fast, versatile, and, above all else, it is *simple*.
> It is a paper clip!
> As with paper clips, simplicity is also a great virtue in graphics. Eliminating the unnecessary and concentrating on the simple, the designer can create graphics that will help readers understand information quickly and clearly.
> Keeping graphics simple makes them easier to produce and print as well.

▼ Effective Design Checklist

- Balance can be obtained through control of size, tone, and position of elements.
- Balance can be upset by nonharmonizing typefaces and too many non-essential elements in a layout.
- A unifying force should hold the layout together—white space, borders, and consistency in shape, size, and tone of elements can unify a layout.
- Equal margins are monotonous. There should be more margin outside a rule than inside, but the rule should not crowd the type within.
- Contrast adds interest. It can be achieved by varying widths of copy blocks, enlarging one in a group of pictures, or using italics or boldface, but sparingly.
- Orderly repetition of some elements can give motion to a layout.
- Long horizontal or vertical elements will cause the reader to follow their direction.
- The space within a layout should be broken up into pleasing proportions.
- Simplicity is important for attractive layouts. It can be achieved by using a few type styles and reducing the number of shapes and sizes of art.
- Use design clichés with caution. These include picture cutouts in odd shapes, tilting art or type, setting vertical lines of type, or using mortises, overlaps, or tint blocks.

▼ Making the Layout

Now it is time to put the principles of design to work. We might approach the construction of a layout by using a simple advertisement to illustrate the points involved. A communication can include a number of parts: headline, art, copy, logo. How these parts are arranged is the layout. The layout is the blueprint. This blueprint could be a *rough* (a working sketch with just enough detail to guide the person who puts the parts together), or it could be a *comp* (for comprehensive), which is a more complete rendering of the layout, or it could be a *mechanical* (a layout complete in every way and ready to be photographed for plate making and printing).

The layout is one of three parts of the package the communicator might pass on to the producer, compositor, or printer if he or she does not create a camera-ready mechanical on a computer. In addition to the layout, the package includes the typed copy and the art (photographs or drawings).

The rough can be used as a guide in creating a comp or mechanical, either by using the pasteup method or doing the job on a computer screen. Many people seeking entry-level positions in the communications industry will find they have a definite advantage over the competition if they have acquired some layout, pasteup, and computer graphics skills. Quite often mechanicals are made combining pasteup and computer-generated design elements.

Let's go to work and create a rather uncomplicated layout. A good place to start is with advertising layout, as it often includes most of the basic elements—art, copy, head or title, logo—that designers devise for a printed communication.

Here is the situation. The manager of the Cotton Shop has asked a designer to plan a layout for an advertisement to appear in the local newspaper to help introduce the arrival of a new type of sports shirt. The manager has the heading, the paragraph of sales copy, the price of the shirt, a piece of art showing the shirt, and the store's logo. The designer has these elements to work with:

Fig. 8-19 This is the shirt illustration. (This problem is adapted from *Newspaper Advertising Handbook*, with thanks to the author, Don Watkins, and publisher, Dynamo, Inc.)

The Shirt.

It's here now . . . The Shirt . . . It's cotton and it's sweeping the country . . . cool, calm, and collectible . . . just right for this season's fashion mood. Long sleeves. Stripes or solids.

$25
Visa & Master Charge
the COTTON shop
(address)

The designer has to arrange this copy in a space 2 columns wide and 8 inches deep (4 by 8 inches). Then the arrangement will be given to the store manager for approval before it is printed.

The steps the designer uses in planning and making the advertisement layout are an effective guide for planning all types of printed communications, whether for print media or graphics for television.

Fig. 8-20 Layout can be accomplished on the computer, eliminating the tools of pasteup. However, many designers make pencil sketches and thumbnails before composing electronically. (Courtesy Xerox Corporation)

Visualization

The first step in developing a layout requires some brainstorming. This is called *visualization*. The communicator's most important job is not done with a drawing pencil, it is done with the brain. The communicator must "see" in the mind's eye how a layout is going to appear. For example, the newspaper publisher wants the paper to look dignified, quiet, and authoritative. How do you perceive such a newspaper's front page in your mind's eye? What sort of headlines, type styles, and nameplate will transmit this image? What do the illustrations look like?

A company wants a brochure for its anniversary picnic. What should be its shape, size, and color? What sort of finished appearance should it have? Brainstorm possibilities and let them bounce around in your head. (Well, it's not quite this simple, as we will see in Chapter 11. Brainstorming in this case must include knowledge of such things as how the brochure will be distributed, but these examples do help illustrate the visualization step.)

Visualization gives direction. It gets us started. It takes an idea or concept and translates it into visual form. Try it and see if this is true. Suppose you have the job of designing an advertisement using this copy:

> The surf is up at Waikiki. Take a winter break in the sun and surf at the Breakers Hotel.

How do you visualize the ad? The possibilities are many: the sun, surf, and sand; palm trees swaying; the hotel and the beach; people surfing. What sort of art would be best, a photo or a drawing? What sort of type style would say it best? What words would you want to stand out? How would you arouse the reader's interest, make that reader stop and want to know more?

That's visualization.

The designer who has to arrange the elements for the Cotton Shop advertisement will not have the opportunity of doing a lot of creative

Fig. 8-21 Thumbnail sketches of possible arrangements for the Cotton Shop advertisement. (Adapted from *Newspaper Advertising Handbook*)

Fig. 8-22 Headlines on roughs can be indicated in this manner. Guidelines should be drawn to the exact size of the type specified in the comprehensive, but the rough lines only need to simulate the type.

Fig. 8-23 A rough scribble of the approximate size and shape of the art is adequate for rough layouts.

Fig. 8-24 Lines can be sketched for roughs to indicate body matter. The size of the copy block and the width of the lines should be exact, however.

visualization because the art and copy have already been determined. However, before the layout process begins, our designer friend will visualize several possible arrangements and type styles.

Thumbnails

The next step in producing a layout is to get something down on paper. Generally three steps are needed to move from the concept to the finished layout. They are the creation of thumbnails, roughs, and comprehensives.

Thumbnail sketches are miniature drawings in which the designer tries out different arrangements of the elements. The goal is to experiment and settle on the most effective placement of the headlines, art, and copy to achieve two goals: (1) to make the most attractive arrangement possible by putting into practice the principles of effective design by applying the basics of good typography, type selection, and arrangement; (2) to construct a working blueprint that can be understood and followed by everyone engaged in the production of the communication.

We don't have to be artists to produce thumbnails, or roughs for that matter. Headlines can be indicated by drawing guidelines to indicate the height of the types to be employed and then filling in between them with up and down strokes.

For easy work and best results, try using a soft pencil for making thumbnails and roughs. The best are either 2H or HB. If softer pencils are used, they are likely to smudge and need to be sharpened more often (pencils numbered 2B to 6B are very soft). Harder pencils are difficult to use, will punch holes in the paper, and the lines are light and difficult to erase. Pencils numbered 3H to 9H are hard. You can obtain 2H or HB pencils at most stationery sections of discount stores or at art supply houses, and they do not cost much.

Art in thumbnails and roughs can be indicated by an outline of the general shape of the illustration, shaded to match its tone.

Body copy or copy blocks can be designated by squares, rectangles, or whatever shape the set type takes. Lines should be drawn in these areas to simulate the width of the type and the leading.

A very rough approximation of the size and shape of the elements is adequate when making thumbnails. Roughs need to be more exact.

Quite often we will need to make several thumbnails before we find an arrangement that we believe to be the best. An easy way to get started is to take a sheet of layout paper the size of the finished layout and fold it into quarters or eighths. This will give four or eight areas for thumbnails in exact proportion in the finished layout.

As an example, the designer working on the Cotton Shop advertisement might make thumbnails in areas 1 by 2 inches or one-fourth the 4 by 8 size of the finished advertisement. Since this is such a small area in which to work, it is more likely that the designer will use a 2 by 4 or a full-size 4 by 8 area.

With a little practice most communicators can produce credible thumbnails. This is the first step in the creation of advertising layouts, and it is the first step in creating layouts for newspapers, magazines, brochures, and other printed communications.

After experimenting with the various possible arrangements, the layout artist selects the one that appears to be the most effective. A rough layout is then produced with the thumbnail serving as a guide.

Producing Rough Layouts

The rough layout is a drawing the actual size of the finished advertisement. All elements are presented fairly clearly and accurately as far as size, style, spacing, and placement are concerned. It is not quite as finished as the comprehensive, however.

Display type is lettered in and art is sketched in the same size and tone as in the final product; there is enough detail so that the art is a close approximation of its final form. Quite often a copy of the original art will be made and pasted in place on the rough.

Text copy, or reading matter, is still indicated by drawn copy blocks, but these are made precisely as the finished product will look, with lines drawn, or "comped," to indicate the type size and leading to be used. Sometimes "greeking" or simulated type is used for copy blocks in roughs.

There are a number of methods available for producing display lines and headlines for beginning layout work. People planning careers as graphic artists will, of course, want to develop lettering ability. One of the procedures used to develop this skill is useful for communicators who want to have some layout skill but who do not necessarily want to go beyond the rough stage in layout work. The technique is *comping type.*

Both reading matter and display type can be comped for rough layouts. The results will be accurate enough to give the precision of the finished product and realistic enough to show how it will appear.

Reading matter is comped by drawing lines to indicate the type size and the leading or space between lines. This is done in two ways. An ordinary pencil can be used to draw two thin lines to indicate the tops and bottoms of the *x* heights of the lines of type.

Or a chisel-point pencil can be used to draw a solid line to represent the lines of type. This kind of pencil can be obtained at most art supply stores. If the point is too wide to represent the size of type desired, it can

Fig. 8-25 The two-line method of indicating reading matter on layouts. The two lines indicate the *x* height of the type chosen.

Fig. 8-26 The solid-line method of indicating reading matter for copy blocks on layouts. The solid line is the same height as the *x* height of the type selected.

Fig. 8-27 A rough layout (*right*) is made from the thumbnail (*left*) selected from those sketched for the Cotton Shop advertisement.

be shaved down with a single-edge razor blade. A double-edged blade should not be used since it isn't sturdy enough.

Pencils for comping body type should not be too soft. They will not hold the point long, and it will be easy to smear the comped lines. Layout artists recommend HB or 2H pencils.

The procedure for comping solid lines of reading matter includes these steps:

1. Tape the paper to be used on a flat surface (a drawing board is best). Use a T square to be sure the paper is lined up square with the edge of the surface so the T square can be used to draw straight lines.
2. Draw a light outline of the area the type will occupy (the copy block) with the T square and a triangle.
3. Select the type for the finished layout. Indicate the proper leading between lines. Now make small dots vertically in the area to be comped to indicate the base of the *x* height of the letters to fill the area.

THUMBNAIL

ROUGH

4. Check to see that the pencil point is the same as the *x* height of the type selected.
5. Use the straight edge of the T square as a guide to draw the lines to represent the lines of type.

If the pairs-of-lines method is used, the same procedure is employed except that dots are placed to represent both the top and bottom of the *x* height of the letters.

After some practice with full lines, try comping with equal paragraph indentations and lines that are ragged right, ragged left, and centered.

Display type can be comped by tracing. The tools required are a complete alphabet of the size and style of type to be used, a drawing board, a T square, a piece of tracing paper, a ruler, and a hard pencil with a sharp point. The steps include:

1. Draw a light baseline for the type on the tracing paper.
2. Place the tracing paper over the alphabet at the point where the word should begin. Be sure to line the drawn baseline up with the baseline of the alphabet.
3. Trace the first letter in the head or display line in outline form.
4. Proceed to trace the rest of the letters in the line. Take care to space the letters properly and to line each up evenly on the baseline. Time spent in doing the job carefully will pay dividends in the end.
5. Fill in the outlines and erase the guidelines as much as possible.

The letters can be filled in with pencil, ink, or felt pen. After a little practice, you will be surprised at the realistic results that can be obtained even with Scripts and Cursives, Black Letter, and ornamental types. If color is needed, the letters can be filled in with colored pencils, felt pens, or ink.

The display lines can be cut out and pasted on the layout. Sometimes the display lines can be traced in outline form and the image transferred to the layout. This is accomplished by rubbing the back of the traced area with a soft pencil. Then the line is placed in position on the layout and the letters traced again. The pencil coating on the back of the tracing paper will act as a carbon to transfer the image to the layout. Again, the letter can be filled in with colored pencils, felt pens, or ink.

There are alphabets of display type that can be transferred to a layout by rubbing. The procedure is the same as when tracing. A baseline is drawn and the letters are lined up for even horizontal and correct vertical spacing. Instead of tracing letters, you transfer them from the master alphabet to the layout by rubbing them with a ballpoint or rounded-point instrument.

Alphabets are also available in printed sheets and in pads of individual letters. The letters can be assembled in display lines and pasted on the layout.

Fig. 8-28 It takes very little practice for the novice to make excellent titles and headlines for layouts by using transfer letters.

The Comprehensive

The final step in constructing the layout is the comprehensive. Often this is followed by creating a mechanical, which is a pasteup or camera-ready layout completed exactly as it will appear when printed. The comprehensive requires careful work, and many graphics students prepare comprehensives for use in their portfolios. In the comprehensive the illustrations and headlines or titles are shown exactly as they will appear in print. If there is extensive body matter it might be indicated by using "greeking" or simulated type.

When the comprehensive is completed it is mounted on illustration or prepared grid boards and covered with a protective flap, as described in Chapter 9.

In the case of our advertisement, proofs of the type and art are pasted into place to make an exact rendering of the advertisement as it will appear in print. This is called a mechanical. It then can be used for making the printing plate.

But if our comprehensive goes to someone else who will make the camera-ready mechanical, we have one more step to accomplish—the *markup.*

Fig. 8-29   The comprehensive layout for the Cotton Shop advertisement.

The Markup

In much communication work, particularly advertisements, brochures, and flyers, the working layout is a rough that includes instructions for the compositor or printer. This is called a *markup*. The markup, illustrations, and type copy are turned over to the printer. As we noted, comprehensives—a further step—also need to be marked up.

All instructions to the printer are written—often in a color to prevent confusion with what is to be set in type—and circled. One method of markup on a comprehensive is to cover it with a protective flap of tracing paper taped to the top of the illustration board or grid and to write instructions on this flap. Abbreviations are used whenever possible and a sort of code is developed based on basic printing terms.

Instructions on a marked-up rough or comprehensive might include:

- The family, series, and size of all type faces.
- The length of all the lines of type, plus the leading between lines.
- The line setting, such as centered, flush left, or flush right.
- Any special instructions for the particular job in question.

Below are the most frequently used terms for indicating instructions on roughs:

CAPS	Set in all capital letters
U&lc or clc	Set in capitals and lowercase letters
lc	Set in all lowercase
pt	Abbreviation for point size of type
BF	Set in boldface
8 on 10	(Also written as a fraction, 8/10.) The top number indicates the point size of the type and the lower number the leading between lines (in this case, 8-point type set on a 10-point base, or 8-point type leaded 2 points)
18	Set copy block 18 picas wide
8/10 × 18	Often the type size, leading, and line length are combined like this
⊐⊏	Center this line
⊏	Set flush left
⊐	Set flush right

The family names of types to be used are often abbreviated: Bod. for Bodoni, Bask. for Baskerville, Cent. for Century, and so on.

Copy blocks are indicated by letters circled on the rough and the same identifying letters are placed at the top of the page of the typed copy: Copy A, Copy B, and so on.

Illustrations can be keyed to the layout with numbers or letters and indicated as Photo 1, Photo 2, or Photo A, Photo B. Different practices may exist in different shops. Always check that the producer of the printing understands the instructions and how they are coded.

The techniques used to lay out effective printed advertisements can be used to advantage by anyone working with the printed word.

Tips for Ordering Typesetting

The following suggestions for ordering typeset matter can save money and help ensure that the finished product will be what you want.

- Specify the typefaces for reading matter and headings and make certain that the printer has these faces and sizes. If this is not certain, the notation "or equivalent" should be included on the purchase order or instructions.
- Specify the type size in points. If you are not certain exactly what size you want, do not hesitate to ask the compositor to recommend a size.
- For reading matter, keep the line measure about 39 characters. Remember that one and a half times the length of the lowercase alphabet is considered an ideal measure.
- Give the line measure (width) in picas, not in inches.
- Specify any copy that is to run ragged. Otherwise the compositor will justify the lines.
- Designate the leading desired (the space between the lines).

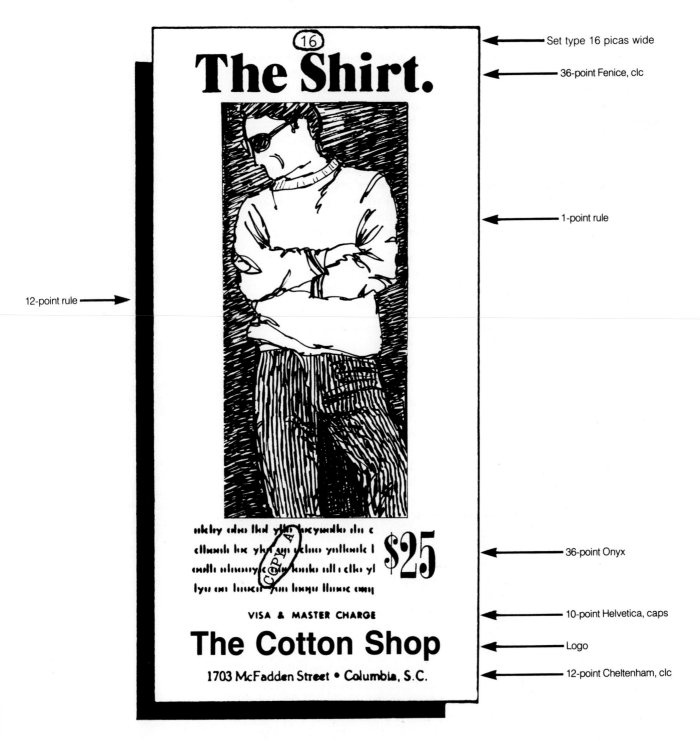

Set type 16 picas wide

36-point Fenice, clc

1-point rule

12-point rule

36-point Onyx

10-point Helvetica, caps

Logo

12-point Cheltenham, clc

Fig. 8-30 The comprehensive layout marked up and ready for the production department. "Greeking," or simulated type, has been used to indicate copy block A.

▼ Effective Design Checklist

- Use visualization. A designer's most important tool is still the brain.
- Integrate the verbal (words) and visual (graphics) elements so they work together to do the job in the most effective way.
- Feature the most important idea or selling point in the layout.
- Check every element in a layout—each should be irreplaceable. If an element can be eliminated or replaced, it should be eliminated or replaced.
- Determine the form and arrangement of the layout by the task it is intended to perform.
- Graphic elements in a layout should facilitate quick and easy comprehension of the message, offer additional information, or set the mood desired.
- Remember that if the layout doesn't get the attention of its intended audience, it will not be read.
- Avoid using Scripts, Black Letter, and decorative types for headlines. If they are used it should be for a very special reason, the line should be very brief, and it should not be in all capitals.

▼ Graphics in Action

1. Find in advertisements, art, headlines, magazine pages, or other printed material one example of each principle of design. Mount them neatly as exhibits and write captions explaining how the exhibit illustrates the principle. Type the captions and paste them on the exhibit adjacent to the illustrations.
2. Study the design of the front page of a newspaper. List the ways you believe the principles of design have been applied (for example, an italic headline may have been used for contrast). Evaluate the page from a design standpoint and point out any changes you might make if you were redesigning the page.
3. Find a two-page spread in a magazine. Clip and mount it on poster board. Assume that you are the editor of the magazine and the art director has asked for your evaluation of the design. Write an evaluation in the form of a memo to the art director.
4. Make four thumbnail sketches, each in an area 3 by 5 inches, using the following elements: a one-line headline about 36 points high; an irregular illustration approximately 1½ by 1½ inches; a copy block approximately 1½ by 1 inch; and a logo approximately 2 by 1½ inches.
5. Draw an area 5 by 7 inches on plain white or layout paper and arrange the following information in the area. This is a cover for a booklet. Sketch the words as carefully as you can to simulate the size and style of type for each line. Remember to apply the principles of design. Use rules or borders if you wish. The copy is:

   A Self-guiding Tour of _____(choose a city:
   Paris, New York, Moscow, Tokyo, your hometown), Walking
   Tours Publishing Company, Your Town, Your State

   (*Hint:* The first step should be thought grouping).

CHAPTER NINE

Preparing for Production

There are a number of steps to follow in preparing copy for production. Many of them have been discussed so far, but several still remain. For example, here is a statement with a headline:

```
Graphic Arts Defined For The Communicator*

Graphic arts is defined by Webster as ''the fine and applied arts
of representation, decoration, and writing or printing on flat
surfaces together with the techniques and crafts associated with
each.'' For the communicator, graphic arts might be defined as the
process of combining all the typographic and artistic materials and
devices available in a way facilitate communication.
   Graphic arts is not the technique of arranging types, illustra-
tions, borders, and other devices to create an unusual visual ex-
perience. Rather, the selection and arrangement of materials should
be to achieve the transfer of messages from a source to a receiver.
```

It is to be printed in a newsletter for communicators. What must be done to transfer the typewritten copy to typeset copy ready to be printed?

Some of the preproduction tasks that must be performed include determining the style and size of type for the headline and the amount of space the piece will occupy in the publication. After it is set in type it will have to be proofed and read for errors.

These steps, along with cropping and sizing art and marking the layout to ensure that the specifications are clearly understood by all who will be working on the job, are part of preparation for production. Pasteup can be a part of the preparation, too.

Some people working in communications do the entire process themselves. In other situations designers and technicians do much of the work. But all who work in the world of printed communications will find a knowledge of the steps in preparing for production worthwhile.

▼ Copy-Fitting without Fits

Old-timers in the industry could squint at a piece of typewritten copy and tell within a fraction of an inch how much space it would occupy when set in a certain size of type. They called it "casting off" type. We call it *copy-fitting*.

Copy-fitting is a pesky business until you get used to it. However, proficiency can come quickly with a little practice. There are some things to remember from the start. For one thing, copy-fitting is an estimation. There is no exact method of fitting copy. The formulas have to be leavened with judgment.

One problem with copy-fitting is that every letter of every font of type, along with the spaces and punctuation marks, is not exactly the same

---

*While standard English usage calls for lowercasing articles, coordinate conjunctions, and prepositions, regardless of length, unless they are the first words in the heading, some newspapers capitalize the first letters of all words in headlines. We will follow this pattern in this chapter.

width. If they were, copy-fitting would be simple. But since this is not the case, we have to use some judgment and make allowances in estimating how much space to designate for copy in layouts.

There are five situations in which copy-fitting might be used in planning printed communications. They include:

1. Estimating how much space a headline or title will take when set in a specific size and style of type.
2. Estimating how much space typed copy will occupy when set in a certain size and style of type on a predetermined width.
3. Finding out what size of type to use if the typed copy must fit a predetermined area in a layout.
4. Finding out how much copy to write to fill a certain space when the size and style of type have been selected.
5. Finding out how much space copy that has already been set in type or printed will occupy when reset in another size.

All of these calculations can be made easily with an understanding of only two copy-fitting methods. One is the unit-count system for display type and the other is the character-count system for reading matter.

Copy-Fitting Display Type

The *unit-count system* used to determine the size of display lines, headlines, and titles for articles is based on the assignment of units for the letters, numerals, punctuation, and space in a head or title line. The units are assigned on the basis of the letters' relative widths.

Of the twenty-six letters in the lowercase alphabet, eighteen are virtually the same width. These eighteen are assigned a unit value of 1.

All of the other letters and punctuation marks are assigned unit values in relation to these eighteen. For instance, *m* is about one and a half times wider than *x*, so while *x* has a unit value of 1, *m* has a unit value of 1½. The lowercase *i* and *l* are about one-half the widths of *x*, so their value is ½.

Below is a typical unit-count system for display type:

1 unit	All lowercase letters (except *f, l, i, t, m, w,* and number 1), all spaces, larger punctuation such as ?, ¢, $, %
½ unit	Lowercase *f, l, i, j, t;* capital letter *I,* spaces between words; most punctuation
1½ units	Lowercase *m, w;* all capital letters except *M, W, I*
2 units	Capital *M* and *W*

In solving a copy-fitting problem for display type, the first step is to determine how many units of the size and style of type will fit into the width selected and then see if the lines of the title or headline fit that limit. We can find this by taking a sample of the type and measuring the units that fit into the width.

For example, the headline noted earlier in this chapter was:

Graphic Arts Defined
For The Communicator

Let's say we want to set it in an 18-point medium Sans Serif type on a line 15 picas wide. Will it fit?

We first have to determine how many units of the type selected will fit into a 15-pica width. We can do this by taking a sample of the type and counting the units in a 15-pica line.

Here is how the units will count out:

Now we know that the maximum number of units of this type that will fit into a 15-pica width is 22½. So we check the lines we want to use to make sure each does not exceed 22½ units. If we count the units according to our system, we will discover that the top line contains 18½ units and the bottom line counts 20½ units. Both are under the maximum so the head should fit nicely.

That's all there is to the unit-count system. Communicators who work with type in planning brochures, newspapers, company magazines, and other publications soon discover that if they count everything as 1 unit and then make allowances if there are many wide or narrow letters in a line, they can estimate the proper fit quickly and easily.

Those who work on newspapers or magazines that use the same type-faces in each issue for headings, titles, and headlines can make up sample sheets of frequently used head forms and determine the maximum counts for each line. The maximum (and minimums) can then be noted on these style sheets, or *head schedules*, as they are called. (Head schedules are discussed in detail in Chapter 15 on newspaper design.)

Copy-Fitting Reading Matter

Several methods of copy-fitting are used in planning the reading matter for printed communications. Two of these methods are not very reliable so we look at them just briefly. It is assumed that the copy to be set will be justified. Designer Edmund C. Arnold points out that type set ragged right or unjustified takes up at least 3 percent more space than justified type. So if copy-fitting estimates are being made on type set unjustified, the estimate should be increased by at least 3 percent and even at that it will not be a tight estimate.

The copy-fitting method used in newspaper work is the word count system. This involves counting the number of words in the manuscript and then determining how much space that number of words will occupy when set in a certain type on a specified width.

First, you obtain a sample of the type style set in the specified width and calculate the average number of words per line. Then you multiply this average number of words per line by the number of lines of type in a column-inch (1 column wide by 1 inch deep), which gives the average number of words in a column-inch. Finally, you divide this number into the number of words in the typed copy, and the result is the number of column-inches the copy will occupy when set in type.

Another method that is seldom used is called the *square-inch method*. The area to be filled with the type is determined, and its square-inch capacity is calculated. Then the average number of words in a square inch is multiplied by the number of square inches designated in the layout. This gives the average number of words that can be fitted into the area.

For example, if an area 24 by 30 picas is designated in a layout, this will equal 4 by 5 inches or 20 square inches. If it has been determined that, say, 21 words will fit into a square inch, then the area will accommodate 420 words.

The square-inch method is used to estimate the number of words or the number of pages needed in books or fairly large pamphlets.

The most accurate method of copy-fitting, however, is the *character-count method*. It is universally accepted and used in communications work. It is based on determining the number of characters in a manuscript and then the number of typeset characters needed when this manuscript is set in type.

It works like this:

First, the number of characters in the manuscript is determined. There are several ways to calculate this, depending on the equipment being used. Some word processors maintain a running count of the total number of characters being written.

Typewriters are equipped with a character counting device of sorts. The scales on the paper guide and the paper bail on typewriters are calibrated for ten characters per running inch for pica (12-point) type and twelve characters per running inch for elite (10-point) type. If the average number of characters per line is multiplied by the number of lines, the average number of characters in the typed copy can be calculated.

In the example shown in Fig. 9-1, the typewriter stops were set at 15 and 65 to average fifty characters per line. The lines were counted and then the number of characters beyond the fifty limit on each line was added. Paragraph endings were counted as full lines, as they would be equally short when set in type. Thus it was determined that this copy contained 665 characters.

Next, the number of characters in a line of type in the designated width is determined. Suppose we wish to set our illustrative copy in 10-point type, with 2 points of space between each line (10/12 or 10 point leaded 2 points) on lines that are 15 picas wide. We must determine how many characters of type will fit into each line.

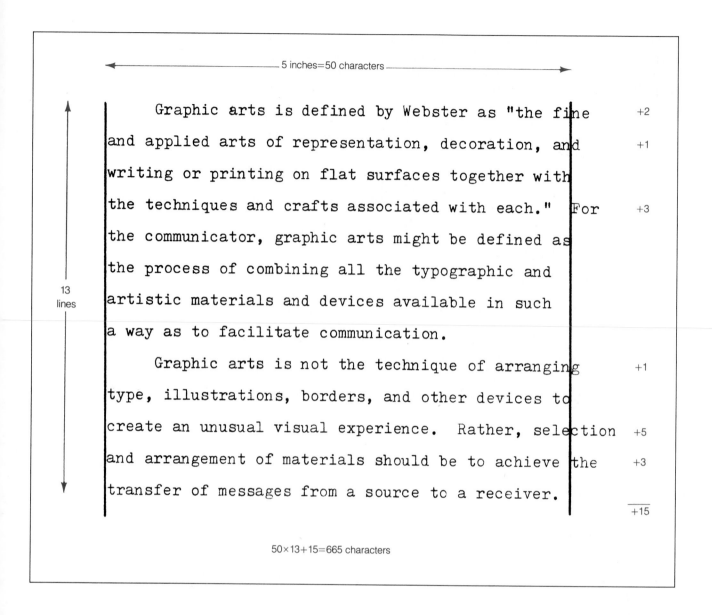

5 inches=50 characters

Graphic arts is defined by Webster as "the fine   +2

and applied arts of representation, decoration, and   +1

writing or printing on flat surfaces together with

the techniques and crafts associated with each." For   +3

the communicator, graphic arts might be defined as

the process of combining all the typographic and

artistic materials and devices available in such

a way as to facilitate communication.

Graphic arts is not the technique of arranging   +1

type, illustrations, borders, and other devices to

create an unusual visual experience. Rather, selection   +5

and arrangement of materials should be to achieve the   +3

transfer of messages from a source to a receiver.

+15

13 lines

50×13+15=665 characters

Fig. 9-1 An example of character counting with typewritten copy.

Manufacturers of types usually supply this information in the form of a certain number of characters per pica. In addition, we can obtain charts to help in the calculation. Here are some examples of typical types and their characters per pica:

Bookman, 10 point	2.60 characters per pica
Bernhard Modern, 10 point	2.99 characters per pica
Century, 10 point	2.70 characters per pica
Futura, 12 point	2.50 characters per pica

For our illustration, let's assume that we have selected Century. In 10 point, this type has 2.7 characters per pica. If we decide to set copy in lines 15 picas wide and multiply this width by the number of characters per pica (15 × 2.7), we will find the number of characters per line will be 40.5.

Now, if we divide this number of characters per line into the number of characters in the manuscript, we will discover the number of lines of type (665 ÷ 40.5 = 16.419 or 16.42). We will have 16.42 or 17 lines of set type. Since we have determined that we will have 2 points of space between lines, each line will occupy 12 points (10 plus 2) or 1 pica. The copy when set will occupy an area 15 picas wide by 17 picas deep.

Here is how it worked out when the copy was set in Century according to our instructions:

> Graphic arts is defined by Webster as "the fine and applied arts of representation, decoration, and writing or printing on flat surfaces together with the techniques and crafts associated with each." For the communicator, graphic arts might be defined as the process of combining all the typographic and artistic materials and devices available in such a way as to facilitate communication.
>
> Graphic arts is not the technique of arranging types, illustrations, borders, and other devices to create an unusual visual experience. Rather, selection and arrangement of materials should be to achieve the transfer of messages from a source to a receiver.

In summary, then, to find out how many lines typed copy will make when set in type we proceed as follows:

1. Determine the size and style of type.
2. Determine the width of typeset lines desired.
3. Calculate the number of characters in the typed manuscript.
4. From tables, or working averages, find out how many characters of the specified type will fit in 1 pica and multiply that by the width of the lines in picas.
5. Divide the number of characters in the typed manuscript by the number of characters in one line of type. (*Note:* 6-point type will have twelve lines per inch; 8-point, nine lines; 9-point, eight lines; 10-point, seven lines and a little over; 12-point, six lines, assuming that all these are set solid.)
6. Multiply the number of lines of type by the point size of the type with the space between lines added to find the total depth of the area in points. To convert to picas divide by 12.

Here are some other common copy-fitting situations you may encounter:

1. What size type should I use to fit the copy into a given area? *Solution:* Scan type specimens and decide on a size that seems to be about right. Calculate the number of lines in this size that you believe will fit into

ALPHABET LENGTH	CHARACTERS PER PICA	8	10	11	12	13	14	15	16	17	18	20	21	22	23	24	25	27	28	30
75	4.25	33	42	46	51	54	59	63	67	71	76	84	88							
80	4.00	32	40	44	48	52	56	60	64	68	72	80	84	88						
85	3.75	29	38	41	44	48	52	55	59	63	67	74	77	82	85					
88	3.65	28	37	39	43	46	50	54	57	61	65	72	75	79	82	87				
90	3.60	28	36	39	43	46	50	54	57	61	65	72	75	79	82	86	90			
93	3.52	28	35	38	42	45	49	52	56	59	64	71	73	77	80	84	87	95		
96	3.40	27	34	37	40	44	47	51	54	57	62	68	71	74	78	82	85	91	95	
100	3.30	26	33	36	39	42	46	49	52	56	59	66	69	72	76	79	82	89	92	99
103	3.22	25	32	35	38	41	45	48	51	54	58	64	67	71	74	77	80	86	90	97
106	3.15	24	31	34	37	40	43	46	49	52	57	62	65	68	71	75	78	83	87	94
110	3.05	24	30	33	36	39	42	45	48	51	54	60	63	66	69	72	75	81	84	91
113	2.95	23	29	32	35	37	41	43	46	49	52	58	61	64	67	70	73	78	82	88
116	2.90	23	29	32	35	38	40	43	46	48	51	58	61	64	67	69	72	76	81	87
120	2.80	22	28	30	33	36	39	42	44	47	50	56	59	62	64	67	70	75	78	84
123	2.75	21	27	29	32	35	37	40	43	46	49	55	57	59	62	65	68	71	76	82
125	2.70	21	27	29	32	35	37	39	42	45	49	54	57	59	62	65	68	70	75	81
127	2.65	20	26	28	31	33	36	39	42	44	47	52	54	57	59	63	65	69	73	79
130	2.60	20	26	28	31	33	36	38	41	44	47	52	54	57	59	62	65	68	73	78
133	2.55	20	26	27	31	32	36	37	41	43	46	51	53	56	58	61	63	67	72	77
136	2.50	20	25	27	30	32	35	37	40	42	45	50	52	55	57	60	62	67	70	75
140	2.45	19	24	26	28	31	33	36	39	41	43	49	51	53	55	58	60	63	68	73
145	2.35	18	23	25	27	29	32	34	37	39	41	47	49	51	54	55	57	62	65	70
150	2.30	18	23	25	27	29	32	34	36	39	41	46	48	51	53	55	57	62	65	69
155	2.25	17	22	24	26	28	31	33	35	38	40	46	48	49	51	53	55	59	62	67
160	2.15	16	21	23	25	27	29	31	33	35	38	43	45	46	48	51	53	56	59	64
165	2.10	16	21	23	25	27	29	31	33	35	38	42	44	46	48	50	53	56	59	63
170	2.05	16	21	23	25	27	29	31	33	35	37	41	43	45	47	49	51	55	57	62
175	2.00	16	20	22	24	26	28	30	32	34	36	40	42	44	46	48	50	54	56	60
180	1.95	15	20	21	23	25	27	29	31	33	35	39	40	42	44	46	48	52	54	58
185	1.90	15	19	20	22	24	26	28	30	32	34	38	39	41	43	45	46	51	53	57
190	1.85	14	19	20	22	23	26	28	29	31	33	37	38	41	42	44	46	49	52	56
195	1.80	14	18	19	22	23	25	27	28	30	32	36	37	40	41	43	45	48	50	54
200	1.75	13	18	19	21	23	24	25	27	28	31	35	36	38	39	41	42	45	48	52

Fig. 9-2 A handy chart to use when copy-fitting by the character-count method. The alphabet-length column gives the length of the lowercase alphabet. The next column gives the approximate characters per pica for the alphabet length. The column headings give the length of a line in picas. To use the chart, measure the lowercase alphabet (a through z) in points. Then find the alphabet length that comes closest to this alphabet length in the alphabet-length column. You can then readily find how many characters of the type selected can be accommodated in the width of line selected by reading across the columns. For example, if the length of the lowercase alphabet selected is 75 points and the line to be set is 10 picas wide, then each line will accommodate 42 characters.

the space. Next, find the number of characters of this type that will fit into a line of the specified width. Multiply your estimated number of lines by the number of characters in each line.

If the total is less than the number of characters in the copy, try a smaller size. If, on the other hand, the total is more than the number of characters in the copy, a larger size will be needed. If one size is too small but the next size is too large, select the smaller size and have it leaded out by adding space between lines as needed.

2. How many words will I have to write to fit into a specified space when I have selected a style and size of type? *Solution:* This crops up most frequently in writing cutlines or short takes of copy for advertisements or brochures, or short takes for boxed type, and so on. First, determine how many characters of the type will fit into the width that has been selected. Next, calculate the number of lines of the type that will be required for the given depth.

Then, set the word processor or typewriter line stops to a length of line that contains the number of characters needed for the typeset lines. Write the required number of lines. These lines will run very nearly line for line with the lines when the copy is set in type.

3. I want to use some copy that has already been set into type. But I want to use it in a different size and width of type. How do I calculate the changeover? *Solution:* First, find the average number of characters in each typeset line of the original. If you know the type style, this can be determined from charts if they are available. If tables are not available, several lines of the original can be counted to determine the average character count per line.

Once the average number of characters per line in the original has been determined, multiply that number by the number of lines in the original to find the total number of characters. Next, calculate the number of characters per line in the new setting. Divide this by the number of characters in the original. The result should give the number of lines in the new setting.

Example: We would like to reprint an item that was originally printed in 6-point Bodoni, 12 picas wide. There are 5 lines in this original. We would like to reprint it in 10-point Bernhard Modern on lines 14 picas wide. How much space must be allowed for this new setting in a layout?

First we must determine that the 6-point Bodoni has an average of 47 characters per 12-pica line. There are 5 lines, so there are 235 characters in the item. Next, we find that there are 2.99 characters of 10-point Bernhard Modern per pica from our charts, or 41.86 characters per 14-pica line. If we divide 41.86 into 235 we will find that the new setting will take 5.6 or 6 lines.

Copy-fitting may appear forbidding, but with a little practice the communicator can master this vital tool for planning printed communications. And accurate copy-fitting will mean dollars saved when the compositor doesn't have to reset copy that was estimated inaccurately.

▼ Preparing Copy for the Press

Use Standard Editing Marks

If hard copy is produced (copy printed on paper as opposed to computer storage of material), it should be typed on standard 8½ by 11 white paper and typed double-spaced. Copy should never be sent to the printer on little slips of paper or small note sheets. This is inviting problems. All copy should be edited carefully to make sure there are no mistakes.

The standard editing marks generally used in marking copy are shown in Fig. 9-3.

Catching the Errors

After the compositor sets the type, or if it is set with computer equipment, the communicator will be supplied with *proofs* (also called *galleys*). These are usually made on a copying machine if the type was set "cold type." In the unlikely event that the type was set with a "hot type" machine, galley proofs will be provided. (The term is a holdover from the days when virtually all type was set from metal and placed in a tray, or galley, from which sheets were pulled for proofing.)

Most proofs, however, are supplied on reproduction paper. This is a high-quality white paper. Such proofs should be marked with a nonreproducing blue liquid or fine felt pen, unless other arrangements have been made with the production department.

There are standard proof marks to be used in marking proofs. They are similar to editing marks, but some are slightly different.

▼ Working with the Production Department

The people who do the actual production work for the designer or editor can provide valuable suggestions, often gleaned from long experience, which can make the communication more effective.

Whenever we work with the production department personnel or commercial printers we should agree on certain conditions and specifications for our communication. Here are the principal accepted "trade customs" of printers in the United States:

1. *Quotation:* A quotation not accepted within sixty days is subject to review. All prices given by a printer are based on material costs at the time of the quotation.
2. *Orders:* Regularly placed orders, verbal or written, cannot be cancelled except on terms that will compensate the printer against loss incurred as a result.
3. *Experimental work:* Experimental or preliminary work performed at the customer's request will be charged at current rates and may not be used until the printer has been reimbursed in full for the amount of the charges billed.

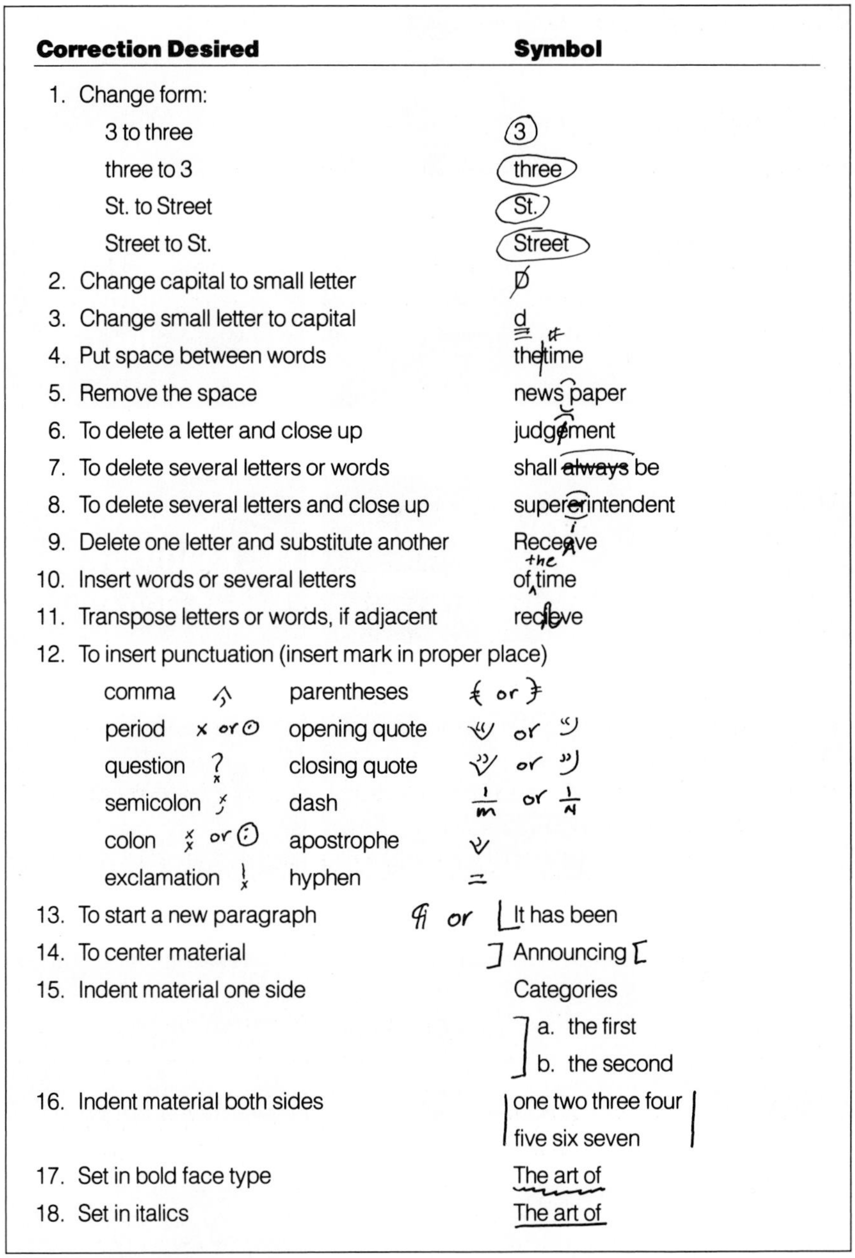

Correction Desired	Symbol
1. Change form:	
3 to three	③
three to 3	(three)
St. to Street	(St.)
Street to St.	(Street)
2. Change capital to small letter	ø
3. Change small letter to capital	d̲̲
4. Put space between words	the⌢time
5. Remove the space	news ⌣paper
6. To delete a letter and close up	judgǿment
7. To delete several letters or words	shall ~~always~~ be
8. To delete several letters and close up	superᵉrintendent
9. Delete one letter and substitute another	Receᵢve
10. Insert words or several letters	of ⌃the time
11. Transpose letters or words, if adjacent	recᴵᵉve
12. To insert punctuation (insert mark in proper place)	

comma	⋏	parentheses	{ or }
period	x or ⊙	opening quote	ⱷ or ⁶⁶
question	?	closing quote	⁹ʸ or ⁹⁾
semicolon	⨍	dash	$\frac{1}{m}$ or $\frac{1}{N}$
colon	x̱ or ⊙	apostrophe	ⱽ
exclamation	!	hyphen	⸗

13. To start a new paragraph	¶ or ⌊It has been
14. To center material	⌉ Announcing ⌈
15. Indent material one side	Categories
	⌉ a. the first
	⌋ b. the second
16. Indent material both sides	| one two three four |
	five six seven |
17. Set in bold face type	<u>The art of</u>
18. Set in italics	<u>The art of</u>

Fig. 9-3 Standard editing marks for correcting copy before it is set into type. These marks differ from proofreading marks.

4. *Creative work:* Creative work, such as sketches, copy, dummies, and all preparatory work developed and furnished by the printer, shall remain the printer's exclusive property and no use of the such work shall be made, nor any ideas obtained from it, except upon compensation to be determined by the printer. This compensation would be in addition to the original agreed upon price.

5. *Condition of copy:* If copy delivered to the printer is different from that which was originally described by the customer and on which a quote was made, the original quotation may not apply and the printer has a right to give a new quote.

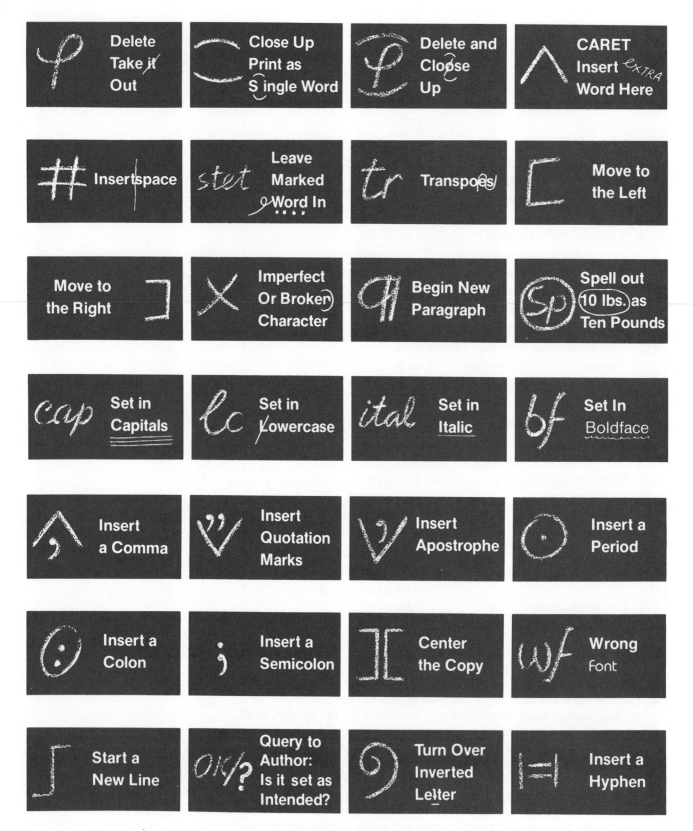

Fig. 9-4 Standard proofreading marks. (Courtesy Metro Associated Services, Inc.)

WEST/Conover Art, P-8064M, Galley—38 of Tape BOOKS4 8064M$$$10

4
6
7
8   Among professional communicators more and more attention is being
9   paid to effective creativity. There are some important reasons for this.
10   We have reached the point in technological advancements where our
11   computers, laser printers, and scanners are capable of producing highly
12   refined work in typesetting, in creating graphics, and in arranging the
13   elements of a graphic communication. We have the tools that have ex-
14   panded our ability to execute what we create. At the same time, we are
15   working in an overcommunicated enviroment. If we want our commu-
16   nication to penetrate the consciousness of our target audience, we have
17   to come up with new and novel and creative ways to get our message to
18   do its job.
19   But, you say, I am just not the creative type. I have to work and struggle.
20   I have a hard time getting started. How do I begin? how did people like
21   Einstein and Galileo and Thomas Edison come up with their ideas? I wish
22   I could get ideas that work, too.
23   So, before we take a look at how we can select and work with type and
24   art to create the most effective communication possible, let's listen to the
25   expert for a minute or two.
26   "Human creativity uses what is already existing and available and changes
27   it in unpredictable ways," explains noted psychiatrist Silvano Arieti.[1] As
28   we shall see, we can take words, experiences, pieces of art, even a scrap
29   of paper in the street, and use them to trigger creative ideas of all sorts.
30   Writer Arthur Koestler/ has an idea to get us started thinking about
31   creativity. Koestler notes that children are instinctively creative. They all
32   create without restraint. They sing, dance, draw, fantasize, play, all in a
33   world of their own creation.[2]
34   But the, as children get older something happens. Their lives become
35   increasingly structured in a mass of rules and regulations imposed by
36   mother, father, and teacher. The creativity of their early days is restricted
37   and repetition replaces creativity.
38   Now, to unleash our abilities and come up with ideas that are unique
39   and different we have to break out of the restraints, if only for a while,
40   that have ordered our lives. We have to bring the right brain into action.
41   Psychologists tell us we have two brains. The left brain operates out
42   talking, walking, and all our mechanical actions. The right brain is the
43   repository of our imagination, intuition, and creative thinking. We use
44   our left brain in much of our day-to-day living; to be creative we need to
45   unlock the right brain—the home of our creative abilities.
46   We need to call up our right brain and let it work with the left to
47   develop creative problem solving. This starts the whole brain operating—
48   using things we have learned about fundamentals of effective commu-
49   nication that are stored in our left brain with the innovative and creative
50   potential of the right brain.

Fig. 9-5 A typeset galley that has been proofread and marked for correction.

Scaling Art

Steps in scaling art using the diagonal line method:

1. Place a tissue overlay over the original. Use a red grease pencil to mark the four corners of the area to be reproduced.

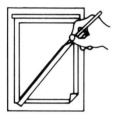

2. Connect the corner marks on the overlay to make an outline of the original image area and then draw a diagonal from one corner to and beyond the opposite corner.

6. *Materials:* Working mechanical art, type, negatives, positives, flats, plates, and other items when supplied by the printer remain the printer's exclusive property unless otherwise agreed in writing.

7. *Alterations:* Work done on changes from the original copy will be quoted at current rates and an accounting will be given to the customer.

8. *Prepress proofs:* The customer should insist on proofs before a job is printed, and the printer should insist that the customer mark the proof with an "O.K." or "O.K. with corrections" and signed. The printer cannot be held responsible for errors in proofs if proofs were marked O.K. Changes should not be made verbally.

9. *Over runs and under runs:* Overruns or underruns not to exceed 10 percent on quantities ordered, or the percentage agreed upon, shall constitute acceptable delivery. (Note: If you need exactly 10,000 copies be sure the printer knows this. Under this custom he could deliver as few as 9,000 copies on a 10,000 order and still be technically correct as far as filling the order is concerned. But he should bill you for 9,000 copies.)

10. *Terms:* Payment shall be whatever is set forth in the quotation or invoice unless you have made previous arrangement with the printer. Claims for defects, damages, or shortages must be made by the customer in writing within fifteen days of delivery.

Both the printer and the customer should understand all conditions concerning the printing and delivery of orders. Most commercial printers will supply their customers with a copy of these accepted trade practices. It is very important that these practices are understood and that the copy and instructions you give to the printer are perfectly clear to all involved. This can't be emphasized enough.

When contacting a printer for the first time it is a good idea to look over samples of work produced in the shop to see the type and quality of printing it has done for its customers.

In addition to understanding the customs of the trade, it will save time and money as well as help you get a satisfactory job if you make sure you have answered the following questions and communicate the information to the production department or printer.

1. When is the job needed? The more lead time the printer has, the more care can be taken in production.
2. What sort of paper is best suited for the job? How durable should it be? If the cheapest paper available is specified, how will this affect the quality of the job?
3. What about color—for the paper and/or the printing?
4. Are the photos or illustrations of satisfactory quality to reproduce well? If there are shortcomings, the printer should point them out so the communicator will know what to expect when the finished job is seen.
5. Are the specifications and instructions on the layout, copy, overlay (mask), and illustrations clearly understood by both the communicator and the printer?

The more details we can give the printer about the job the better the chance that our expectations will be fulfilled. Printers are skilled craftspeople—or they should be—and they can provide many helpful suggestions if they know just what we want done.

Instructions for compositors and printers must be clear, concise, and legible. There are some universally accepted abbreviations for marking copy. However, there can be variations from place to place, and it is important that both the communicator and the printer agree on those being used. Instructions generally can be grouped under three categories: (1) instructions for cropping and sizing art, (2) for marking layouts, and (3) for marking reading matter. Art instructions are considered in Chapter 5.

Instructions for layouts and reading matter should include:

- Type size in points. However, for very large display type, inches and centimeters are used in some shops.
- The type family by name, such as Caslon, Stymie, Times Roman.
- The family series, which could be condensed, light, medium, bold, extra bold, light italic, and so.
- The posture of the letters, whether all capitals, all lowercase, capitals and lowercase, small capitals, swash letters, or other special characters.
- The width of the lines in picas.
- The leading between lines.
- The justification—whether the lines are to be set flush left, centered, flush right, or justified. Usually, the justification is obvious by the comp lines on the layout.

In addition, it might be necessary to indicate if parts of the copy are to be set indented, in italics, boldface, or in some other special way.

This list may seem formidable, but it can be mastered quickly. There are a number of shortcuts for marking copy and they can be understood with a little practice. For instance,

8/10 Caslon Bd. itl clc × 15
or
8/10 Caslon Bd. itl U&lc × 15
(either U&lc or clc mean capital and lower case)

means "Set this in 8-point Caslon bold italic capitals and lowercase letters with 2 points of leading in a line 15 picas wide."

The designation 8/10 comes to us from the days of linotype composition when the term "8 on 10" meant to cast 8-point type on a 10-point slug. This was the equivalent of putting 2 points of leading between the lines.

When a brochure, advertisement, or general printing layout is made, most of the instructions are written on the layout sheet outside the actual layout. Any instructions written within the layout should be circled to indicate that they are not to be set in type. All instructions should be clear and accurate. After all, the specifications are like those an architect would put on a blueprint. The builder is expected to follow them precisely. The printer is expected to follow the layout's specifications precisely.

▼ The Designer's Tool Kit

So far several tools of the trade have been mentioned. Below is a list of basic tools and materials that you will need to complete the projects—or similar ones—suggested in this book.

Scaling Art, continued

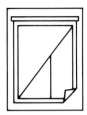

3. Mark the desired width on the bottom line of the original area. Example: If you want to reduce the original to 5 inches from 8 inches, measure across the width from left to right to 5 inches. Then draw a line from this new width to extend through the diagonal.

4. Mark the final size of the width and new depth (where the vertical and diagonal lines intersect) on the overlay and retape it over the original as well as show the cropping and final size.

Spell It Out

Whether we set type in house or order it set by a compositor, it is important to write out all instructions and be sure the instructions are perfectly clear. Typesetting costs will be reduced and better quality will result.

It is an old axiom in the printing trade that the compositor follows copy, mistakes and all, "even if it flies out the window."

Before ordering typesetting:

- Read the copy and correct errors in spelling, punctuation, capitalization, and grammar.
- Check for paragraph numbering, if any, as well as indentation, page numbering, and general format.
- Copy should be double spaced, on one side of a sheet, and paginated in sequence.
- Long tables and other matter that is to be set in a special way should be typed on separate pages.

- *Paper:* Two kinds of paper will prove useful for drawing thumbnails, roughs, and comprehensives. One is *transparent tracing* and the other is *layout paper.* Layout paper is white and not as transparent as tracing paper but it still can be used for tracing. Both can be obtained in pad form in sizes from 8½ by 11 to 19 by 24. Paper weights of 16 or 24 pound work well.
- *Rulers:* Two types of rulers will come in handy. One is a 12- or 18-inch etched steel and the other is a *pica rule* or *graphic arts rule* with both inch and pica scales.
- *T-square and triangle:* A good 24-inch T square and an 8- or 12-inch triangle are musts. A T square with a plastic edge should never be used for cutting. Obtain a T square with a steel arm for cutting. A plastic triangle should also not be used for cutting. It could become nicked and useless for drawing straight lines. (It is a good idea to have a steel T square, triangle, and straight-edge for making cuts with a knife or razor blade.)
- *Pencils and pens:* The most useful pencils for layout work are 2H or HB. Felt pens come in handy too.
- *Erasers:* Magic Rub pencil and art gum work well.
- *Masking tape and rubber cement:* Both will come in handy for tacking down art on grid or layout paper and for many other uses. Rubber cement thinner is needed to thin out rubber cement when it thickens in the jar. It also can be used to loosen something that has been pasted down with rubber cement.
- *Scissors and art knives:* A good pair of scissors and an X-acto knife or single-edge razor knife will take care of the cutting needs for starting in layout work.

Only a few additional items are needed to add pasteup capability to the tools required for layout. These include:

Nonreproducing blue pencils and pens
A waxer: There is an inexpensive hand-held model that works fine.
A burnisher: Burnishers are either in roller or stick form. If much pasteup work is done, the roller is most satisfactory.
White correction fluid or graphic white paint: These fill in shadow areas or cover blemishes in the pasteup.
Preruled grids or plain white bristol, about 4 ply.

▼ Steps in Basic Pasteup

Once the layout has been completed and all the images to be used assembled, the next step is to paste them into position as shown in the layout. Then this pasted-together form is photographed for plate making. The process is called *pasteup.* It can be accomplished by following ten steps:

1. Prepare the grid or base sheet.
2. Double check all elements for errors and compare them with the layout to see if they are all available.

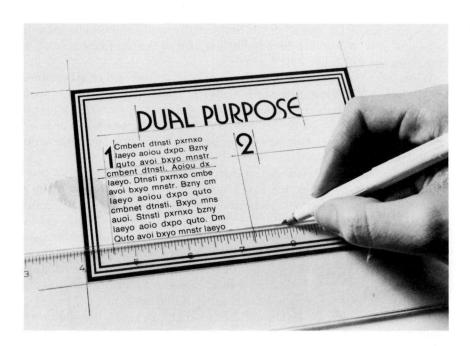

Fig. 9-6 The pasteup process. Guidelines are drawn with a nonreproducing blue pen or pencil. Then the elements are affixed in position, usually by using a waxer to coat their backs so they will stick to the grid. The body type in the illustration is an example of greeking (simulated type that makes no sense). (Illustrations used in Fig. 9-6 through Fig. 9-10 courtesy Graphic Products Corporation, Rolling Meadows, Illinois 60008)

3. Trim the excess paper from the type and art forms.
4. Apply wax to the backs of the type and art forms.
5. Position the elements on the grid or base sheet.
6. Check all elements to be sure they are square.
7. Burnish the form to affix the elements to the grid.
8. If halftone negatives are to be stripped in, affix a window (usually red plastic) the exact size of the halftone.
9. Add borders and rules, usually in tape form.
10. Make a copy on a copying machine and proof the copy. Then add an overlay to protect the pasteup.

In step 1 a piece of illustration board or a preruled grid (in nonreproducing blue, which will not show up in printing) is fastened to a drawing board or flat, perfectly smooth surface. It is critical that everything be square and true. The grid should be lined up on the surface with a T square and then held in place with small pieces of tape in the corners.

Next, the nonreproducing blue lines on the grid are used to position the heads, art, body copy, and other elements. Nonreproducing blue pencils or pens can be used to mark the spots where the elements are to be placed. If art is to be bled (printed to the edge of the page) it should extend about ¼ inch beyond the outside dimensions of the page margins for trimming after printing.

Rubber cement or melted wax is applied to the backs of the assembled images. They should be trimmed as closely as possible to the print area but care should be taken not to cut into any images. It may prove more satisfactory to wax after trimming, depending on the size of the elements.

The various elements with adhesive applied are placed in position on the grid. Great care is taken to make sure everything is straight and square.

A T square is used to line elements horizontally, and a right-angle triangle with one edge placed on the edge of the arm of the T square is used to line them up vertically.

When adjacent columns of reading matter are lined up or when lines are added to a column of reading matter, a Haber rule is a handy tool. A *Haber rule* is a plastic rule marked with type sizes and various leadings so that, for instance, two lines of 8-point type leaded 2 points can be lined up evenly. The point of an art knife is handy for moving small elements on the grid for alignment.

If a halftone negative is to be stripped in later, the exact area is drawn on the grid with a nonreproducing blue pen and the area is coded (such as "photo A"). The photograph is similarly identified and sized to fit the area exactly.

Once all the elements are in place, the entire pasteup is burnished to ensure that everything is firmly in place. There are several types of burnishers but the rubber or ceramic roller is the most frequently used. A sheet of clean white paper is placed on the images for burnishing to help keep the pasteup clean and prevent smearing during the burnishing process.

If the pasteup is to be produced in more than one color, the additional color areas are placed on an *acetate overlay sheet*. The entire pasteup is often protected with a cover sheet, or *frisket*, taped as a flap over the completed grid.

Fig. 9-7 (a) In pasteup the elements are placed into position on grids with pre-printed guidelines in nonreproducing blue. (b) The completed pasteup is photographed and a printing plate is produced. The grid lines will not show in the final printed product.

(a)

(b)

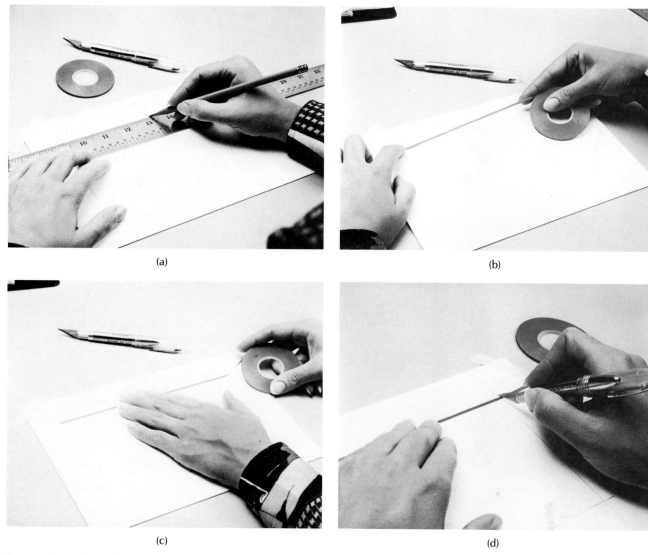

(a)

(b)

(c)

(d)

Fig. 9-8 To apply border tape: (a) Draw a faint pencil line, nonreproducing blue guidelines, or use a guideline on the paste-up grid for aligning the border tape on the layout. (b) Press the end of the tape into position and unroll enough tape to go just beyond the length required. (c) Carefully lower the tape into position along the guideline and gently smooth it into place. (d) Trim the tape to the desired length and firmly burnish it into place.

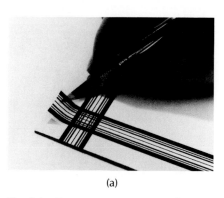

(a)

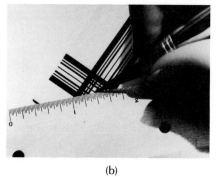

(b)

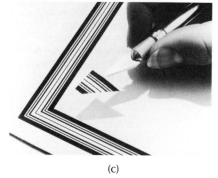

(c)

Fig. 9-9 To create attractive mitred corners on a pasteup: (a) Place the border tape in position but overlap the ends of the tape beyond the corner of the box. (b) Lay a straight edge on the overlapped area and cut carefully on the diagonal. (c) Remove the excess material and press the border into position.

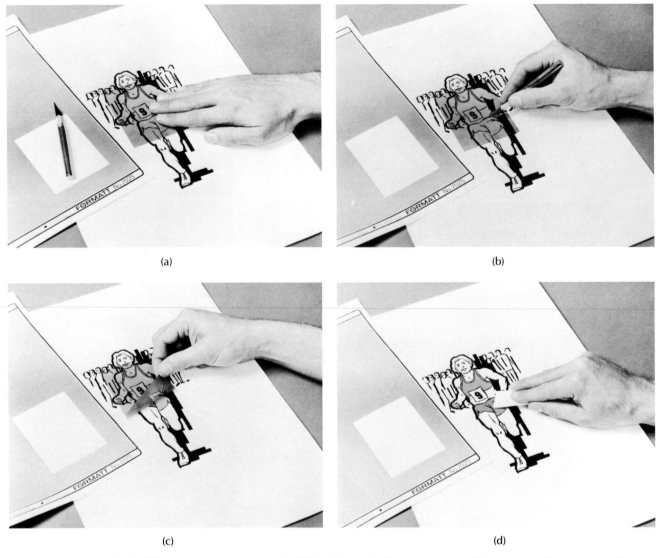

Fig. 9-10 To apply shading film to art: (a) Cut and remove enough shading film from a backing sheet to cover an area slightly larger than the drawing. Smooth out any air bubbles. (b) Cut lightly around the outline of the area to be shaded on the artwork. (c) Remove all excess shading film. (d) Burnish the film until it appears as if it were actually printed on the artwork.

Fig. 9-11 Borders can be customized and their widths narrowed if desired. (a) First cut the border to the desired width. Next, slide an art knife under the border and remove the excess. (b) Finish the border by adding rules to create different effects and press it into position.

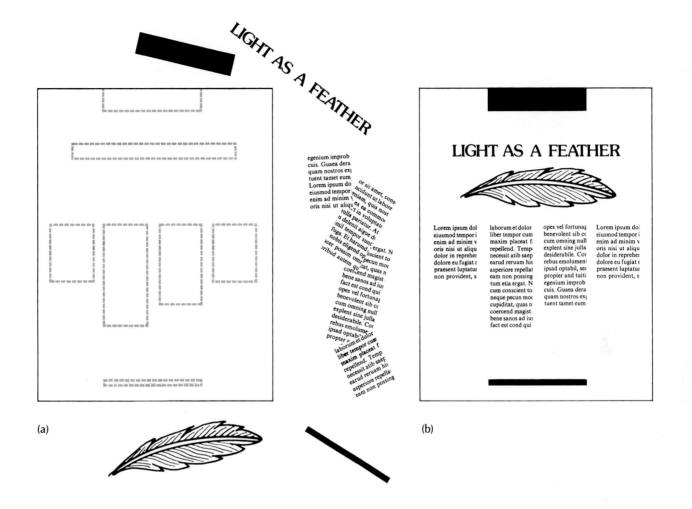

(a)

(b)

▼ Effective Design Checklist

- Border tapes used with care can add a pleasing element to your pasteup (see Fig. 9-8).
- Mitre border corners for neater, more professional-looking pasteups (see Fig. 9-9).
- Shade art for a more realistic effect. Use shading films for shading or for color overlays (see Fig. 9-10.)
- Masking film can be used as in the first three steps for shading to create *overlays* for color printing. Follow these steps:

  1. Cut a piece of masking film (red or amber is used most frequently) large enough to cover the area to be printed in color and tape it firmly at the top to the layout.
  2. Cut around the outline of the area to be printed in color as in step 2 in the shading technique.
  3. Peel off the taped large piece of masking film. The outline form remaining can be affixed to a clear polyester overlay ready for the camera.

Fig. 9-12 (a) The individual pieces of art and copy and their positions on the grid or pasteup board (often a lightweight white bristol cardboard). (b) The completed pasteup. A protective flap of lightweight paper or tracing paper can be taped to the back and folded over the face of the pasteup. (Courtesy A. B. Dick)

4. Care must be taken to place the red masking film cutout over the original layout to obtain a perfect register.

• Use creativity and customize borders (see Figs. 9-9 and 9-11).

▼ Graphics in Action

1. Take the front page of a daily newspaper. Measure the width of the columns in picas. Determine the maximum unit counts of the headline types used on the page for the various column widths.
2. Use the copy that defined graphic arts at the start of this chapter. Assume you want to set it in 10-point Century (2.70 characters per pica) leaded 2 points, in a column 12 picas wide. How deep would the space be when the copy is set in this size and style of type in lines 12 picas wide with 2 points of leading?
3. Find an advertisement in a newspaper or magazine about 4 by 5 inches. Select one that has few display lines and only one or two copy blocks. Mount it on plain white paper. Mark it for publication. Attempt to identify the type families or identify them as closely as possible by comparing them with type specimen sheets or the type specimens in the appendix.
4. Assume you want to print the copy used to demonstrate copy-fitting in this chapter as an envelope stuffer for your business. Your business is "(Your Name) and Associates" Graphic Designers; use your address. You want 5,000 flyers, 6 by 9. Make a rough layout for such a flyer, mark it, and write all the information you would want to tell the printer.
5. Paste up a prototype for the flyer in number 4. Or practice pasteup by creating prototype advertisements by using display lines, blocks of copy, art, and logos clipped from magazines (try to find magazines with heavy paper). Some practice in pasteup can be done with a minimum of equipment: Rubber cement, art knife and scissors, nonreproducing blue pencil, T square and ruler, and a border tape or two are about all that is needed to get started. Try to devise shadow boxes and other special effects. Examples:

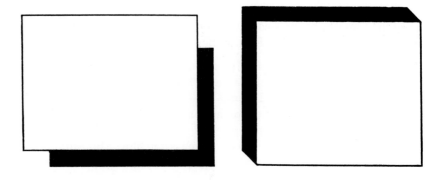

CHAPTER TEN

The World of Desktop

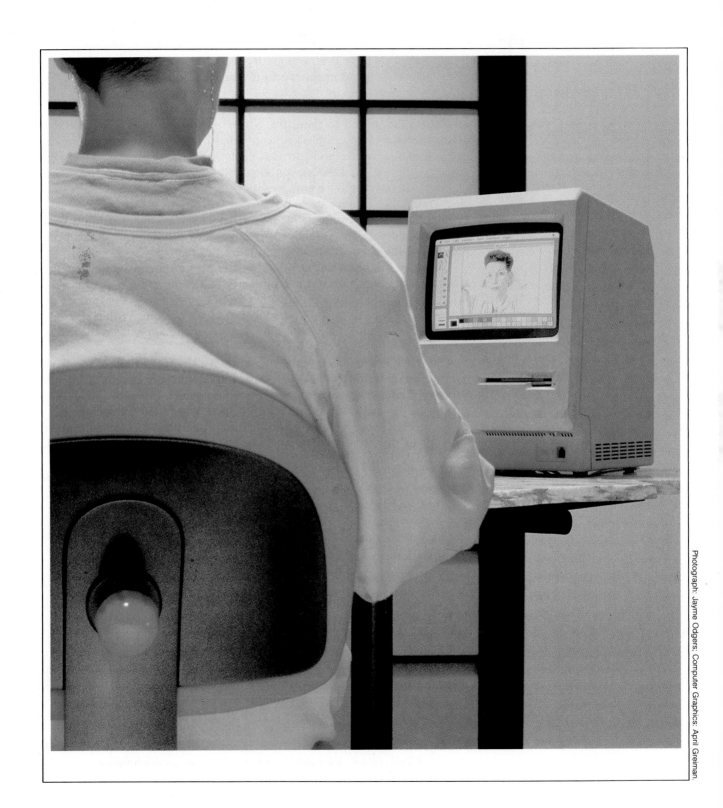

Photograph: Jayme Odgers; Computer Graphics: April Greiman

Life on the Mississippi

Life on the Mississippi

Life on the Mississippi

Fig. 10-1 Over-tight kerning, as in the top example, is seen all too frequently in the world of desktop publishing. The middle example shows proper kerning, while the bottom line is the type set in normal letterspacing. The middle and bottom examples are acceptable, the top line is not.

Words crammed together. Words spaced so loosely there are gaps and rivers of white running all over the page. Letters kerned so tightly the distinctive designs that have led to their uniqueness to communicate has been virtually destroyed. Arrows here and bullseyes there. Heavy rules that build walls the eye must climb over to get at the words. Reverses used with abandon. Welcome to the world of desktop publishing.

These words have been echoed frequently since "desktop publishing" became a buzz word in the world of graphics and communication.

They were echoed by designer Jeffrey Parnau, who published his book *Desktop Publishing: The Awful Truth* in the spring of 1989.[2]

Parnau attempted to bring the reality of desktop publishing to those who believe desktop publishing is a magic process by which anyone can sit before a computer and turn out professional-quality work. The awful truth is the reality. Desktop publishing is a Pandora's box from which ineffectual and illegible communications can—and do—pour at astounding speed. But it is also a tool that can free the designer from the constraints of the past and help him or her create excellent printed communications that are attractive, effective, and economical. It can bring control of the product into the hands of the creator.

Consider this scenario. Dennis Chesnel, technical consultant for Time Incorporated, sits before a Macintosh II. Watching are a skeptical managing editor and a doubting art director. Chesnel places a red bar at the top of the computer screen, a yellow bar at the bottom. Within seconds, he blends colors to create a rainbow. Next a square appears on the screen and dances into a variety of shapes.

Chesnel fills the screen with type, gives it a shadow, changes the shadow's color, and transforms it to look as if the letters are caught in a gentle breeze, then a strong wind, then a howling gale.[3]

The editor and art director cannot believe what they are seeing. They are so enthusiastic that they go ahead and order more desktop equipment to expand the Time Incorporated's move, already under way, into the world of desktop publishing. Time is preparing to launch a new magazine, *Sports Illustrated for Kids,* to appear early in 1990. Desktop publishing is expected to play a major role in its production.

▼ What Is Desktop Publishing?

What is desktop publishing? Is it a device for amateurs that will turn them into professional publishers, or is it a tool for professionals to produce better and more effective work?

The answer lies at both ends of the spectrum, and it was summed up by Parnau: "Professional designers and writers have always been able to work with or without sophisticated tools. Certain desktop publishing equipment and software make the professional's job easier, but does not make the professional. Design is in the head. Creativity is in the soul. And like an X-acto knife or scissors, desktop publishing is on the desk."[4]

Roger Black, designer of many magazines and editorial art director of the *New York Times,* sees desktop publishing as an enchanting world in

Fig. 10-2 A desktop publisher at his workstation. He has just finished scanning an illustration into the cover design for a magazine. (Courtesy Xerox Corporation)

which communicators can "sit down, write, layout, and print whatever they want whether it's poetry, recipes, a book, or a newsletter."

"You are in control," he adds. "Completely. It's also immediate. You can proof and refine and correct things instantly. It's like having your own printing press and type house. In fact, it's not *like* that at all. That's exactly what you have."

If you are planning to work in communications, and most likely you wouldn't be reading this book if you weren't, you are sure to become involved in desktop publishing, and the graphic skills you are acquiring can help.

You already know better than to produce a communication that contains the shortcomings enumerated in the first paragraph of this chapter. Those words were used to describe the winner of the grand prize in the Gaggie Awards presented by a San Francisco design firm for "the most outstanding examples of graphics illiteracy and/or blinding bad taste." [5]

But do you know the reality of desktop publishing? That is the purpose of our discussion in this chapter. In desktop publishing "you just don't pop in the software and away you go," says Robert Chisholm, art director for a Hewlett Packard facility in San Diego. "In the hands of the right person desktop publishing can be incredible. In the wrong hands, it can become a nightmare." [6]

So let us take a trip through the world of desktop publishing. Let's start off by defining just what it is and how it came about. Then we will consider the equipment needed in the process. We will take a look at how we can learn to operate this equipment and how we can put it to work for us. Throughout we will attempt to provide hints and pointers so that those who enter the desktop world can enter it as quickly and easily as possible.

Fig. 10-3 Desktop publishing is accomplished with a computer (the larger the screen the better), keyboard, and a mouse. This plus a printer and you are in business.

How Does Desktop Publishing Work?

How did desktop get a toehold in the world of communication? It was not too long ago that reporters or copywriters wrote stories and copy on their typewriters or word processors. Their typed pages went to an editor or copy chief who would correct, alter, add to, and change things around as they saw fit. Then the copy went to a compositor who would set it into type for publishing. The compositor might set the type styles and sizes, the leading between the lines, the kerning of the space between letters, and the line widths.

Once the copy was typeset, it was combined with art, rules, borders, and so on on the pasteup table. Then it was ready to be photographed and a plate made for the press.

With desktop publishing, the copy is set at a computer keyboard. The communicator can create, proofread, make corrections, designate type styles, leading, and line widths, set the type, and position the text on the layout without getting out of his or her chair.

One person now can do what formerly would take several people to accomplish. Desktop publishing systems can save time and money, provide better control of accuracy before production, and, ultimately, produce finished, camera-ready ads and pages.[7]

It was in 1984 that "desktop" entered the world of graphic communication. Paul Brainard, president of the Aldus Corporation, created Aldus with the specific purpose of creating a program that would combine the functions of low-cost microcomputers and laser printers to produce camera-ready mechanicals.

Brainard coined the word "desktop," and Aldus produced a software program—PageMaker—that brought writing, editing, designing, and production into the world of personal computers. Since then newspapers, books, newsletters, manuals, reports, and even magazines have been designed with PageMaker. Since then, too, many more desktop publishing programs have entered the market.

Brainard did not claim this new method of producing printed communications would eliminate the skills of the professional designer. "The need for people who understand visual design isn't diminished," he said. The software is simply another tool of the designer, an addition to the toolbox already loaded with T-squares, triangles, rules, and rubber cement. PageMaker might be a substitute for some of those tools, but it is not a substitute for creativity.[8]

Designer Roger Black adds an optimistic and encouraging note:

"Personally, I think almost anybody can do some design and put out a decent looking document. It doesn't have to be heavily designed, but a document that has been desktop published, if it's done thoughtfully, is going to command so much more attention.

"This technology is a force that is going to move everybody up to where we will begin to expect and receive a higher level of sophistication in all areas of our printed communication."

▼ Equipment for Desktop Publishing

A desktop publishing system consists of hardware and software. Hardware includes the machinery—basically a computer, printer, and scanner. Software includes those things needed to run the hardware such as disks which contain a variety of programs. A program contains a sequence of instructions that tells the computer how to do a task. It operates the system.

A word processing program would be used to compose copy. Another program might turn the computer into a page layout system, and another might enable the operator to draw pictures on the screen.

So the equipment needed for a typical desktop publishing system usually includes a personal computer, a printer, a scanner and the software needed to create copy and graphics and blend them to make a camera-ready mechanical.

The Computer

In the pasteup days of design and layout, the communicator used scissors, X-acto knives, waxers, T-squares, right angles, cropping Ls, and so on to create mechanicals. The computer contains all these tools in one convenient machine. Although there are many computers on the market, the choice for desktop publishing narrows down to one of two types. These are the Macintosh by Apple and the IBM PC and its compatibles. (Compatible means that the computer and its hard- and software are interchangeable with those of the IBM PC. Everything that can be used with the IBM can be used with the compatible.)

The Macintosh became the desktop publishing leader when it entered the market in 1984. Before the appearance of the Mac, as it is called, computer design was limited to firms that could afford to invest as much as $100,000 or more for equipment.

Not only did the Mac enable communicators with limited budgets to become desktop publishers but also it was "user friendly," that is, relatively easy to learn.

In addition, the Mac included a mouse. This little hand-held device enabled the communicator to move a cursor, or pointer, in the form of an arrow around the computer screen to point to the function the operator wanted the computer to perform. The mouse could activate everything the computer contained to set type, create art, and combine elements in a page layout.

Apple's innovations made it the leader in producing computers for desktop in the early years. Today the competition is producing computers that are equally satisfactory. But the Mac continues to be a leader in the field.

Those who have worked in desktop publishing have a few suggestions to help you select the right computer. They suggest reading books and magazines on the subject, talking to computer dealers, and attending workshops and seminars.

A computer should be selected that can produce the types of communications you need. It should have enough power and memory to do page makeup. The minimum amount of power for desktop publishing is considered to be 640K. The K stands for kilobyte.

The computer should have a hard drive, or hard disk. A hard disk is a rigid disk used for storing data.

The graphic ability of the computer is an important consideration. Flexibility is important, as well. This includes the scope of communications it is capable of handling. For instance, you would not need a computer with the capacity and ability to compose multipage manuals or books if you are going to use it mainly for flyers and newsletters.

The Printer

The second key piece of equipment for desktop publishing is the printer. Actually, a computer and a printer are all the hardware you need to get started in desktop publishing.

There are two basic types of printers. One operates with a dot matrix system and the other uses a laser.

A dot matrix printer forms images by arranging series of dots. The resulting print is not considered suitable for quality work.

Laser printers can produce type and art with more detail and resolution. Most provide 300 or more dots per inch (dpi). The Apple LaserWriter was introduced in 1985, and it opened the door to quality desktop production.

Fig. 10-4 A comparison of resolutions. The illustration on the left was digitized at 72 dots per inch. Such a low resolution causes dithering and jagged edges. The same illustration, on the right, was digitized at 300 dots per inch. (Courtesy Metro ImageBase, Inc.)

It proved to be suitable for newsletters, small publications, and reports, although not for quality commercial work.

Generally, commercial-quality work requires a much higher resolution such as that produced by phototypesetting machines that provide resolutions of up to 2,540 dpi. This compares with all but the very highest-quality professional typesetting equipment.

While the cost of advanced equipment may be beyond the reach of many communicators, capabilities of such equipment can be obtained through the use of a service bureau. A service bureau is an independent business that provides help in layout, graphics, and production. Often material created on economy equipment can be taken to a service bureau where it can be made camera ready for a reasonable cost. The service bureaus also will provide instruction and use of their equipment on an hourly basis.

Fig. 10-5 Essential for professional quality desktop production is a laser printer. This printer can produce direct output from Macintosh, IBM PC, or compatibles. It is capable of printing ten pages per minute at 600 dots per inch. It can generate type from 4 points to 999 points. (Courtesy Verityper)

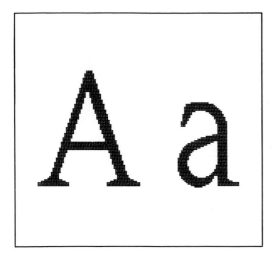

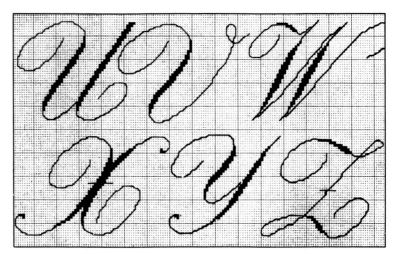

Fig. 10-6 An example of digitized type, above left. It is made up of dot patterns similar to an arrangement for a mosaic. Digitized type could be compared to letters formed in needlepoint, as in this example from a turn-of-the-century textbook for cross-stitching, above right.

When discussing desktop equipment and costs, we must realize that the technology is changing constantly. And, cost-efficient manufacturing methods are being developed to reduce costs, so the situation can change almost overnight.

An example is the development of outline type fonts announced late in 1989. These new fonts will allow the desktop publisher who is limited to digital type to produce type that rivals that generated on expensive laser printers.

Companies are offering software incorporating outline fonts that enable a computer to display or print letters in a wide variety of type styles in virtually any size. Instead of bit maps in which a character is represented by a letter design made up of dots, the outline font approach uses mathematical curves and lines to produce the letter. This eliminates the rough edges of the letters that characterize bit-mapped letters (see Figure 10-7).

The Scanner

A device that enables you to copy a photograph or line art or text and transfer it to a page layout on a computer is becoming increasingly useful

Fig. 10-7 For years personal computers have stored type as a pattern of dots, as in the diagram on the left. In the outline font system, the curves and lines that outline the letter are stored as mathematical formulas.

and economical. That device is a scanner, which reads, or scans, the image and converts it into a series of dots that can be accepted by the computer.

There are economical small hand scanners, but they are limited to scanning areas about two to four inches wide. They are also limited in their resolution capabilities. Most hand scanners are 400-dpi models, which are useful for scanning small images, such as signatures and corporate logos. They also provide a good low-cost introduction to scanner applications.[9]

Moving up the scanner ladder, the designer has a wide choice of sheet-fed scanners that are rated by their gray-scale capabilities: no gray-scale, four-, six-, or eight-bit gray scale. Four-bit scanners can reproduce 16 levels of gray compared with 64 levels for six-bit scanners and 256 levels for eight-bit devices.

The technology in desktop is advancing so rapidly it is assumed that scanners will be marketed soon that will capture all the highlights and shadows in continuous tone art and make possible the incorporation of publication-quality photographs into page layouts on personal computers.

The Modem

A modem is an electronic device that permits one computer to talk to another over telephone lines. This enables the desktop publisher to send files to service bureaus, other offices, and clients who have modems to receive them. Modems are rated by *bauds*. A baud is a measurement of the rate of speed at which modems can transmit data. Most data sent to a service bureau is sent at 1,200 or 2,400 bauds. A modem of this quality costs about $500 in 1990.

Fig. 10-8 This fully automated scanner can produce color separations in 65- to 150-line screens. It is capable of 9,000 color tint possibilities. A layout is placed on the digitizing tablet (left) and the desired art is scanned with a mouse. Everything else is automatic in the scanner (right). (Courtesy Itek Graphix)

▼ Software For Desktop Publishing

An instructor of introductory computer courses likes to build the confidence of her beginning students by impressing on them the fact that computers are "dumb." They can only do what you tell them to do. And you tell them what to do by selecting the right software.

There are a great number of programs on the market for desktop publishing and new programs are being introduced constantly. With such an array it may appear difficult to select the right programs for your publishing goals. However, only three are needed to get started. They are designed for word processing, graphics and page layout.

Word processing is used to enter and edit written copy which will be incorporated with graphics to form the basis of a layout created with the page layout program. Word processing programs give the communicators the capability to write and edit on the screen. Such programs have features that enable them to search through copy and replace words, cut out copy, bring in copy and insert it where desired, and to put headers and footers on each page automatically.

Most word processing programs can check spelling, justify columns of type, and adjust space between lines. Just about anything needed to be done with copy before it is sent to the printer can be accomplished. These programs should be evaluated in terms of the features they contain that would be useful for the type of work you wish to do.

Graphic programs are used to prepare and edit illustrations and decorative graphics and to produce diagrams, drawings, charts and graphs, and digitized line art and photographs. Many graphics programs are bit-mapped. A bit-mapped graphic is composed of a pattern of individual pixels which form the images on the screen, as is done in composing bit-mapped letters.

Some graphics programs are object-oriented, which means the resolution of the output will match the capabilities of the output device. The MacPaint program for the Apple Macintosh, for instance, is bit-mapped so that the image is created at the fixed resolution of the Macintosh screen and does not change when printed on the Apple LaserWriter.

By contrast, the MacDraw program is object-oriented so that each individual object you draw on the screen is redrawn at the resolution of the output device being used. This means that an object drawn on a screen with a limited resolution could be printed on a laser printer and come out looking much better than it did on the computer screen.

It is important to see graphics programs demonstrated, as well as to know if the software and hardware are compatible, before making a purchase decision.

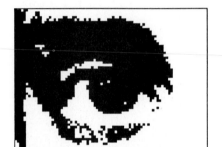

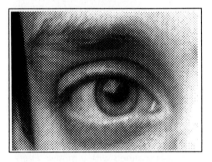

Fig. 10-9 The quality of computer-generated art depends on whether a single bit or a multi-bit scanner is used to copy the original. The example (above) was produced with a single-bit scanner. The same image (below) was produced by a multi-bit scanner which can create a grayscale format that compares well with the continuous tone of the original. (Courtesy Aldus Corporation)

Page Layout Software

Page layout software will let the communicator combine the text from the word processing programs with the images produced by the graphics program. A page layout program can eliminate the old pasteup process and enable you to produce complete comprehensives from the laser printer. A good quality laser printer is capable of producing mechanicals that can

be taken to the printer for plate making and printing. Sometimes short runs of copies can be made on the laser printer or other duplicating equipment.

These programs vary greatly in price and in the features they contain. Some programs are better suited for certain types of desktop publishing than others. In selecting a page layout program the communicator should consider the results he or she is hoping to achieve and go from there. Constant monitoring of changes in programs is important. As we have emphasized, the field is one of constant change. For instance, *Personal Publisher* magazine, which monitors changes in software packages, lists in its September 1989 update of popular programs thirteen changes in one thirty-day period.[10]

Some features to look for in selecting a page layout program include:

- The page size on the computer screen, the bigger the better.
- The ability to create grids that are standardized from one page to the next for use in designing publications.
- The ability to select and bring up to the screen grids with a standard set of column widths per page.
- The ability to run copy through a number of pages and adjust it automatically throughout the newsletter or other type of publication.
- The assortment of tools contained in the program such as fill patterns, screens, rules, and tools that let you create ovals, boxes, and so on.

Fig. 10-10 Note the realistic detail and depth that can be obtained with a multi-bit scanner. A multi-bit scanner breaks down an illustration into dots and sometimes lines of varying sizes while a single-bit scanner uses only one size of dot to copy an illustration. Multi-bit scanners can record the many levels of gray tones in the original art. This illustration was scanned and then printed with a Linotronic 300 laser printer. (Courtesy Aldus Corporation)

Fig. 10-11 This page was written and designed with the Ready, Set, Go program on a Macintosh SE computer. It was printed with an Apple Laserwriter Plus. The large ampersand was made by typing a 250-point Goudy Old Style ampersand into a text block, selecting the text block with the selection arrow, and cutting and pasting it into a picture block, scaled to 123 percent horizontally and 182 percent vertically.

②

This caption and the box, right, are both good ways to add supplemental information to the page. They stand out from the rest of the text because of their different shapes, placement, and especially textures. Texture in type is the result of a combination of size, weight, and line spacing. You can make a block of type look light and airy, like the type in the box, or dark and dense, like the Franklin Gothic Heavy in this caption.

24-Pt Franklin Bold
and a few other indispensable faces

Goudy *or Galliard or Garamond—an "old style" serif face for readable body text*

Franklin *or Helvetica, Futura, or Univers—a "contemporary" sans serif face for weight contrast in heads and captions*

Century *or Times or Baskerville— a refined "transitional period" face substantial enough for either text or display type*

Bodoni—*an elegant "modern" face for decorative accents or classic-looking text*

Layout: tactic two

Layout is the art of placing text, art, and other elements on the page. You should base your newsletter layout on an underlying grid to provide structure and consistent placement from page to page. The illustration at right shows how the grid for this page appeared on screen as the page was being designed. The rows and columns serve as guidelines for placing the text, art elements, and captions.

Grid structure

Grid structure can consist of simple vertical columns or can incorporate more complex combinations of both vertical columns and horizontal rows. At first glance, this newsletter seems to have a simple three-column structure, but that structure is actually based on a six-by-ten modular grid. The three text columns are each two modules wide, and the half-column captions are one module wide. The horizontal rows help provide guidelines for placing the art elements.

Ready Set Go, the program used for these pages,

bold weight and strong contrast to the text. Effective typography comes not so much from the *variety* of typefaces used as from the imagination employed in their use. Using fewer faces is simpler, costs less, and, frankly, usually makes for better design.

So what faces should make up your basic type library? The chart above lists four categories of typefaces and their most common uses. Owning one or two faces from each category will give you a good selection to start with; more will expand the range of looks you can achieve.

With a repertoire of classics like these, and by using the following basic design techniques of *layout*, *contrast*, and *details*, you'll be able to use just a few good typefaces over and over again while keeping your newsletter fresh and exciting looking.

Type illustrations like this   **can add punch to a page**

Typographic forms, when exaggerated in scale, can provide interesting effects and can even become art elements. This ampersand is being used as a "typographic illustration," an inexpensive graphic alternative to photography or illustration.

&

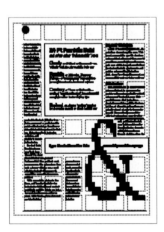

Fig. 10-12 This is how the page in Fig. 10-11 appeared on the computer screen. The printed page seems to have been constructed on a three-column grid, but actually a six-column grid was used. The six-column grid gives more flexibility in laying out the pages. (Used by permission of *Publish!*, the How-to Magazine of Desktop Publishing, Vol. 4, no. 3, March 1989, published at 501 Second St., San Francisco, CA 94107.)

- The cropping and scaling capabilities for processing art.
- The type of kerning program. Some are automatic and some are manual.
- Hyphenation and justification ability.
- Ability to import text from the word processing program and graphics from the graphic program.

Supporting Programs

Many supporting programs can be obtained to augment the basic word processing, graphics, and page layout process. An example is a hyphenation program. One way to destroy readability in desktop publishing is to write copy and then simply tell the computer to justify this copy. You could end up with something like this:

> Family planning providers have won the battle with the Reagan administration over counseling and referring for abortion while retaining Title X funding. Unfortunately, thisfundingtoPPNNhasbeencutby2.5%, after remaining static since 1982. During this time, our SERVICES GREW BY 47%.

Now there are hyphenation programs available that can help you turn out copy that is more realistically word spaced. These programs can read text and check all the words against a comprehensive dictionary. Words of as little as five letters are matched with the dictionary and broken into proper divisions for hyphenation. The result is a neat appearance that eliminates rivers of white and unsightly gaps between words.

Here is an example of lines properly spaced. Compare them with the improperly spaced lines in the preceding example.

> Family planning providers have won the battle with the Reagan administration over counseling and referring for abortion while retaining Title X funding. Unfortunately, this funding to PPNN has been cut by 2.5%, after remaining static since 1982. During this time, our SERVICES GREW BY 47%.

Hyphenation programs can be obtained for less than $50. No newsletter, magazine, newspaper, or brochure should be produced without such a program to maintain uniform spacing between words. It is better to set copy ragged right than allow copy with unsightly word spacing to appear in print.

A wide variety of clip art software is also coming onto the desktop publishing market. Most of it is not looked on too favorably by many professional designers who would prefer to originate the art themselves. However, there are some excellent clip art programs. Metro ImageBase collection, for instance, offers high quality clip art for newspapers and other publications.

Type fonts for laser printers as well as fonts for the computer screen can be purchased very reasonably. In selecting fonts for desktop publishing the same criteria should be used as discussed in Chapters 3 and 4.

▼ The Complete Desktop Publishing Shop

What do you need to get into desktop publishing and how much will it cost? When you get right down to it, that's what most communicators want to know right off the bat.

The most economical entree to desktop publishing can cost as little as $3,000. It can create documents of one to several columns on a page plus crude posters and newsletters. It does not produce quality printing that would be acceptable for any but internal use.

Such a system would consist of a personal computer with limited power, a monochrome monitor, and a graphics adapter. The printer would be a 24-pin dot matrix model plus programs for word processing and layout. The quality of the work produced would depend a lot on the skill of the operator.

The next step up would be a desktop publishing setup capable of producing somewhat better quality work for newsletters and reports. It would consist of a more powerful computer with a monochrome graphics monitor and adapter card which would cost, at 1990 prices, between $2,000 and $2,500 plus a LaserJet printer and a rather sophisticated program such as PageMaker. It would need a mouse, and it would produce much better quality printing than the economy package just described. Such a system would cost in the neighborhood of $5,000.

If you wish to produce good quality newsletters and small publications as well as reports and flyers, you should consider an investment of, say, $10,000. The result would be a product that was good but not quite tops. The final product could be improved if mechanicals were produced and sent out to a professional typesetting facility that was equipped with a typesetter such as a Linotronic or Postscript.

It would take about $25,000 in hardware and software to produce quality newsletters with art and graphics; long, complicated reports; quality publications; and even books. Now the desktop publishing office would include a computer with considerable power, a laser printer, a program with many features such as PageMaker or Ventura Publisher, and a scanner. It would still be necessary to send out mechanicals to a commercial shop to make color images and four-color separations. The finished product would have to be printed by a commercial printing plant.[11]

The technology is changing so fast that it is virtually impossible to predict from one day to the next what new products will be available, how much they will cost, and what changes in existing software programs will be made. Anyone considering becoming a desktop publisher should monitor the field constantly. Regular reading of at least one of the desktop publishing magazines would be a good way to keep abreast of changes.

▼ Learning to Be a Desktop Publisher

The first step in publishing on a desktop is to turn on the computer. This may seem obvious, but at least one student in a course in desktop publishing could not get output from the computer at her workstation because she had neglected the first step.

Once you turn on the computer you have entered a magic world. The electronic magic is capable of helping the communicator create limitless possibilities, but the computer is still just a machine. It is the creativity and skill of the person at the keyboard and the mouse that determines the effectiveness of the output.

(a)

(b)

THEIXCHANGE   THE EXCHANGE   THEIXCHANGE

(c)

**PHELAN
&SMITH   PHELAN
& SMITH   PHELAN
&SMITH**

Fig. 10-13 Logo design made easy. These examples were created by the firm of Galarneau & Sinn, using Adobe Illustrator 88. (a) Once the type was set in Italia Bold, the house and roof were drawn with the pen and box tools. To cover the serifs in the typeface, they used the box tool to create white shapes. They also elongated some of the strokes in the letterforms using the same tools.
(b) In this example, they set the words in Futura Extra Bold Condensed. White rules were placed over the *E*, which was effortlessly flopped using the reflection tool and placed in front of *xchange*.
(c) After the type was set in Futura Bold Condensed, they used the pen tool to draw a wrench. Then it was a simple matter of placing the illustration over the ampersand and coloring it white.

Learning to be a desktop publisher is simplified because the manufacturers of the hardware and software use everyday terms to designate commands and functions. Page makeup programs such as PageMaker and Ventura Publisher use terms such as "cut," "paste," "clipboard," and "scrapbook" to make the programs understandable. For instance, the term "cut and paste" means just what it says: You "cut" copy and "paste" it in place on a page just as you would in the traditional method, but you do it on a computer monitor instead of a drawing board.

In working with a page design system, the designer operates the computer which takes information in digital form that has been generated with word processing software and stored on disks and converts it into type. This type is displayed on a screen in the format determined by the designer using a page makeup program.

It can be helpful to do some preliminary work, such as creating thumbnails and roughs with pencil and paper, to visualize how the page will look when assembled on the screen.

Various kinds of illustrations, obtained from a clip art program stored in the computer or introduced with a floppy disk or created by the designer on the screen with a draw or paint program, can be incorporated in the layout.

All this is done on a grid that has been designed on the screen before the type and art are brought in.

Once the information is assembled, the computer allows the designer to manipulate the material on the screen in almost every imaginable way, with far more speed and accuracy than any traditional method. You can

Fig. 10-14 Examples of electronic art from a computer art program. These were created and separated using Arts and Letters. (© 1989 Computer Support Corporation)

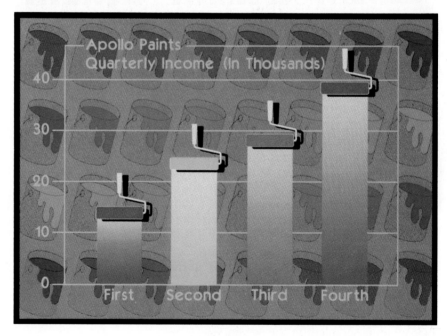

change page size or typeface size and posture—regular to bold or italic—at the press of a key. Line length and leading, position of columns and margins, and location, size, and shape of illustrations can be manipulated with ease. You can even reverse type in seconds.

After the designer is satisfied with the layout, it is printed on a laser printer. Laser output may be used for the mechanical. It can then be handled just as a mechanical is handled in the traditional way. A flap can be attached and instructions for the printer added.

One of the advantages of desktop publishing at this stage of the production process is that last-minute changes can be made simply and quickly. Quite often, once the mechanical is ready for the printer it can be sent on a disk or transferred by a modem to a service bureau or a commercial printer. There the digital information from the designer's computer can be translated by a typesetting machine that produces either film or paper output.

In many association and business offices these days, the product of the laser printer is printed in house; that is, the organization has its own production facilities.

But before all this can be accomplished, those who have not learned desktop techniques have to turn on the machine and go from there. It seems that a big barrier in the path of the neophite desktop student is the manual that is produced by the manufacturers of the hardware and software. Many times such manuals are written by technicians and contain a maze of information that is baffling and frightening.

Myrick Land at the University of Nevada in Reno has found a solution to this problem. He encourages his beginning desktop students to write their own manuals in language that they can understand and use.

Here is a twenty-two step manual written by Derron Inskip, a student in Professor Land's class, for the PageMaker publishing program on the Macintosh computer. (The computers used in this class also had MacPaint and MacDraw programs for the creation of illustrations, Hypercard with an assortment of clip art, and Microsoft Word for creating copy.)

▼ Introduction To PageMaker

Step 1: To get into PageMaker go to the hard drive, then place the arrow on the PageMaker symbol and click twice.

Step 2: When PageMaker 3.01 appears, place arrow on the symbol and click twice again. Move the arrow to "File" and hold the button down, then pull down to "New" to open a new file.

Step 3: The page setup will appear. Here you can select the page size, number of pages, options, and margins. After you have done this, click once on "O.K."

Step 4: Your screen should show a blank PageMaker page.

Step 5: Move the arrow to the apple symbol, hold down the button on the mouse, and pull down to "Scrapbook." Here you can select an image by placing the arrow on the gray bar below the image and pressing the button to select an image saved in the scrapbook.

Step 6: After you select your image, place the arrow on "Edit," hold the button, and pull down to "Copy."

Step 7: Place the arrow on the page number at the bottom left corner of the screen that you wish to put your image on and click once.

Step 8: Place arrow on "Edit," hold the button, and pull down to "Paste." The image you selected will appear on the page.

Step 9: To enlarge or reduce the image, place the arrow on one of the small black dots that appear around the image, hold the button down, and pull the image in the direction you wish to enlarge or reduce.

Fig. 10-15 A graphic created by a student for his desktop publishing manual. The student moved the arrow to "file" and clicked the button on the mouse. The various symbols and names of the programs appeared on the menu screen of the Macintosh. If the student wishes to work in a program, he will move the arrow to the proper symbol and click the button on the mouse twice. The program will appear on the screen.

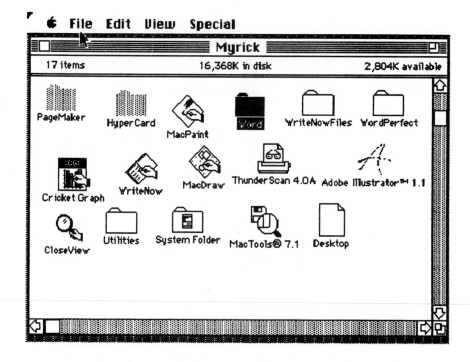

Step 10: To place the image in the layout, move the arrow on top of the image, hold the button down, and you can move the image to where you wish to place it. Once you place the image, move the arrow outside the image and click once.

Step 11: If you wish to set column guides (the same as the column areas that are created on a paper grid) move arrow to "Options" and pull down to "Column Guides." Select the number of columns you wish on the grid and click once on "O.K."

Step 12: To protect the image from type being placed over it, move the arrow to "Options," hold the button, and pull down to "Text Wrap." Place the arrow on the wrap option you prefer and click once. The Scrapbook image should appear surrounded with dots.

Step 13: To type on the PageMaker page, place the arrow on the "A" in the toolbox and click once, then place the symbol where you wish to begin typing and click once. You are ready to type.

Step 14: To see what you have typed, move the arrow to "Page" and hold down the button; then pull down to "Actual Size" or "75% Size." To return to the PageMaker page, move the arrow to "Page" again and pull down to "Fit in Window."

Step 15: To select type size, move arrow to "Type" and pull down to "Size"; then pull to the right and select the type point size.

Step 16: Next, move arrow to "Type" and pull down to "Font"; then pull to the left to select the font. (Repeat Step 11 if you wish to set different column guides.)

Step 17: To place text that you have previously typed into a word processing program such as Microsoft Word onto the PageMaker page, place the arrow on "File" and pull down to "Place." A list of documents will appear on the screen. Choose the document you wish to place on the screen. Place arrow on "O.K." and click once.

Step 18: Move arrow to "Options" and pull down to "Autoflow." This option will feed the text around the illustration.

Step 19: Place text symbol where you want your text to begin and click once. The text will appear.

Step 20: To remove the text, if you wish to do so, place the arrow on the text to be removed and click once. Move the arrow to "Edit" and pull down to "Cut."

Step 21: If you wish to print the page you have created, move the arrow to "File" and pull down to "Print." That will activate the laser printer.

Step 22: If you wish to save the pages you have created, move the arrow to "File" and pull down to "Save as." Type a name you want to use to identify the pages and then place the arrow on "O.K." and click once. Your pages are saved in the file.

A number of design programs include so many features it is a temptation to let the computer dictate how your layouts will appear. For instance, the PageMaker Portfolio contains a collection of design templates. It is a simple matter to select a template and let the machine fit your design into this predetermined form. Do not take this easy way out unless the resulting mechanical will be just what you want.

Sometimes best results can be obtained by combining what you can get out of your computer with elements created by other methods. These might include conventionally screened halftones and professionally rendered original art or logos, as well as color separations. All the elements can be brought together and a mechanical constructed in the traditional way—with scissors, X-acto knife, waxer, and grid.

▼ The Future of Desktop Publishing

"It is fun being part of the revolution in electronic graphic design and publishing. If I had to give up my computer and software, I think I'd have to find something else to do for a living. I don't even want to ponder it. I'd never go back to using a drawing table, waxer, and T-square. I'm completely spoiled," is the way Fletcher H. Maffett, Jr., a professional designer, explains his enthusiasm for the computer as a design tool.[12]

Desktop publishing has changed graphic communication and it will continue to change how things are done in the future. The hardware and software necessary to create spot color and color separations on the computer are now available. Magazines are discovering that desktop publishing can be professional publishing when a competent designer is at the computer controls.

Desktop publishing "has made people rethink the way they put words on paper," says Serge Timacheff, associate editor of *Infoworld*. "The problem comes with the realization that you can give someone a palette and a brush, but that doesn't guarantee they can paint the Mona Lisa."[13]

The successful desktop publisher is the person who can take advantage of the capabilities of electronic publishing equipment and combine them with skillful use of graphics and typography.

Two people who have done just that are William E. Ryan, associate professor, visual communications and advertising design, at the University of Oregon, and David Cundy, designer and principal of David Cundy, Inc., a Connecticut design firm.

Ryan has this to say about the future of desktop publishing:

"Today, more than ever before, it is important—even critical—that the person working in communications has a basic understanding of type, design, composition and production.

"Desktop publishing has imposed responsibilities in these areas on those who may know very little about them. And so along with writing, writers need to be able to make intelligent decisions about design and type. Editors, copywriters, public relations directors and other communication professionals also must be educated beyond their fields because computers and desktop publishing are central to their work. Like it or not, that is today's technology, and if you expect to work effectively (or even land a job, for that matter), you'd better know it."

Fig. 10-16 Liz Eselius, University of Oregon advertising major, uses David Ogilvy's recommended layout format of visual/headline/copy for this clever 1989 Nissan—Chiat/Day award-winning design. Layout, typography (including the text wrap) and logotype were executed on a Macintosh SE.

Fig. 10-17 After creating over a dozen thumbnails on the Macintosh, University of Oregon advertising student, Andrea McHone, selected this format for her art and copy. She then scanned in the artwork—two separate photographs—placing the second image, the 240SX, atop the background to suggest a 3-D effect. Finally, the headlines, copy and logo were carefully positioned. Notice how the art shows the headline and vice versa. Note, too, how the subhead, NISSAN 240SX, is framed and set off by the product. This piece was also a national award-winning student advertisement for Nissan.

Fig. 10-18 Jarrett Jester designed this striking piece and co-authored it with Michele Tarnow. Designing this advertisement began on the computer via a series of thumbnails. Eventually all the typography (including construction of the logotype) was carefully figured, set and positioned in the ad on a Macintosh. Jester's airbrushed art was painted to the ad's specifications and the typography was then overlaid atop the art and shot.

Fig. 10-19 Debbie Hardy worked as the designer/artist for this piece and Leslie Jones the typographer/copywriter. In this instance, the students designed black borders for the art and reversed copy to give it a dramatic flair. Care was taken to improve the readout of the reverse block by increased point size and leading: 14/17. Helvetica, also 14/17, was used for the signature and identification of Clean Ocean Action because it nicely contrasted the serif face used for the headline and copy. The tag, "Until the coast is clear," was set in a bold, oblique Helvetica. Artwork was executed to fit the designed format exactly and dropped in one-to-one.

Figs. 10-18 & 10-19 Both of these student advertisements competed in the 1989 Annual One Show awards. They were among the finalists for that competition; the "Catch of the Day" piece was among the One Show student winners. The client was Clean Ocean Action and the idea was to make a general audience aware of the ever-increasing pollution problem caused by the dumping of sewage, sludge, medical and industrial wastes into our oceans. (The ads above and on the facing page were created by students of William E. Ryan at the University of Oregon.)

Ryan continues, "However, along with all the additional responsibilities and expected skills come tremendous advantages. Desktop provides an inherent immediacy. It also brings a vast array of designing and editing tools that are capable of rendering immaculate layouts with razor sharp graphics. What's more, desktop will only get faster, more sophisticated and powerful in the years to come. In many ways, its future is now because as you read these words, its evolution continues to snowball. What was considered incredible four or five years ago is mundane today. Yesterday's design and production fantasies are available as standard features in today's desktop and computer technology."

Cundy, who specializes in corporate identity and communication design, says the computer is "creating a revolution in the visual communications profession." He sees it affecting architects, designers specializing in environmental, graphic, industrial and interior design, illustrators, photographers, printers, and typographers. He says they all will feel the impact.

"A design student must comprehend this new phenomenon in order to survive in an increasingly competitive employment environment."

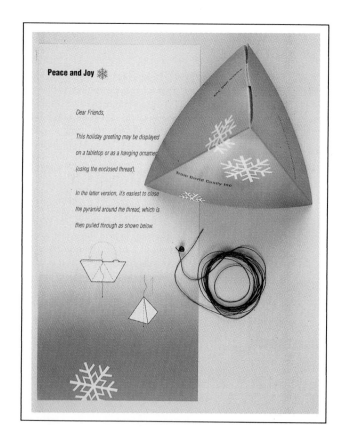

Fig. 10-20 *David Cundy Inc. 1988 holiday card:* The computer's mathematical capabilities streamline creation of geometrical objects. Computers are being used in most two- and three-dimensional visual arts disciplines today.

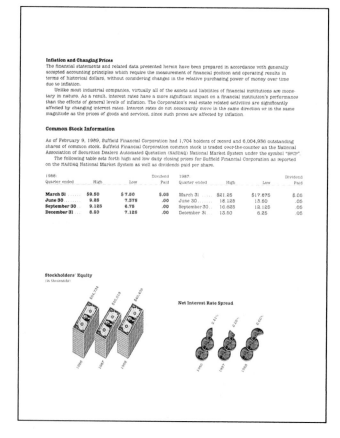

Fig. 10-21 *Suffield Financial Corporation 1988 Annual Report:* Distinctive, theme-related graphics were created with an illustration program. The computer provides a far higher degree of accuracy for data presentation than previously achievable.

Fig. 10-22 *NovaCare corporate identity:* The computer permits quick, clear visualization of many permutations, streamlining the identity development process. Final artwork can be stored on disk for reproduction.

Fig. 10-23 *AG/ENA poster:* The computer is well-suited to system- and grid-oriented projects, where repeated elements and precision alignment are required. This poster of an acclaimed architectural firm utilizes modules suggestive of the firm's work.

Fig. 10-24 *Great Northern Nekoosa 1988 Annual Report and 10-K:* Illustrations were created on-screen and transmitted to a service bureau for film separations. Software is currently under development which will give designers a direct link to prepress, eliminating conventional artwork and mechanicals.

Figs. 10-20 through 10-26 The examples shown here and on the following page were designed by David Cundy, of David Cundy, Inc., a Connecticut design firm that was established in 1985. The firm's pioneering work with electronic design tools has been featured in *Print* and Adobe's *Font & Function.*

Fig. 10-25 *Eastern Press October No-vember December 1987 Calendar:* Using the right tool: This poster was designed and produced conventionally, with type prepared by the computer. The computer is not always the best tool for visualization or production.

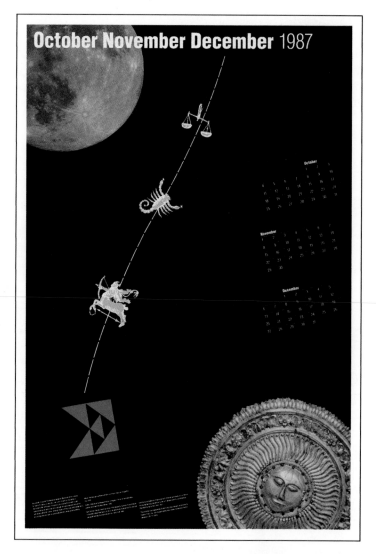

Fig. 10-26 *Chartwell Group Ltd. Annual Report 1987:* This report was designed and produced with the aid of a page lay-out program. The complex financial ty-pography used in annual reports requires advanced typographic skills.

Cundy adds a word of advice and caution.

"But the computer, like the paintbrush or pen, is only a tool, and must be regarded as such. Design education must continue to emphasize aesthetics (imagery), typography (structure), communication skills (content), and experimentation (objectivity/risk). Students with a broad understanding of art, craft and culture, as well as an eagerness to explore, will find it easier to adapt to the methods—and mindset—the future will surely require."

▼ A New Vocabulary

Desktop publishing uses a new vocabulary for a new world of design. Here, trimmed to the bones, are the essential words needed for working with electronic design tools.

- *bit (binary digit):* the smallest unit of information making up the digital or dot image of a character or graphic; small parts of a letter; just little dots.
- *bit-map graphic:* a graphic image document formed by a series of dots, with a specific number of dots per inch. Also called a "paint-type" graphic.
- *boot:* getting your computer going; getting it started up and into the program you're going to use.
- *byte:* eight bits make a byte and a byte can store one character; a unit of measurement.
- *card:* a printed circuit board; computer systems are made up of these boards.
- *clipboard:* a temporary holding place for material, in the computer. You can store text, graphics, or a group selection on a clipboard for later use.
- *cursor:* a blinking, movable marker or position indicator on the screen to show you where you are.
- *data base:* a collection of information that is organized and stored so that an application program can access individual items.
- *download:* to transfer data from one electronic device to another. You could download information from one computer to another with a modem, for instance, or you could download information from a hard disk to a floppy disk.
- *downloadable fonts:* fonts that you can buy separately and install so as to expand the variety of fonts available on your printer.
- *dpi:* dots per inch.
- *DTP:* desktop publishing.
- *file:* a collection of stored information with matching formats, the computer version of filing cabinets.
- *finder:* the file that manages all the other files; the finder is like an index; it saves, names, renames and deletes things in a file.
- *footer:* one or more lines of text that appear at the bottom of every page, similar to folio lines or running feet.
- *header:* same as a footer, but at the top of each page; like a running head.

- *H&J:* hyphenation and justification; there are programs that will do the hyphenation of text for you following a standard dictionary.
- *icon:* a small graphic image that identifies a tool, file, or command displayed on a computer screen.
- *kerning:* to decrease the space between letters, by moving them closer together; if you are not careful you may move them so they overlap and become illegible.
- *kilobyte:* 1,024 bytes or 1K; a 3½ inch floppy disk holds 800 kilobytes (800K) or about 400 double-spaced typewritten pages (a rough estimate).
- *menu:* a list of commands that appears when you point to and press the menu title in the menu bar.
- *menu bar:* the area at the top of the publication window that lists the menus.
- *modem:* a telecommunications device that translates computer signals into electronic signals that can be sent over a telephone line; a way to get information from one computer to another or from your computer to a print shop and so on.
- *object-oriented graphic:* an illustration created in an object-oriented, or draw-type application. An object-oriented graphic is created with geometric elements. Also called a "draw-type" graphic.
- *pixel (picture element):* the smallest part of a graphic that can be controlled through a program. You could think of it as a building block used to construct type and images. The resolution of text and graphics on your screen depends on the density of your screen's pixels.
- *RAM:* random access memory; the temporary memory inside the computer that allows you to find stored text and graphics.
- *resolution:* the number of dots per inch (dpi) used to represent a character or graphic image. The higher the resolution the more dots per inch and the clearer the image looks.
- *scanner:* a hardware device that reads information from a photograph, image, or text, converting it into a bit-map graphic.
- *text wrap:* to run text around an illustration on a page layout. Some programs have an automatic text-wrap feature that will shorten lines of text when a graphic is encountered; in other systems you need to change the length of lines to go around a graphic.
- *vertical justification:* automatic adjustment of leading or the space between lines, in very small amount so columns on a page can all be made the same depth.
- *WYSIWYG:* an acronym for "what you see is what you get."

▼ Graphics In Action

1. This exercise is intended for those who are learning desktop publishing. You will need the three basic software programs: word processing, graphics, and page layout. Find a news story in a newspaper. The story should be about 10 to 15 column inches long. Copy the story with your word processor. Write a head for the story to fit in a three-column width on a page grid of your page layout program. Create a

drawing or find a graphic in your graphics clip art program, if you have one. Combine all three elements into a layout with the page layout program and print the results.

(*Note:* A similar project has been devised by William Ryan, University of Oregon, whose students attempt to duplicate information graphics found in newspapers. Many students are surprised to discover they can reproduce graphics very similar to the originals.)

2. Assume you are going to produce a monthly newsletter for an organization or business. Plan a desktop publishing workstation to produce the newsletter. Obtain literature and visit computer stores to obtain information and prices of hardware and software. Your goal is to create a camera-ready four-page newsletter. Work out a prospectus for the proposed newsletter that would include a budget for the equipment purchases.

3. If you have a draw program, practice by attempting to duplicate a simple illustration similar to the one included here. You will often be surprised at the results, especially if you can print your drawing with a laser printer.

4. Write your own manual for your particular needs (see the manual written by a student in this chapter). Use a looseleaf notebook and organize it as your permanent reference book. For instance, one section could contain an alphabetical listing of all the material you have filed with the identifying file name. Another could contain ideas for graphics and layouts clipped from various sources. The most important part of your manual would likely be step-by-step instructions in your own words for various procedures.

5. Plan a newsletter as in exercise 2, but build your plan around the capabilities of the hardware and software already available to you rather than the purchase of new equipment. It might include a "wish list" of hardware and software you would like to add in order of the priority of purchase when funds would be available.

Notes

[1]"Pandora's Desktop," *Adweek Special Report*, October 3, 1988, p. 4.

[2]"Doc, will I be able to play the piano?" *Typeworld*, Second March Issue 1989, pp. 6, 12, 23.

[3]"Time for a change," *Publish!* February 1989, p. 77.

[4]"Doc, will I be able . . .," *Typeworld* previous citation, p. 23.

[5]"Pandora's Desktop," p. 5.

[6]Ibid.

[7]"A Primer on Desktop Publishing," *Plus Business*, July 1988, pp. 5, 10.

[8]"PageMaker Is Still Choice for Desktop Publishing," *Magazine Design & Production*, December 1988, p. 14.

[9]"More Than Meets the Eye: A Scanner Review," *Electronic Publishing & Printing*, January/February 1989, pp. 40–42.

[10]"Updates," *Personal Publishing*, September 1989, p. 70.

[11]"Power Publishing," *PC Magazine*, December 27, 1988, p. 91.

[12]"Desktop Publishing: An Ode to Joy," *Magazine Design & Production*, January 1989, p. 22.

[13]"Pandora's Desktop," p. 5.

Designing Printed Communications

Photograph: Mark Jenkinson

Every time an image is impressed on a piece of paper, the ultimate effect is communication. No matter whether the production is an elaborate magazine, a metropolitan newspaper, or a letterhead, the same care should be taken to ensure that it is doing all it can to perform this function to the fullest.

Every printed communication should exhibit the attributes of good design principles—and be legible and readable. It is now time to put to work the principles of layout and design, the use of type and illustrations, and the role of paper and ink as well as the various methods of production in creating printed communications.

In this chapter we consider some of the various types of printed communications all of us working in the profession will encounter at times during our careers. Then, in succeeding chapters we examine the design of advertising, magazines, newspapers, and newsletters. By developing a basic understanding of the graphics of communication we can go from there to the area of the industry in which we might wish to specialize.

First we consider brochures since they are so closely related to newspapers, magazines, and newsletters. Then we examine stationery, programs, books and pamphlets.

▼ Designing the Brochure

They are called brochures, folders, flyers. Regardless, the basic approach to their design is the same. A folder can be thought of as a finished brochure that is folded, and a flyer as a single-sheet handbill. We will refer to them as brochures to prevent confusion.

Let's say we are to design a brochure that will be part of an organization's communications mix. The brochure will serve as an informational piece, or an introduction to prospective members. Where should we begin? How do we move from inception to the completed project?

Planning

While adjustments must be made to fit the particular situation at hand, there are some steps that can serve as a guide in handling any brochure or folder design execution from start to finish. *Planning* is the first step. The more details that can be worked out in the early stages of the project, the more effective and cost-efficient the end product will be. Planning should include (1) forming a statement of purpose, (2) determining the audience and its characteristics, (3) making a checklist of the essential information to be included, (4) listing the benefits to the audience of the information, and (5) making a timetable for execution and distribution.

The planning could be organized around the "three Fs" of communication—function, form, and format. Consider the brochure's *function*. What is its purpose? What are some possibilities? Usually these purposes will fall into one of three or four alternatives. It might announce a workshop or program of some sort. It might be used mainly to provide information. (In this case, we need to determine if the brochure is an end

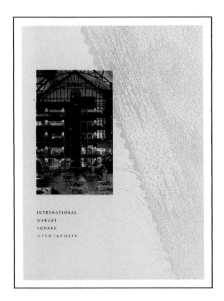

Fig. 11-1 Attention-getting art combined with an appropriate type style were used in designing the cover page of a brochure for the International Market Square in Minneapolis.

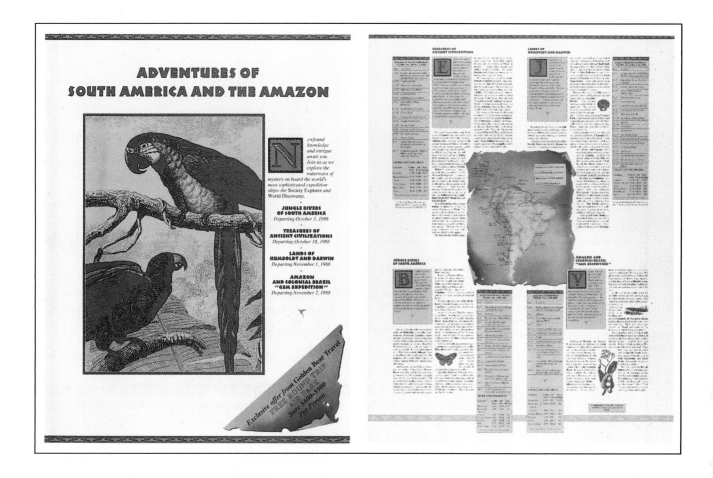

in itself or a supplement to other communications such as an advertising campaign or a lecture series.)

Perhaps the brochure is to serve several functions. It might be a mail-out announcement and at the same time a bulletin board notice. It might be part of a series of brochures that the audience will be encouraged to keep in a binder for future use or reference.

The distribution method and the life expectancy of the brochure should play a part in design planning. Will the brochure be mailed, distributed in an information rack, or handed out at a meeting? Will it be used just to announce an event, or will it have more permanent use?

It will also, of course, be necessary to consider budget factors. How much money can be spent on the production? Money can be a limiting factor on the size, number of illustrations, use of color, quality of paper, and so on.

As this preliminary visualization of the situation progresses, a form may begin to evolve since form follows function. What should the physical size be? What shape should the brochure have? There are a number of possibilities. It could be simply a flat sheet; it could be folded in any number of arrangements.

Now might be a good time to start considering a rough dummy. Try this. Settle on a size and take a blank piece of paper and see what can be done with it. For example, suppose we decide our brochure will be

Fig. 11-2 The designer can create the mood for a brochure by selecting type, art, color, border that is fitting for the subject. Note how the elements used in this brochure give the feel of South America and the Amazon. Note, too, the placement of the teaser in the lower right pointing towards the inside pages.

Fig. 11-3 Paper is available that has partial color added. This paper can add a spark of color to a brochure at a minimum cost. It can also be used in a laser printer.

printed on an 8½ by 11 standard typewriter-size sheet of paper. We have determined that the brochure will have a multi-informational purpose and that it should be easy to mail, place on information racks, and pass out at meetings.

What can we do with this piece of paper to make the most effective brochure?

There are a number of alternatives. The first is to make the brochure a flat sheet, printed on both sides. This could be punched to be kept in a ring binder. Or it could be designed as a combination flyer and bulletin board notice. One side would be designed as a poster and the other could contain general information, or be left blank, or have an address box for mailing.

Other possibilities might include folding the sheet of paper in certain ways. But before we can explore those possibilities, we should become acquainted with paper finishing and folding operations.

Folding

If we are going to produce a brochure or pamphlet that requires folding, that operation should be a part of the planning. Michael Blum, a printing instructor, wrote in *In-Plant Printer* that "approximately 35 to 40 percent of the labor cost of the average printing job is in finishing." [1]

Modern equipment is capable of producing many different types of folds. But all folds are basically either parallel or right angle. *Parallel folds* are used for letters where two folds are required to fit the letter into the envelope. This same fold, when the sheet is held horizontally, becomes a six-page standard—also called *regular*—fold.

An *accordion fold* is another parallel fold. It, too, is popular for brochures and envelope stuffers. The most basic parallel fold is a simple fold to make a four-page brochure. The two-fold accordion made from an 8½ by 11 sheet is popular for brochures as it creates a handy size that mails easily in a number 9 or 10 commercial envelope. It also has good design possibilities. Accordion folds can be made with additional folds to create eight, ten, or more pages.

A popular fold for invitations and greeting cards is the *French fold* This is an example of a right-angle fold, which is made with two or more folds at right angles to each other. The French fold is also used to create an eight-page publication that can be saddle stitched if desired and trimmed off at the head, or top, of the folded sheet.

Folding machines can handle signatures of many pages to create booklets and publications of twelve, sixteen, twenty, twenty-four, or more pages. Some machines are equipped with pasting attachments to produce paste-bind booklets (booklets with the pages pasted rather than stapled).

There are many other options for folding depending on the equipment available. These options should be considered when brochures and other printed communications are planned. Thus it can be quite worthwhile to discuss all possible options with the printer or binder operator. Sometimes the capability to make certain folds can spark ideas for a new and unique way to design the brochure.

Now take the blank sheet of paper and experiment with the fold possibilities for your particular brochure. How many alternatives can you devise considering all the factors that went into the planning so far? Keep in mind that you would want to preserve the basic dimensions of the format you have in mind for your brochure in selecting folding alternatives.

Fig. 11-4 The possibilities presented by various types of folds should be considered in brochure planning. Here are some alternatives: 1, vertical parallel fold; 2, horizontal; 3, book fold; 4, short fold vertical; 5, short fold horizontal; 6, tent fold; 7, gate fold, 8, z or concertina fold.

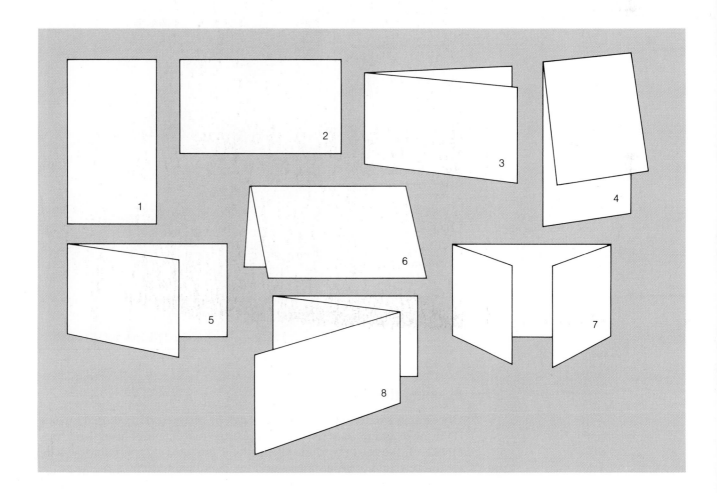

Fig. 11-5 In designing a brochure, the designer considers the entire area and does not always confine the layout to rigid horizontal or vertical grids. Note the unifying features of this brochure such as the angled lines and use of tint blocks. Questions readers might have should be anticipated and answered such as was done here with the map. (Courtesy Reno Printing, Inc.)

Planning the Format

Once a size is determined, the format can be planned. This should include (1) the size and form of the margins, (2) the placement of heads and copy blocks, (3) the use of borders, illustrations, and other typographic devices, (4) the selection of type styles—and the overall determination of how the principles of design will be applied.

Now we can begin making rough layouts on paper. A good idea is to take several sheets of paper, fold them into the final brochure form, and sketch roughs of possible arrangements. Some designers like to make thumbnails in smaller but exact proportions to the final layout. Others prefer to work with full-size roughs. Whichever we choose, once we have selected a general arrangement, we would produce a full-size rough or comprehensive, following a procedure similar to designing an advertisement. Then we would mark the rough and submit it plus typed copy and illustrations to a printer. Or we might have the type set and then do the pasteups for comprehensives or completed mechanicals. These would be photographed and the negatives used to make plates for printing.

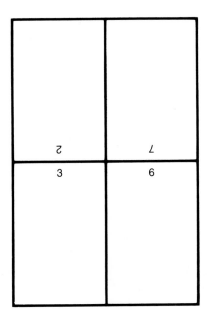 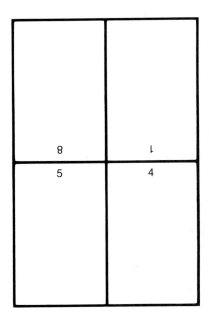

Fig. 11-6 In preparing a mechanical for a brochure, the "lay," or arrangement of pages on the grid, should be such as to ensure proper order for printing. Usually, several pages are printed on one sheet which is folded and trimmed after printing. This is the arrangement of pages on a mechanical for a typical eight-page brochure.

Two Ways to Design Effective Brochures

Although the design for the brochure may evolve out of the preliminary planning, if we find it difficult getting started, there are other ways to approach the overall design. One might be called the "headline method," and the other the "attention to action" approach.

In the *headline method* we devise a brochure outline around the headlines to be used. First is the feature head, or the title for the brochure. This head stresses the most compelling reason for the target audience to read the brochure. It should set the theme for the brochure by answering the question, Why should I bother to read this brochure?

This feature head can become the title line for the brochure, and it can be amplified by *main support heads.* These can number two or three to six or more, and they become the heads for the copy blocks in the brochure. Often each copy block and head cover one topic or point to be made.

If the copy block under the main support head is long and involved, it can be broken down by "reinforce heads" or subheads, each with a brief copy block. This approach can be used as an outline for writing the brochure copy as well.

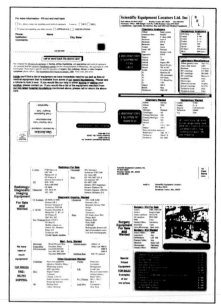

(Before)

(After)

Fig. 11-7 Scientific Equipment Locators wants their brochure redesigned. SEL wants emphasis on their desire to buy used scientific equipment and they want their name to be remembered. What would you do? Here's how designers Donna Kelley and Kelly Archer handled the problem. They noted the original's shortcomings: lack of organization, too cluttered, little incentive for the target audience to plow through the material, tries to emphasize buy and sell rather than zeroing in on one theme. The new design emphasizes SEL's desire to buy equipment. A separate brochure for selling equipment was recommended. Goals for the new design included: get the target audience to read the brochure, present the information in a logical order, provide memorability for the company name, make it easy for the target audience to take action. The mechanical of the new brochure was created desktop and printed on a laserjet printer. Adobe Garamond and Helvetica Black types were used. Compare the Before and After brochures. How do you think the goals were achieved? (Courtesy Scientific Equipment Locators and D. Kelley, Zetatype, San Francisco)

The *A-I-D-C-A (attention, interest, desire, conviction, action) formula* some advertising copywriters use, as explained in Chapter 12, can be a handy guide in planning brochures. The parts of the formula that are relevant to the situation can be used to check the copy and to plan the layout. This method also helps to organize the material in an orderly and forceful manner.

▼ Effective Design Checklist for Brochures

- Work with the pages as the units readers will see. For example, if the brochure is four pages, the two inside pages should be designed together. If the brochure has two parallel folds, to make six pages, the three inside pages should be considered as a unit.
- In designing facing pages, cross the gutter with the same care as if designing magazine spreads.
- Apply all the basic design principles—balance, proportion, unity, contrast, harmony, rhythm—when laying out a brochure.
- Define the margins for the pages first and work within them when making layouts. Use ample margins and avoid a jammed-up look.
- Stress simplicity and careful organization in layouts.
- If the rough layout is to be examined and approved by another person before production and the roughs are on thin paper, cut the spreads apart. Paste them on heavier backing for a more impressive presentation. A protective flap or cover taped at the top will help, too.
- Check to see if all possible readers' questions are answered.
- If a mail-in coupon or registration form is included in the design, place it where it can be detached easily and where it will not destroy pertinent information such as a program listing when it is removed.
- Use big, bold display when the brochure is to double as a bulletin board announcement.
- Have the purpose and content dictate the design: The announcement of an event calls for stronger display than an informational brochure does.
- Don't try to achieve too much in one brochure. Designers often need to limit the quantity of information to be included in a single brochure.

▼ Choosing the Right Paper

Choosing the right paper for a brochure or any other printed communication requires making two basic decisions. One is *selecting the proper finish and weight*. The other is *planning the size* of the printed piece to obtain the most paper for the least amount of money.

Once the format of a job has been determined the next step is to select the best paper. This should be done after deciding the mood of the mes-

A Mini Course in Division

Quite often when designing brochures we want to create a format of three pages on each side of a sheet of paper. This requires dividing the sides into thirds when making a grid or template for the layouts.

Here's a quick way to divide an area into thirds.

Just multiply its width in inches by 2 and the result will be one-third of the area in picas.

Say you have an 8½ by 11 sheet of paper and you are planning accordion or parallel folds to create a six-page folder with each page 8½ inches by one-third of the width. Multiply 11 by 2 and the result in picas will be 22. Each one-third segment of the sheet will be 22 picas wide by 8½ inches deep.

Some other examples:

- An 8½ inch width times 2 equals 17 picas wide for each third.
- A sheet 14 inches wide will have three 28-pica segments.
- A sheet 9 inches wide will have three 18-pica segments.

sage, the type styles, and the kinds of art being used. All will play a part in determining which papers will work best.

In addition to antique finishes, machine finishes, bonds, and coated papers, there are papers of bold colors, pastel colors, iridescent colors. Nearly every color imaginable is available to call attention, set a mood, or add distinction. There are even "duplex" papers with a different color on each side.

The printing process and mailing costs will play a part in the paper choice. Offset and bond papers are closely related and are finished to take the water involved in offset printing and writing with ink. Screen printing can be accomplished on almost any kind of paper or other material.

In selecting papers, their reflectance or brightness and opacity should be considered. The more light the paper reflects, the brighter it will appear. This is its *reflective* quality. A glossy paper will appear bright, will show the contrasts in art to their maximum, and will bring out the tones. These papers will print colors vividly. However, glossy papers can be tiring to the eye if used for large amounts of reading matter.

The *opacity* is the ability of a paper to help prevent printing on one side from showing through to the other side. Opacity is an important quality to consider when selecting a paper that will be printed on both sides.

Since the paper has such an important effect on both the design and the quality of the printing, it is helpful to obtain samples of possible papers to use, preferably with printing on them, from the printer or paper company before starting a design project.

Another aspect in selecting paper is its size in relation to the size of the printed piece. As was noted in Chapter 7, unless papers are manufactured in rolls, they are produced in standard-size sheets.

This creates no problem where letterheads are concerned. Since the standard letterhead is 8½ by 11 or 5½ by 8½ inches, and the standard bond paper sheet is 17 by 22, four 8½ by 11 sheets or eight 5½ by 8½ sheets can be cut from each full sheet with no waste.

But it is worthwhile to understand how the printer prices and cuts paper stock. Sometimes a slight adjustment in the format of a project can create significant savings in paper costs.

Here is how it works. Say we are planning a brochure. It will be a simple, four-page folder with each page measuring 8 by 10 inches. The folder will be created by using a parallel fold on a 10 by 16 sheet. We have decided that the folder will look best if printed on machine-finish book paper. The standard size is 25 by 38. We require 5,000 brochures.

The printer will calculate the number of 25 by 38 sheets needed and cut them to make 5,000 10 by 16 sheets. This is the formula he or she will use:

$$\frac{\text{Paper width}}{\text{Brochure width}} \times \frac{\text{Paper length}}{\text{Brochure length}} = \begin{array}{l} \text{Number of 10 by 16s} \\ \text{that can be cut} \\ \text{from a 25 by 38 sheet} \end{array}$$

That is, the printer will divide the widths and depths to get the most cuts out of a sheet. This may be affected by the grain of the sheets if the brochure is to be folded, especially if printed on heavy stock. However, in this case, we get

$$\frac{25}{10} \times \frac{38}{16} = 2 \times 2 = 4 \text{ 10 by 16 sheets from each 25 by 38 sheet}$$

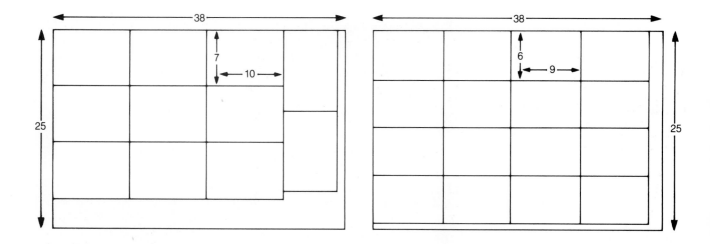

If the 4 is then divided into 5,000, we find that 1,250 full-size sheets will be needed for the job. We will be billed for that number plus about 5 percent for spoilage in printing and processing.

However, if the brochure is designed for a 6 by 9 page rather than 8 by 10, it can be printed on a 9 by 12 sheet cut out of the 25 by 38 size.

$$\frac{25}{12} \times \frac{38}{9} = 2 \times 4 = 8 \text{ 9 by 12 sheets from each 25 by 38 sheet}$$

Fig. 11-8 Careful planning that considers paper sizes can be cost efficient. On the left, six cuts, possibly eight if paper grain is not a factor, produce 7 by 10 inch sheets from a 25 by 38 sheet. By reducing the size just 1 inch for both width and depth to 6 by 9 inches, sixteen cuts can be obtained. The paper cost would be decreased by about 50 percent.

Now the job will require only 625 sheets, cutting the paper cost in half even though the brochure page size is only reduced by 1 inch in one direction and 2 inches in the other.

The economies of choosing sizes for the final printed products that correspond to standard paper sizes cannot be overemphasized. A designer can always choose a unique size, but that designer must be prepared to pay for higher levels of waste. Also remember that sizes that can be cut efficiently from bond papers will not cut efficiently from book papers because of the different standard sizes. One designer has pointed out that beginners often design a communication on standard office (8½ by 11) paper and then specify book papers for production. This causes confusion and increased costs.

In planning printing needs, quite often paper can be saved by printing more than one job on the same sheet of paper. Money-saving procedures such as this can be worked out if the communicator and the printer have developed good rapport.

▼ Designing Effective Stationery

Letterheads, statements, invoices, envelopes, business cards, and other printed materials needed to help us function in our work environment may not appear to offer much challenge. However, even something as deceivingly simple as a business card should receive thoughtful

Fig. 11-9 Recognition and memorability as well as the desired image projection can be reinforced by coordinating design elements in all communications. The Chermayeff & Geismar design firm created these coordinated stationery pieces for the Carousel Center (see pages 242 and 243).

consideration. The principles of good design apply here just as they do for all printed communications.

The design of these items should be coordinated so they all work together to create the desired impression. Often it is effective to use the same layout but perhaps in a smaller size for statements and invoices, envelopes and cards, as is used for the letterhead. And even though a different design may be chosen for each, depending on its use, it is worthwhile to use the same logotype or typeface for the name of the organization on all printed pieces. This will help build recognition and memorability.

Creating A Logo

Designers keep two basic considerations in mind when creating a logo. The first is the characteristics that will make it suitable for printed communications. The second is its adaptability to all the identity requirements of the organization.

An effective logo should identify the organization when it stands alone. It should be simple enough to reproduce well on office copiers and more sophisticated printing equipment. It should reduce or enlarge without losing its design subtleties. It should reflect the tone of the organization, and it should not become outdated as times and styles change.

The logo should be executed with consideration of the possible expansion of the organization or company into new areas or activities. And the design should be suitable for use on vehicles, work clothes, uniforms, and so on.

One of the country's most prestigious design firms is Chermayeff & Geismar Associates, which has created corporate identity programs for such firms as Xerox and Chase Manhattan Bank. It also devised the official American Revolution Bicentennial symbol.

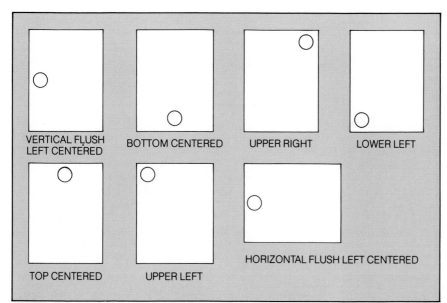

VERTICAL FLUSH LEFT CENTERED

BOTTOM CENTERED

UPPER RIGHT

LOWER LEFT

TOP CENTERED

UPPER LEFT

HORIZONTAL FLUSH LEFT CENTERED

Fig. 11-10 "Where to position the logo and the type, which typeface works best visually and best represents the client, and what size and shape the letterhead should take are some of the things that make every letterhead job an exciting challenge," according to "A Letterhead Production Guide" by the Gilbert Paper Company. Here are some possibilities for locating the logo.

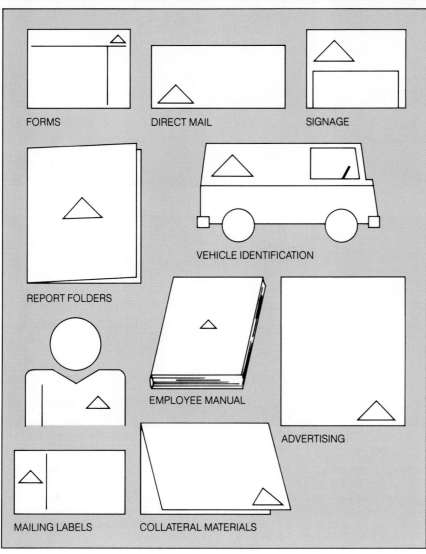

FORMS

DIRECT MAIL

SIGNAGE

VEHICLE IDENTIFICATION

REPORT FOLDERS

EMPLOYEE MANUAL

ADVERTISING

MAILING LABELS

COLLATERAL MATERIALS

Fig. 11-11 The design can be incorporated into a total identity and marketing program for a business or organization.

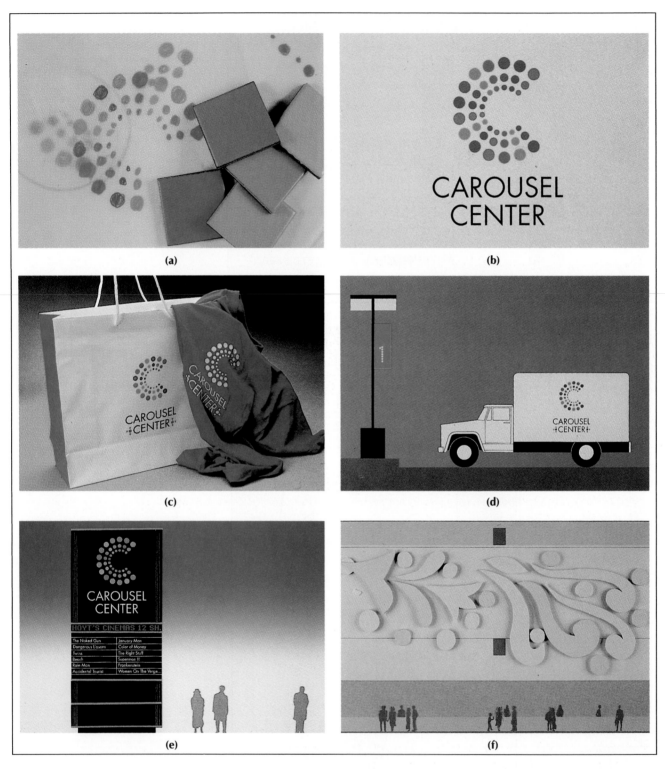

(a)　(b)　(c)　(d)　(e)　(f)

Fig. 11-12 (a) In its original, schematic design work for the Carousel Center, Chermayeff & Geismar's design team first made a palette of pastel colors to coordinate with the tiles used on the shopping center floor. (b) The completed logo was combined with a Futura Light typeface. (c) For the initial presentation to the client, the logo was applied to shopping bags, T-shirts, and other promotional materials. (d) The logo as it would appear on trucks. Note how the number on the parking lot standard is coordinated with the logo palette. (e) The logo as it would appear on large information signs at the Carousel Center. (f) A proposed interpretation of traditional carousel designs used in the Center.

The firm follows the same philosophy for each project regardless of its size. This philosophy is stated in its promotional brochure: "Design is the solution of problems, incorporating ideas in relation to the given problem, rather than the arbitrary application of fashionable styles."[2]

How Chermayeff & Geismar handles a project might provide helpful insights for the neophyte designer. Their work on the Carousel Center is a good example.

Carousel Center is a large shopping center in Syracuse, New York. It was developed by Pyramid Companies as a flagship shopping and convention complex. The Center was created to be a major destination for movie goers, shoppers, and conventioneers.

A late eighteenth century restored carousel was used to set the theme for the festive atmosphere of the Center. A six-story atrium and convention facility creates an observatory for vistas over the New York state lake region.

The design process for Carousel Center drew on many talents—an architect, an interior designer, lighting consultants, engineers, and graphic consultants. Chermayeff & Geismar was hired to create a logo, marketing graphics, and environmental artwork.

The first step was research. The design team investigated carousel motifs but decided the logo should reinforce the modern spirit of the Center's architecture and the festive atmosphere of the shopping arcades.

For the logo, a color palette of pastel colors was selected to coordinate with decorative tile patterns used on flooring throughout the Center. This was incorporated with an abstract "C" rendered on a Macintosh computer coupled with a Futura Light typeface.

The logo and graphic identity device were applied to shopping bags, T-shirts, stationery, etched glass mockups, and promotional materials for the initial presentation to the client. After the completion of their preliminary work, Chermayeff & Geismar's design underwent numerous changes. By early 1990, the final, revised Carousel Center design program was being implemented by the client.

The Letterhead

Often the letterhead is the initial contact the receiver of a message has with the sender. Not only that, but many times the letter is the only contact made between an organization and its prospective clients. The letterhead, then, carries the weight of creating an impression as well as transmitting a message.

The letterhead should make a statement about its originator. The type styles selected and their arrangement help set the stage for the message. Effective letterhead design generally should be neat, dignified, and orderly. It should have character but not be obtrusive. It should be unique but restrained in terms of type sizes, tonal values, and use of space.

Letterhead designers suggest that the most effective results are obtained by skillfully accentuating the name of the firm, product, or logotype in relation to the other less important type elements while maintaining harmony, balance, and tone. In seeking this goal, designers usually consider two approaches: traditional or modern.

These are also referred to as formal and informal letterheads.

In the *traditional design,* type groups are usually arranged in either a square or inverted pyramid. If a trademark or symbol is used, it is centered. Sometimes rules are included but if they are, the formal symmetry is maintained. One feature of the inverted pyramid arrangement is that it forms a downward direction motion toward the message.

The *modern approach* is to create a basically asymmetrical arrangement while maintaining the principles of good design. The type can be arranged on an imaginary vertical line or it can be counterbalanced. If larger type is used for the company name on the left of the page, a smaller two or three-line address may be placed on the right for balance.

Rules are sometimes used to add motion to the letterhead. Decorations can give a letterhead individuality, but they should be selected and used with discretion. Color should also be considered. Remember, though, that the typographic embellishments should not draw attention away from the type.

Papers are available in a variety of finishes—linen, wove, pebble—to add individuality and distinction to a letterhead. If a paper with a textured finish is selected, it's a good idea to examine the inking and reproduction abilities of the paper.

Fig. 11-13 Traditional layout in letterheads with symmetrical balance. (Courtesy Woodbury and Company, Inc., Worcester, Mass.)

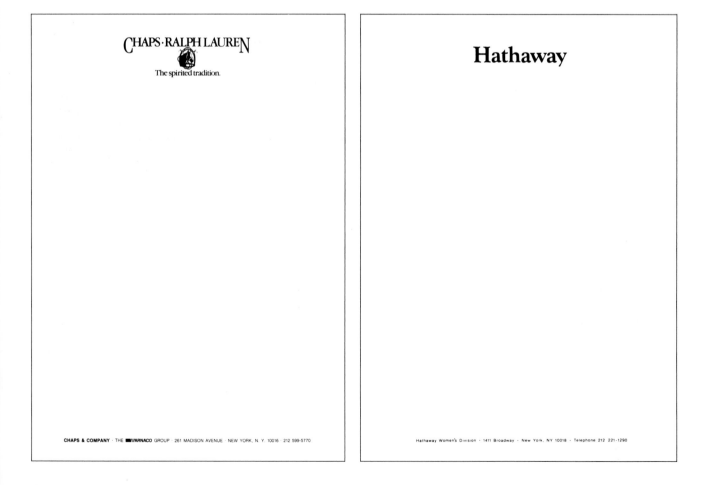

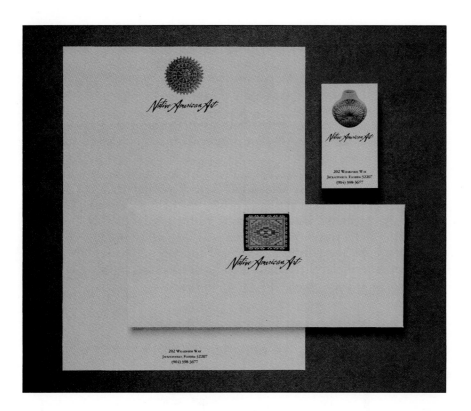

Fig. 11-14 Calligraphy and photography were used to create this coordinated combination of letterhead, envelope, and business card for Native American Art. Earth tones reinforce the appropriateness of the designs. Photographer was Tom Schifanella of the Robin Shepherd Studios. Calligrapher was Pamela Stanholtzer.

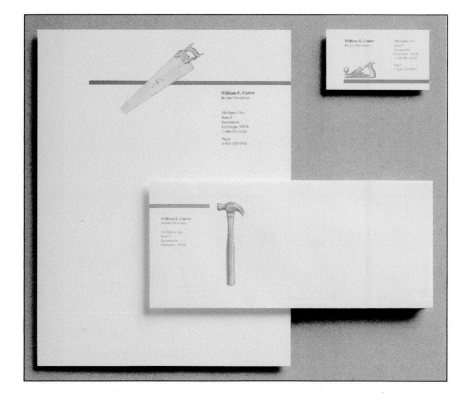

Fig. 11-15 Designer Michael Kennedy used clip art to create these coordinated letterhead, envelope, and business card designs. Note the appropriateness for a construction company. Even though an illustration for a different tool was used for each item the result is a harmonious trio.

Margins are important in letterhead design regardless of the arrangement selected. Side margins of type lines should approximate fairly closely the usual line width of the typed message. Sometimes elements such as the name and position of the person sending the messages are placed on the right of the page, to balance the typed name and address of the recipient of the letter.

Some organizations insist that long lists of officers be included on a letterhead. This can cause design problems, but they can be solved. If only a few names are involved, they can be listed across the bottom of the sheet. The other obvious alternatives are to list the names down the left or right margins. These lists should be kept as unobtrusive as possible. Sometimes printing them in a lighter color will help. Of course, placement of all elements should always be made with consideration of the format of the letter to be typed on the sheet.

A typographically effective letterhead printed on a carefully selected stock that reflects the character of the organization can create a favorable impression and help project the desired image.

Fig. 11-16 The top of the sheet for this letterhead was die cut to resemble trimming by pinking shears to provide identity, memorability, and distinction. (Courtesy Hammermill Papers Group)

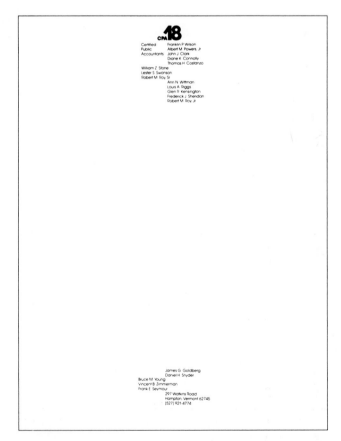

Fig. 11-17 The problem was to list eighteen names on the letterhead. It was solved creatively rather than following the outdated impulse to list them on the side and clutter the message. (Courtesy Hammermill Papers Group)

Fig. 11-18 Good design does not have to be complicated. Note the simplicity of the illustration used to create this distinctive and memorable trio of business stationary items. It was designed by Rubin Cordaro Design, and the client was the Minneapolis chapter of American Institute of Architects.

Special Processes

There are several processes that can be used in producing letterheads and other printed communications, which can add effectiveness and distinction. They include engraving, embossing, hot stamping, thermography, and die cutting.

Engraving   Engraving is an excellent method for projecting a high-quality, prestigious image. The sharp lines and crispness of an engraved announcement or letterhead project elegance, strength, and dignity. It reproduces fine lines and small type well. And it brings out the subtleties of shadings and patterns in a design.

A chromium-coated copper or steel plate with an etched-in design plus a smooth counterplate are used on a special engraving press. The plate is covered with ink, then wiped dry. The ink remains in the etched portions. Paper is fed into the press, and the impression transfers the ink to the paper.

Keep in mind that most engraving plates are limited to 4 by 8 inches. Also, paper selection is critical because of the stress exerted by the press. Paper lighter than 20 pounds should never be used.

Embossing   This technique is enjoying great popularity. It uses heat and pressing paper between dies to produce a raised effect. It provides a distinctive element to a printed communication, and introduces a third dimension for added memorability. Often new life can be added to an old logo by using embossing.

Embossing can be done in three styles: *blind, deboss,* and *foil-embossed.* Blind embossing uses the process without ink or foil. Debossing is embossing in reverse. The image is pressed down in the paper rather than

Logo Design

In designing an effective logo, you have to stretch your imagination and be creative. These are international symbols. Can you match them with their meanings?

1

2

3

A Lost Child   **D** Florist
B Keep Frozen   **E** Go This
C Snack Bar   Way

(*Answers: 1-C, 2-B, 3-D*)

raised. Foil embossing uses a thin material faced with very thin metal or pigment.

Usually the embossed effect works best alone. The visual impact is impaired when printing or color are added to the embossed image. When using embossing remember that the paper is formed, or molded, and this will use more of the paper than the layout might indicate. So type and other elements should not be too close to the embossed area.

Hot stamping   Hot stamping uses the same heat-pressure process as embossing, but it goes one step further by transferring an opaque foil material to the surface of the paper. A variety of foils and designs are available, but the most common hot-stamping techniques use gold or silver foils.

Hot stamping is used for greeting cards, ribbons, paper napkins, and so on. With the introduction of new hot-stamping presses it is also becoming a popular technique for business cards and letterheads.

In hot stamping a very thin ribbon of foil is fed into a press and releases its pigment onto the paper when pressed between a die and a hard, flat surface as heat is applied.

Hot stamping is expensive. It can cost as much as 10 to 15 percent more than engraving or embossing. Also, the process can discolor certain colors of paper stock, especially browns, yellows, and oranges.

Thermography   Thermography produces a raised printed surface by dusting the wet ink printed on the surface with a resinous material. This material is fused to the ink with an application of heat. The image is permanent and chip-proof and crack-proof. It is considered by some to be similar in appearance to engraving, but it can be added anywhere on a sheet of paper and is not limited by a plate size, as is the case with engraving. It can also be used with any color ink.

Die cutting   Die cutting is rather like cookie cutting, except paper instead of dough is used. It can be an effective and dramatic attention-getting device. The cuts can be straight, circular, square, rectangular, or a variety of special shapes. Lasers are now being used to cut dies.

Delicate or lacy patterns, which tear easily, should be avoided. Also, since it is difficult to maintain a tight register in the die press, elements should not be designed close to the cuts.

Die cutting is not expensive compared with some of the other special-effect processes, but care is needed in selecting paper with sufficient strength to take sharp, clear-cut lines.

▼ Effective Design Checklist for Letterheads

- Letterhead design should never interfere with the purpose of the letterhead, which is to convey a message.
- The *monarch size* (7¼ by 10½) can add dignity for professionals such as doctors, designers, architects, executives.
- When color stock is used the color should fit the character of the organization or the service it renders. It should not interfere with the typed letter.

- Half-sheets (5½ by 8½) can save money, and they can be folded twice to fit a 6¾ envelope.
- The logo should be original, stimulating, imaginative, and straightforward. It should be adaptable as well.
- The weight, color, and texture of the letterhead paper should be compatible with its envelope.
- Always keep in mind that a letterhead is a platform for words.

▼ The Envelope—A Tale of Diversity

The part played by the envelope in printed communications begins with the vast size and diversity of materials available. Here is another tool that if used properly can make communications effective. The communicator should be able to sort out the different styles of envelopes and select the best one for the job.

But there are so many to choose from! The Old Colony Envelope Company, which produces the most extensive line of envelopes in the United States, offers more than 1,700 different styles and sizes. Other converters (which is what the envelope people call themselves because they take flat sheets of paper and cut and fold and glue them to create envelopes) also offer envelopes in hundreds of sizes and styles for thousands of uses.

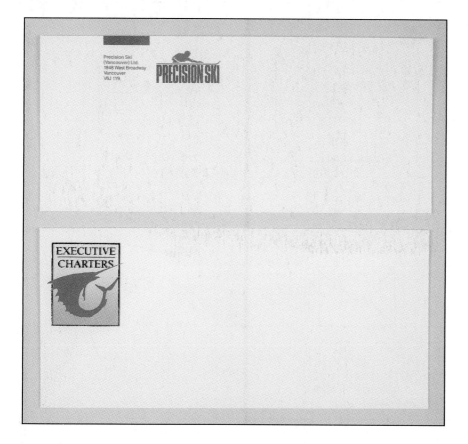

Fig. 11-19 Envelope graphics are important. The envelope is often the initial contact with a target audience. (Courtesy Gilbert Paper.)

ENVELOPES

STANDARD STYLES AND SIZES

The envelope manufacturers and the graphic arts industry have settled on a variety of standard sizes and styles. Although a number of others are available, these are the most common:

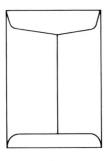

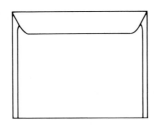

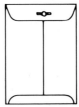

CLASP

3-1/8 × 5-1/2	7-1/2 × 10-1/2
3-3/8 × 6	8-1/4 × 11-1/4
4 × 6-3/8	8-3/4 × 11-1/4
4-1/2 × 10-3/8	9 × 12
4-5/8 × 6-3/4	9-1/2 × 12-1/2
5 × 7-1/2	9-1/4 × 14-1/2
5 × 11-1/2	10 × 12
5-1/2 × 8-1/4	10 × 13
6 × 9	10 × 15
6-1/2 × 10-1/2	11-1/4 × 14-1/4
7 × 10	12 × 15-1/2

STRING and BUTTON
Same Sizes as Clasp

OPEN END

3-7/8 × 7-1/2	7 × 10
4 × 6-3/8	7-1/2 × 10-1/2
4-5/8 × 6-3/4	8-1/4 × 11-1/4
5 × 7-1/2	8-3.4 × 11-1/4
5-1/2 × 7-1/2	9 × 12
5-1/2 × 8-1/4	9-1/2 × 12-1/2
6 × 9	10 × 13
6-1/2 × 9-1/2	11-1/2 × 14-1/2

BOOKLET

3-3/4 × 6-3/4	6 × 9
4 × 5-5/8	6-1/4 × 9-5/8
4-1/2 × 5-7/8	7-1/2 × 10-1/2
4-3/4 × 6-1/2	8 × 11-1/8
4-1/4 × 9-5/8	8-3/4 × 11-1/2
5-1/2 × 8-1/8	9-1/2 × 12-5/8
5-3/4 × 8-7/8	

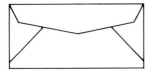

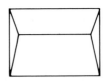

OFFICIAL

7	—3-3/4 × 6-3/4
7-3/4	—3-7/8 × 7-1/2
*Monach	3-7/8 × 7-1/2
9	—3-7/8 × 8-7/8
10	—4-1/8 × 9-1/2
11	—4-1/2 × 10-3/8
12	—4-3/4 × 11
14	—5 × 11-1/2

*Deep Pointed Flap.

COMMERCIAL

5	—3-1/16 × 5-1/2
6-1/4	—3-5/8 × 6
6-3/4	—3-5/8 × 6-1/2

ANNOUNCEMENTS & INVITATIONS

A-2	—4-3/8 × 5-3/4
A-6	—4-3/4 × 6-1/2
A-7	—5-1/4 × 7-1/4
A-8	—5-1/2 × 8-1/8
A-10	—6 × 9-1/2
Slimline	—3-7/8 × 8-7/8

BARONIAL

4	—3-5/8 × 4-11/16
5	—4-1/8 × 5-1/8
5-1/2	—4-3/8 × 5-5/8
6	—5 × 6

Fig. 11-20

Fortunately, though, they have settled on some basic sizes and grades that are easy to sort out and that will take care of most of the designer's and communicator's needs. Selecting and using just the right envelope need not be difficult. There are four main points to consider: (1) sizes and styles, (2) paper weight and texture, (3) graphic design, and (4) Postal Service regulations.

The sizes of envelopes are given in inches with the shortest dimension first. A 6 by 9 envelope is 6 inches wide and 9 inches deep. Designations also include the location of the opening and the styles of the flap and seam. There are several devices for closing the envelope, and each should be considered.

For example, most standard stationery envelopes are "open side" with the seal flap and opening on the long dimension. "Open end" envelopes have the seal flap and opening on the short dimension. Flap styles are called pointed, square, wallet, and mail-point. Seam styles are determined by the construction and location of the parts of the paper folded and glued to form the finished envelope. These styles include a diagonal seam that is used most commonly in business correspondence, a pointed flap used for announcements, and side seams used for mailing programs and booklets. A center seam is used in making envelopes that must be rugged to withstand heavy duty.

Envelopes can also be classified by the way they close. There are gummed flaps, flaps with metal clasps for added security, and "button-and-string" tied-down flaps, which were designed for envelopes to be used over and over again. Some converters have other patented closures such as self-sealing adhesives.

We can design the printed material and then seek the proper envelope. However, time and money can be saved and a much more effective communications package produced if we consult a listing of styles and sizes of envelopes first. See figure 11-20 for such a listing.

When selecting an envelope, we need to consider the size and bulk of the material to be inserted and how it is to be stuffed. An envelope for inserting by hand should be from one-eighth to one-fourth of an inch wider and one-fourth to three-eighths of an inch longer than the material it contains. If a machine at the printing plant or at a mailing firm will be used, the inserter should be consulted. The whole package must be compatible with the mechanical system.

Once the size and style have been selected, thought should be given to the envelope paper stock. There are impressive envelopes, envelopes that attract attention, and envelopes that harmonize in texture and color with the messages they contain.

All envelopes, however, need to conform to Postal Service regulations. The Postal Service classifies envelopes as nonmailable, mailable with no surcharge, and mailable with possible surcharge. Since the regulations are subject to change, they should be checked before printing is designed to be sent through the mails.

Graphics for Envelopes

The envelope can introduce the contents. It can help create the stage setting for the message. Often this communications bonus is overlooked and little attention is paid to envelope graphics. Good envelope graphics

---

Size Standards for Domestic Mail

Minimum Size

Pieces that *do not* meet the following requirements are prohibited from the mails:

1. All pieces must be at least .007 inch thick, and
2. All pieces (except keys and identification devices) that are ¼ inch or less in thickness must be:
   - Rectangular in shape,
   - At least 3½ inches high, and
   - At least 5 inches long.
   Note: Pieces greater than ¼ inch thick can be mailed even if they measure less than 3½ by 5 inches.

Nonstandard Mail*

First-Class Mail, except Presort First-Class and carrier route First-Class, weighing one ounce or less, and all single-piece rate third-class mail weighing one ounce or less is nonstandard if:

1. Any of the following dimensions are exceeded:
   Length—11½ inches
   Height—6⅛ inches,
   Thickness—¼ inch, or
2. The length divided by the height is not between 1.3 and 2.5, inclusive.

*Nonstandard mail is subject to a surcharge in addition to the applicable postage. Please check with your post office for rates.

Fig. 11-21 Modern design for business cards emphasizes symbolism and bold display for recognition and memorability. (Courtesy Gilbert Paper and Reno Printing, Inc.)

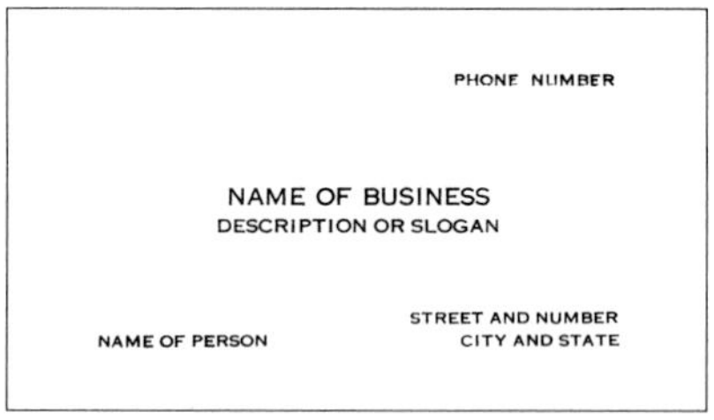

Fig. 11-22 Traditional layout for a formal business card.

produced with some thought can aid in getting the container opened. Direct-mail advertisers know this, and they do all they can to design envelopes that will get the prospect to look inside.

Envelope graphics should be determined by the nature of the message and the sender. The types selected and their arrangements should harmonize with those used on the message.

There are three principal categories of envelope graphics: those designed for direct-mail advertising, those to accompany letterheads, and those for pamphlets or publications. Direct-mail envelopes are designed to use every device possible and proper to attract attention and get the prospect to open and read the contents, just as art or headlines are used to lure a reader to read a newspaper or magazine advertisement. The envelope is an integral part of the whole sales plan.

The graphics of envelopes that accompany letterheads should extend the basic letterhead design. The same types, logos, and symbols used on the letterhead but in suitable smaller sizes can help create the tone and recognition impact of the message.

The graphics of envelopes used for publications or pamphlets should reflect the contents and project the same type and tone harmony.

In other words, whatever the purpose of the envelope, there should be a strong visual relationship between it and the contents. This includes the graphics as well as the color and texture of the paper.

Sometimes business reply cards or envelopes accompany a mailing. Here again, certain Postal Service regulations govern the graphics, sizes, and information that can be included. The designer should become familiar with these regulations.

Fig. 11-23 The typography of these business cards creates just the right feeling for shops that cater to a young, urban, trendy audience.

▼ The Business of Cards

Business cards are like letterheads—they can be formal or informal depending on the person or organization they identify. Most *formal cards* are carefully arranged to preserve balance and dignity. The copy should identify, explain, and locate. That is, it should emphasize the name of the person or firm, tell the nature of the business or service, and give the address and telephone number. In the formally balanced card all elements are balanced along a visual vertical line down the center of the card with attention paid to the optical center.

Informal cards still retain the attributes of good design and typography, but are less rigid in arrangement. The designer has greater freedom in selecting type styles and arranging them on the card. Scripts, Romans, Sans Serifs, or Square Serifs can be used along with rules and small symbols or logos as long as they reflect the nature of the organization or service. Of course, all the essential information should be included in legible typefaces.

There are several standard sizes for cards. The generally accepted business card is identified as a number 88. It is 2 by 3½ inches. Resist the temptation to be different by using a different size. Odd-sized cards are often thrown away, and they will not fit the standard desktop file systems for business cards that serve as an excellent reference source for busy people.

In preparing business card layouts, don't place the type lines too close to the edge of the card. A margin of 12 to 18 points should be allowed on smaller cards and at least 18 points on larger cards. If a card is a number 88, 3½ inches wide, the type should be designed in an area 18 picas wide.

EXHIBITION
OF PAINTINGS
By
EDWIN MORRIS

*Monday, January 9th
to Saturday, January 28th
inclusive*

At the
PEMBROKE GALLERIES
ARLINGTON AVENUE, AT 60TH STREET
NEW YORK CITY

Fig. 11-24 Title page for a program, illustrating proper word grouping, proper type choice for the subject, and good balance and unity. (From U&lc, publication of the International Typeface Corporation.)

▼ Programs

Programs for plays, concerts, and other events do not necessarily follow a standard size or style. Quite often, too, the limitations imposed by budget restraints may create an interesting design challenge. Of course, the same basic criteria for good layout and typography used in all graphics work still apply.

The simplest plan for a program is a single sheet. The dimensions should follow the principle of good proportion while cutting most economically from the paper size chosen.

There are several common formats for programs, depending on number of pages and content. Most four-page programs with printing on two pages are designed with a title page on page 1 and the program itself on page 3. If three pages are printed, the program copy occupies the second and third pages. If a menu and a program are included, the usual format is the title page, the menu on page 2, and the program on page 3.

The typography should be consistent on all pages. A single family, with italic or oblique, if contrast is needed, is usually best for harmony and a pleasing appearance. If other type styles are used, they should be used sparingly, for heads or for contrast, as long as they harmonize in design and tone.

Lines of dots or dashes are used to separate items on many programs. These devices are known as *leaders*. They are also used for setting what is called tabular material such as financial statements. Hyphens or periods can be used in place of leaders, with a standard separation between each period or hyphen. Most designers find that 1 or 2 ems is about right.

▼ Books and Pamphlets

Communicators often become involved in writing and designing books and pamphlets. It is thus helpful to know a few principles and practices concerning the format and design of these publications. They are treated very briefly here as, once again, the basic tenets of good design and typography apply.

Let us begin by considering the standard arrangement used in the book industry and see how it can be helpful to us. This standard arrangement is followed for an average book and followed or modified depending on the size and nature of the book or pamphlet.

The order of arrangement of the contents of a typical book includes, from front to back, the following pages and sections:

Cover
Half-title
Title page
Copyright information
Dedication
Preface
Acknowledgments
Contents
List of illustrations
Introduction
Text
Appendix
Glossary
Bibliography
Index

All of the segments except the copyright information, which often includes the printer's imprint, begin on right-hand pages. The copyright information and printer's imprint usually appear on the back of the title page. This order of contents can be used as a guide for orderly arrangement. Of course, items can be eliminated. For instance, on booklets the cover can also serve as the title page. Even many full-sized books do not contain all of the sections.

The half-title is a page containing only the title of the book usually placed at the optical center. The title page gives greater prominence to the title and usually includes the author's name. The publisher's name and address and the date are often included on the title page. These are arranged and designed to harmonize with the content and to have proper balance on the page.

A colophon can be included, usually at the end of the book. It describes the technical aspects such as typefaces, paper, and printing techniques.

Harmony is an important design element, and the types used should work together throughout the book. The preface and acknowledgments are usually set in the same style and size of type as the text matter. Other material is often set in a smaller type size.

Some typographical features of book design that apply to booklet planning include folios, running heads, and margins. *Folios* are the page numbers of a book. The standard practice is to use Roman numbers (xii or XII) to number pages of the sections preceding the text, called the *preliminaries*, and Arabic numbers for the text and ending sections, called the *back pages*.

The numbering of the preliminary pages begins with the half-title. It is important that the numbers are placed consistently throughout the book.

Running heads are the lines at the top of pages that identify the book or chapter and often contain the page number. These heads usually consist of the title of the book on the left-hand page and the title of the chapter on the right. Running heads on preliminary pages ordinarily identify the contents of these pages.

Running heads offer the opportunity of adding a little variety and contrast to the page. But they still should harmonize with the title page

Creative Communication

"Eureka, I've found it!"

What a happy moment when the editor or designer "sees the light" and solves the problem.

But psychologists tell us that after we have collected information about the problem, clearly defined it, and started on a solution, it can be helpful to forget it. Well, not forget it entirely. Just withdraw attention from it for a while.

Often we can then return to the problem with a fresh attitude and a new approach that can lead to a better solution than if we continued working without interruption.

Psychologists call this the "Eureka syndrome"—the "sudden" emergence of an idea or solution to a problem.

Fig. 11-25 Creative design ideas can be obtained from many sources. These cover pages for books illustrate current trends in design. The top pair were the work of Louise Fili and the bottom two were designed by Lorraine Louie.

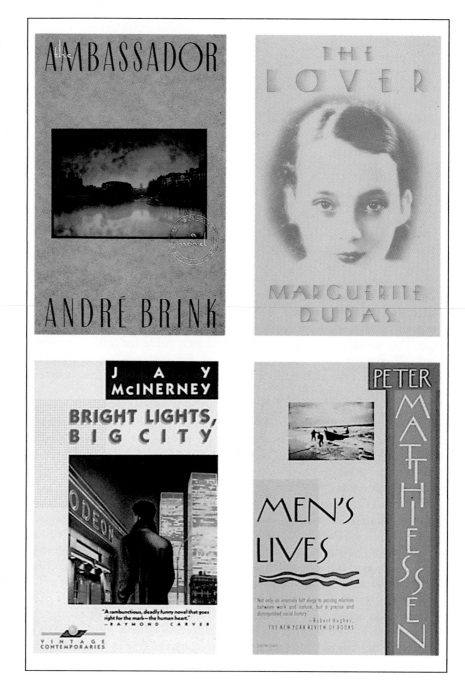

and the body matter. Here are some examples of running heads:

Graphic Communications Today

Graphic Communications Today

Graphic Communications Today

Margins play an important role in book design. They frame the type much like a mat frames a work of art. They help the eye focus on the type area and create a pleasing appearance to the page.

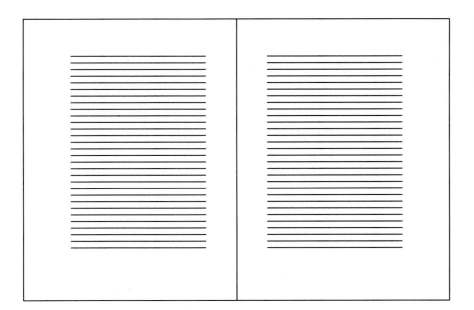

Fig. 11-26 Margins are important elements in book design. Progressive margins (top) and progressive margins with hanging shoulder notes (bottom) are shown.

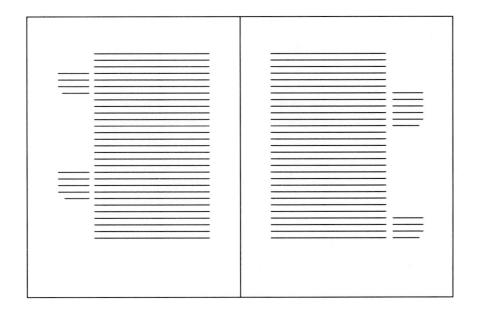

Book designers recommend that there be more margin at the bottom of the page than the top. This will help prevent the appearance of the type falling out of the page. The inner margin should be smaller than the outer margins. This will help make two facing pages appear as a single, unified whole. Careful attention to margins will not only enhance the appearance of a book but will help legibility as well.

Much care is taken in setting type and arranging elements for books. If some of the practices of page layout that are standard in book production were followed closely by others in the printed communications field, much of what we read would be improved.

Fig. 11-27 Two problems encountered in preparing reading (or text) matter for attractive design and pleasant reading. In the top example, the righthand column contains a widow, the last line of a paragraph carried to the top of the adjoining column. The lower example contains rivers of white, vertical gaps in successive lines caused by excessive spacing between words. Both situations should be eliminated.

Further costs accrued if the person injured was transferred out of one county's hospital to a larger facility. If the bill became uncollectible from the patient, the county where the accident occurred was billed. The study defined what accidents occurred and what number of those resulted in uncollectible bills.

Harris said the long standing problem was partially addressed when the Nevada Legislature passed Assembly Bill No. 218. The bill created a fund provided through an ad valorem tax to help cover the cost of transient indigent care.

But the research showed that because of the volatile nature of the mining and agriculture industry of some communities, many more indigent accidents happen to in-state rather than transient patients. Harris thinks those financial burdens will have to addressed some day.

The economic impact of the agricultural school research is obvious in the case of the EMS study. It is harder to assess in such studies as the one being conducted in Lander County.

Another matter, closely connected with even spacing and complementary to it, is the question of close spacing. We have become accustomed to wide gaps between words, not so much because wide spacing makes for legibility as because the Procrustean Bed called the Compositor's Stick has made wide spacing

▼ Effective Design Checklist for Books and Pamphlets

- Eliminate widows. *Widows* are the final words of a paragraph carried over from one column to the top of the next, or from one page to the next. They often contain only a word or two. This practice breaks the unity of a paragraph and creates an uneven contour to the column or page.
- Do not end a column or page with the first line of a paragraph.
- Do not divide a word from one column or page to the next.
- Avoid more than three consecutive lines ending with hyphenated words.
- Prevent "rivers of white" from flowing down columns. *Rivers of white* are obvious gaps between words that run down the columns. They are the result of poor and uneven spacing in the lines of type. To avoid rivers of white use a smaller typeface on the same measure; use the same typeface but in a longer measure; set the copy ragged right; use the same measure but a more condensed typeface.

▼ A Practical Project in Design and Production

A Sacramento, California, graphic designer, who interviewed and hired many young designers right out of school and found them lacking in understanding the production end of the business, decided to do something about it. He created a course in which students engage in hands-on designing and the actual production of what they design.

The designer, Michael Kennedy, who owns Michael Kennedy Associates in Sacramento, has taught the course at California State University, Chico.

The course begins with students examining a variety of projects produced by working designers, from the creation of thumbnails to finished printing. The main task of the course is to develop a design project that is printed in process color.

The assignment is the same for each student—designing a business card. All the cards are combined on a grid to create a poster. The poster is printed in full color, and some of the posters are trimmed so that each student receives 100 of his or her own business cards. The students are involved in all the steps of the production process. They use a variety of materials from color pencils to torn paper.

The class meets once a week and it follows this general schedule:

- *Week 1:* The procedure is first to work on the full-color art, then on the typeset copy. This week rough ideas are worked out in pencil sketches. Students contemplate how best to use the color materials that are available.
- *Week 2:* Students now narrow down their ideas to one or two and create a full-size sketch. They write an explanation of how the final art will be reproduced.
- *Week 3:* The final art for the color portion of the card is prepared. The art for each card is trimmed to exactly twice the size of the poster to fit the grid format.
- *Week 4:* The artwork for each card is mounted on the grid. It is photographed and an 8 by 10-inch color transparency is made of the entire layout. Color separations are made by a laser scanner. The cards, which have been gang-produced, are now 2 by 3 ½ inches with about ¹⁄₁₆ inch all around for trim. They are separated and the students add typeset material to their individual cards on an overlay. They can have the type printed in black or reversed to white. The final step is to specify the instructions to the printer on a tissue overlay.
- *Week 5:* The artwork is sent to the printer. The printer produces two proofs, one in full color and the other a blueline. The students check

Fig. 11-28 The California State University, Chico, design students prepare their poster for platemaking and printing.

Fig. 11-29 The design students spent the day in the printing plant observing and assisting in the production of their project.

Fig. 11-30 One of the posters produced by the design students. The individual business cards were cut from the printed posters.

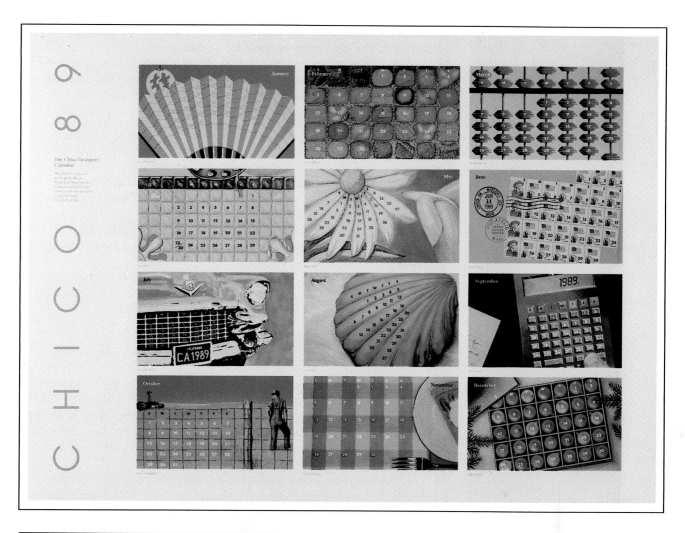

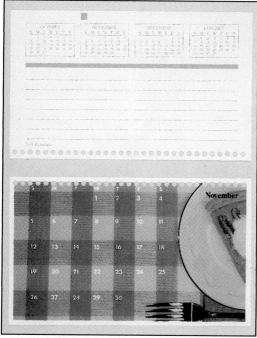

Fig. 11-31 (Above) Another approach to hands-on experience in design and production devised by Michael Kennedy follows a procedure similar to the business card project. A calendar is created with each student designing the graphic for each month. The calendar is reproduced as a wall calendar. Individual month designs are cut and spiral-bound to create desk calendars. The calendars are sold to businesses and others to help defray the cost.

Fig. 11-32 (Left) A page from the desk calendar created from designs for each month on the wall calendar.

the proofs for color, registration, and accuracy and the proofs are returned to the printer to prepare for printing.

- *Week 6:* The class spends the day at the printing plant and follows all the steps in the production process. When the posters roll off the press each student receives 25 posters plus 100 business cards cut from the full sheets.

The cost is covered by a small course fee and sizable donations from companies involved in the project. The major contributor has been the printer Georges + Shapiro Lithograph of Sacramento.

"This is a rare opportunity for students to understand the relationship between design and the printing process," says Joel Shapiro, one of the owners of the printing firm.[3]

The project benefits the university as well. Posters are mailed to high schools and junior colleges in the area to acquaint students with the university and its design program.

▼ Graphics in Action

1. Select the types to use to design a coordinated 8½ by 11 letterhead, number 10 commercial envelope, and number 88 business card for your own use. If you are associated with an existing business or service, use that. If not, devise a fictional name for a firm in your major field of interest. The types and arrangement should reflect the characteristics of the business or service.

2. Outline the procedures you would follow to create an informational brochure for an organization with which you are associated. This brochure would be printed on an 8½ by 11 sheet that would be mailed to prospective members. Its objective is to convince prospects that they should consider joining the organization.

3. Prepare a rough dummy for the brochure planned above.

4. Design one letterhead in the traditional format and one in modern design for a company, either existing or fictional. Use the same copy for each. Use an ornament or devise a symbol. Consider using a rule for the modern format. Analyze the two designs and decide which would be most appropriate for the company.

5. Plan and design the title page for your autobiography. Select types and ornaments or borders that you believe would best fit the subject and enhance the page. The use of ornaments or borders is your decision. The page size is 6 by 9 inches.

Notes

1. *In-Plant Printer*, April 1982, p. 48.
2. "Design By Number," *HOW*, January/February 1988, p. 71.
3. "Hands-On Education," *HOW*, November/December 1988, p. 66.

Advertising Design

Photograph: Mark Jenkinson

After examining all the parts of a complete graphic communication, the time arrives for putting these parts together. One approach to the design of magazines, brochures, newspapers, and all other forms of printing the communicator might be responsible for is to study advertising design.

There are several good reasons to approach the application of graphic design from an advertising base. Some of the best talent in the communication field works in the world of advertising. Advertising is a pacesetter in the use of art, type styles, and the arrangement of elements on the printed page.

In addition, advertising makes a good starting place because all the steps, from conceptualization to the finished comprehensive or mechanical layout, can be followed quite easily. All the principles of design can be seen in action in one comparatively small area. A well-designed advertisement will contain, often on one page or less, balance, proportion, unity, contrast, and rhythm. These principles are put to work to create a communication that does a specific job.

There is another benefit. Quite often communicators who specialize in editorial functions or public relations, for instance, have to coordinate their efforts with those of the advertising personnel. The common task might be a special supplement for a newspaper, a special section for a magazine, or an advertising campaign that requires support from public relations professionals. As with most areas of endeavor, when mutual understanding exists among the various professionals engaged in an effort, the work goes more smoothly and the result can be more effective.

The study of advertisements can help stimulate creative ideas for brochure and magazine page layouts. Sometimes the arrangements of elements in an advertisement can be adapted to editorial content. For example, does the placement of the elements in the advertisement produced here in Figure 12-1 trigger ideas for placement of elements on a magazine page? A possibility is shown in Figure 12-2.

Of course, the creation of a layout is only one small part of the advertising process, just as it is only one part of any communication effort. It cannot be completely isolated from the other steps in effective communication. But even this brief look at advertising design should help us understand something about the process of advertising communication. In addition, since most advertising is communication that forcefully attempts to motivate people, examining the techniques used by advertising professionals can help those in other areas of the profession.

▼ Design and Advertising Communication

Advertising communication differs from most other communication in two ways. Here the communicator pays to have the message circulated. As a result, the communicator has more control over the message than, say, the public relations professional who distributes a press release. The source pays, so the source can specify when and where the advertisement will appear. The source decides the size and content of the message as well.

America's best export is America.

InterNorth and several other large American corporations have been engaged in a unique business-to-business relationship for five years in nations of the Caribbean Basin. Called Caribbean-Central American Action, it is the initiative of an international partnership to encourage development of strong, market-oriented economies.

The freedom and stability of the Western Hemisphere will depend largely upon the economic models its nations choose to adopt. The cooperative role of U.S. businesses can do much to ensure that those models provide productive freedom and a better way of life for the people of the region.

Balance-of-trade figures tell only part of the story of America's success in international business. The rest of the story is the incalculable benefit in sharing the best of America with other nations.

Americans don't export just goods and services. We also send abroad the compelling evidence of how well our system works, and the inspiring model of what free people can achieve for themselves in a free system.

That's why our InterNorth International company and other international companies take pride in sharing the best of America with the rest of the world. Our most valuable export is our nation's 200-year-old success story.

InterNorth is a diversified, energy-based corporation involved in natural gas, liquid fuels, petrochemicals, and exploration and production of gas and oil.

INTERNORTH
We work for America.

International Headquarters, Omaha, Nebraska 68102
© 1983 InterNorth, Inc.

Fig. 12-1 This advertisement could be used as a guide in making a layout for a magazine page (see Fig. 12-2).

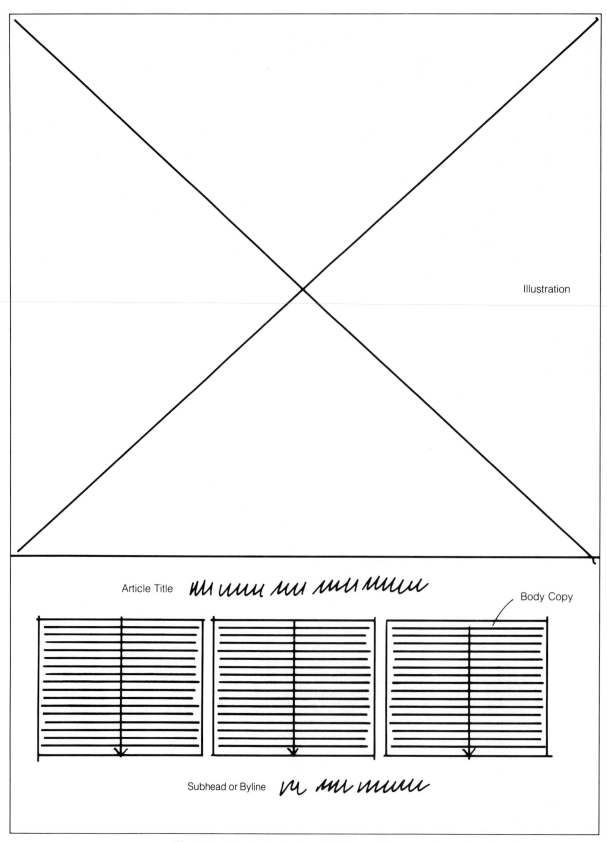

Fig. 12-2   Rough sketch of possible layout for a magazine or brochure based on the Internorth advertisement. Advertisements can be good idea sources for communicators.

Although the source has final say in the message, the actual advertising layout can be created by designers who work for an advertising agency, those who work for the medium that delivers the message, such as the newspaper, radio station, magazine, or television station, or the staff of the advertising department of the source—a business or organization, for instance. Quite often the strategy and its execution is accomplished by a team with representatives from an advertising agency and the client in consultation with representatives of the advertising department of the particular medium.

Advertising communication is aimed basically at getting people to do something or accept something, often against their will or initial inclination. Therefore advertising communication must use all the attributes of the communication process to attain maximum effectiveness. This includes typography and graphics.

Advertising communication is *persuasive communication*. So is most public relations communication and much newspaper and magazine editorial content. This makes them sort of cousins under the skin. Better newspapers, brochures, and other printed materials can be designed by studying the techniques of the advertising designer. Since typography and layout cannot be separated from the message they present, the most effective design cannot be created without understanding how that design is linked to the message.

A quick trip through the advertising copywriting process will help us understand the importance of linking the layout with the message.

The Strategy Platform

The first step taken by most advertising copywriters is to form a "strategy platform" as a guide for the actual writing. Usually more than one person is involved in devising the strategy platform. The account executive, copywriters, and layout people work together as a team. (In a one-person shop, of course, it will be a one-person project.)

The strategy platform is based on extensive research of the source and its product, the media to be used, and the target audience. It is a written statement that answers questions such as the following:

1. Who is the target audience?
2. What is the most important idea in this whole project?
3. What are the most important selling features?
4. What other important features of the product or idea should be considered?
5. What action do we want from the target audience?

The job of the copywriter and the layout artist is to integrate the verbal and visual elements they believe will be most effective. The goal is to get the desired "action" out of the target audience in the most effective and least expensive way.

Quite often copywriters and layout artists work together to accomplish this. Sometimes one person, if that person knows typography and graphics as well as the techniques of effective copywriting, does the whole job.

Fig. 12-3 This advertisement illustrates several design principles and techniques including application of the A-I-D-C-A formula. It has direction and motion through the placement of the grains plus the angle of the background card (which includes the product name in blind embossing). There is unity in the headline, copy, and logo arrangement. The golden tone emphasizes the ''golden touch'' message of the headline. And the art and copy are placed to create a well-balanced layout.

Arranging Information: A Useful Formula

There is an old tried-and-true formula that advertising copywriters often use to arrange the information they believe most effective to create a selling message. This formula plots the message from start to finish, from attention to action, in a series of five steps. This has been dubbed the A-I-D-C-A approach.

The A-I-D-C-A approach is effective in planning all sorts of communications. But few communicators outside the field of advertising are aware of it.

Let's see how the formula is applied to the communications process. Before any communication can take place, we must make an initial contact with the target audience. The audience will not listen if we do not get them to look up from what they are doing or stop them from turning the page.

This process of *getting attention* is the A of the formula. Whether an advertisement or a layout for a magazine article is being planned, the communicator must first capture the attention of the audience. Typography can do it. Effective words presented in the right type style can do it. Effective art can do it too. The designer must blend the words of the copywriter with the skills of the typographer.

The A for attention is followed by I, which stands for *interest*. Something written in the copy and arranged in the layout must stimulate the reader's interest quickly.

Well, we have attention, and we have aroused interest. What happens next? The target must be given a strong shot of *desire*. This means desire to acquire the product, to know more about it, or to endorse the idea or whatever the objective of the message might be. In writing a narrative, for instance, we must build up the reader's interest to make that reader want to continue through to the end. The D is for desire.

Depending on the purpose of the communication, the C may or may not be pertinent. If an advertising layout is trying to sell the reader a product, now is the time for *conviction*. However, if this is a brochure explaining an activity, C may not be needed. It all depends on the type of communication. But C in an advertising message, or in an editorial, is the clincher, when the message has sold the prospect. Here is where the communicator closes the pitch and has the prospect ready to put money, or support, on the line.

Finally, it's back to A, but this time the word is *action*. Action is the means for accomplishing the communication's purpose in the first place. If a communication does not provide the target audience with a way to take action or let them know action has been or will take place, very little will happen. In advertising, action can be generated by a number of devices—limited time offers, send in that coupon, get in on this special deal, price good until the end of the week, and so on.

The A-I-D-C-A formula provides a plan to keep things moving, to establish the rhythm and motion needed for a dynamic layout. The communicator can use this formula to help in placing elements on a page for a dynamic, alive layout such as the one illustrated in Fig. 12-3.

Graphic Elements

Whether the advertisement designer uses the A-I-D-C-A formula or has another approach to the task, the graphic elements of an advertisement facilitate quick and easy comprehension of the printed word. They can supply additional information that the written word cannot convey well, and they can help set the desired mood.

Let us take a quick look at how these elements can be put together into an effective advertisement. Most of these techniques are applicable to all types of printed communications—they aren't just the private domain of the advertising world.

Art can be used in advertising for many different purposes. The most obvious and most frequent use is to simply show what a product looks like. Product art is most effective, however, when readers are already aware of their need for the type of product being pictured.

Art can be very effective by showing a product being used. Art of this sort can stir to consciousness a need or desire for the product that the

---

Creative Communication

The creative communicator can be frustrated by accumulating either too little or too much information.

Effective creative solutions require adequate information. Superficial research results in superficial design.

On the other hand, too much information that is not relevant to the goal of the layout can muddy the waters and cause confusion.

The creative communicator should make an effort to gather enough information to understand the design problem thoroughly. However, the irrelevant information should be eliminated.

Creativity is aided by concentration on information pertinent to the solution of the problem.

Fig. 12-4 A typical advertisement includes: a headline, an illustration, body copy, and a logo or other form of corporate identity. The designer must arrange these elements in the most effective way possible. (Courtesy Brooks Shoe Division of Wolverine World Wide, Inc.)

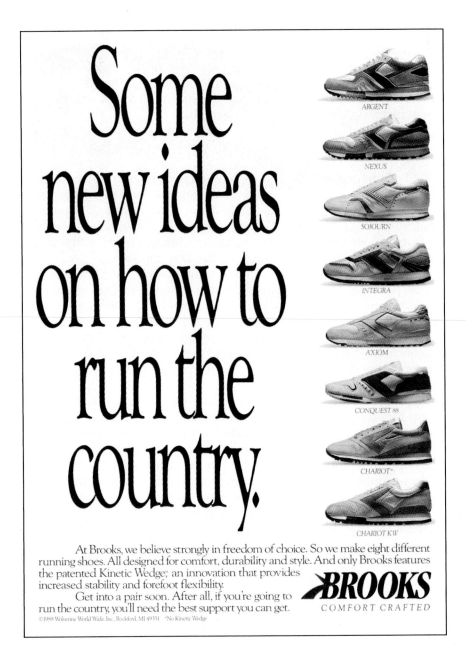

reader had not felt before. It can also reinforce a headline designed to attract a particular target market or make the wording that does this unnecessary in the headline, leaving room in the reader's mind for something else that might make the complete ad message more effective. Art showing a product being used should include people with whom the intended readers can or want to identify.

Art can also demonstrate the happy results of owning or using the product or service or patronizing a particular establishment. This may be shown either realistically or symbolically. Usually this type of illustration is associated with emotional, imaginative copy. This kind of art is especially helpful if the advertiser wants to reach persons who have not felt a need for the product. It attracts the interest of hard-to-motivate prospects by making them want to experience the same joy and satisfaction as the pictured users. It arouses their interest in the product or service, hopefully enough so they will read the rest of the ad to find out more.

Such art can take a positive or negative approach, or two pictures can be used, one negative and the other positive. For instance, they can show the predicament and the solution or some other sort of before-and-after situation. Advertisers have found this format to be very effective.

Closely related to "happy results" illustrations are those that imply psychological relationships between the product and its users. This type of art emphasizes the background and environment of the product's users or portrays people with whom readers would want to be associated. This art must be reinforced by imaginative, emotional copy.

Art used in advertising copy can show real products and real people in actual situations; it can also be abstract and symbolic. Symbolic art can reflect various attributes of a store or product—stylishness, reliability, durability, convenience, femininity, masculinity, gaiety, seriousness. Symbolic art is used to reinforce an element of the product's image. This type of art is difficult to create but it can have great impact and memorability. It is a real challenge to the communicator's creativity to come up with ideas for symbolic art that are not trite or too obvious.

Quite often art can be used much more effectively than words to demonstrate features of a product and how it is made and works. Cutaway techniques or greatly magnified illustrations of product details that may never be seen by the user in normal circumstances can be effective in certain situations and for certain products. This kind of art goes well with factual copy and is a graphic way of presenting tangible evidence that a product can perform as promised.

Illustrations can connect the use of a product with a national or local current event. A picture of a blizzard can remind people of the need for all sorts of bad weather gear, especially if a blizzard is raging outside at the time the advertisement appears.

Fig. 12-5 An approach to advertising layout. Note the use of a flush left arrangement. (Courtesy S. D. Warren Company, a subsidiary of Scott Paper Company)

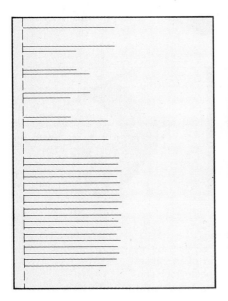

Fig. 12-6 Here is a newspaper advertisement created by using a number of art sources: computer-generated clip art, a line art diagram, and a continuous-tone photograph converted to line art by the halftone process. (Courtesy American Newspaper Publishers Association)

Art can also show "news" about the product or where it may be obtained. Art can show new merchandise being unloaded, a crowd attracted by a special event, or progress on new construction.

Art can also be used to make people laugh. Some of the most memorable communications are those that make us relax and chuckle. People tend to respond well to cartoons and illustrations that have a humorous theme. However, the theme should tie in with the story line of the copy or the object of the advertisement.

Art can be used to attract attention, arouse curiosity, or simply to decorate an advertisement. All art used in a layout, however, should have a job to do. An inappropriate piece of art in no way related to the ad message can do more harm than good. It can detract from the impact. It may even make a reader angry if she or he feels tricked by the irrelevancy of the art to the message. A purely sexist appeal, for instance, such as a pretty woman in a bikini standing beside a product, can be particularly offensive.

Fig. 12-7 In creating an advertisement the designer will test the layout possibilities by making a number of thumbnail sketches. Then a full-size rough is created prior to completing a more detailed "comp" or a "ready" for plate-making mechanical. (Courtesy American Newspaper Publishers Association)

Fig. 12-8 This advertisement was considered very successful even though the headline was only two words and the art very small. Both emphasized the theme "think small." Note the effective use of white space to gain attention and reinforce the idea behind the advertisement. (© Volkswagon of America, Inc. Reproduced with permission.)

Think small.

Our little car isn't so much of a novelty any more.
  A couple of dozen college kids don't try to squeeze inside it.
  The guy at the gas station doesn't ask where the gas goes.
  Nobody even stares at our shape.
  In fact, some people who drive our little

flivver don't even think 32 miles to the gallon is going any great guns.
  Or using five pints of oil instead of five quarts.
  Or never needing anti-freeze.
  Or racking up 40,000 miles on a set of tires.
  That's because once you get used to

some of our economies, you don't even think about them any more.
  Except when you squeeze into a small parking spot. Or renew your small insurance. Or pay a small repair bill.
  Or trade in your old VW for a new one.
  Think it over.

Headlines for Advertisements

The headline is the most important single typographic element in an advertisement. Its primary function is A—attention. In addition to attracting attention, an effective headline states or implies a benefit. It should contain a verb if possible, and active verbs far outperform passive verbs. The good headline identifies the target audience and gets the reader involved.

The headline for an advertisement might be developed from the copy, as it is for a news story or magazine article. It can also tie in with the illustration.

Some ad copywriters say if the headline doesn't get the reader's attention, the rest of the ad will never be read. If it attracts the wrong kind of readers, those who don't want, don't need, or can't afford the product, the ad won't do its job.

How long should a headline be? Basically a headline should say enough to attract the attention of the target audience and make them want to read the rest of the ad. But keep it short. There have been successful Volkswagen ad heads with just two words:

Think Small

Nobody's Perfect

However, use as many words as necessary to do the job. One of the most famous heads in all advertising history had eighteen words:

At Sixty Miles an Hour the Loudest Noise

In This New Rolls Royce Comes from the

Electric Clock

The communicator has many choices when deciding on the type of headline for an advertisement. There is the "news" headline:

Wammo Now Has Flamastan

The news headline should be set in a type style that makes it look like a newspaper headline.

The selective headline emphasizes appeal to the target audience and helps sift out the target from the mob. A headline for Haggar slacks zoomed in on the target like this:

The Slacks for the Untamed Young Man

Benefit headlines stress advantages. The most effective benefit headlines do not boast, they stress benefits to the target audience:

You'll Have No Maintenance Costs for Five Years

Promise headlines, those that offer a reward for use of the product or adoption of the idea, should be followed by proof in the copy that the promise will be kept. Clearly related to the promise head is the "how to" or advice headline:

How To Get More Interest on Your Savings

Here, again, proof should be offered in the copy. The command headline gets readers involved because they are urged to act. Tactfully and subtly written, it can be very effective, but if the command is too strong the prospect might take offense at your arrogance. "Do Yourself a Favor, Buy from Us" is a much more effective approach than "Buy from Us, We're the Best."

Fig. 12-9 This headline identifies the target audience and holds out a promise to lure the target audience into the advertisement. Also note the correct use of negative leading between the lines. The ascenders and descenders do not touch.

To improve your accounts receivable, shift your point of view.

Label headlines in advertisements are weak, just as weak as label heads on news stories or articles. A label head simply states a title or obvious fact. "Arrow Shirts" would be an advertising label head; "Lions Club Meets" would be a label head on a news story.

There are many other types of advertising headlines. Some pose a question ("Do You Suffer from Headaches?"); others attempt to arouse curiosity ("How Many Beans in This Jar?"), and still others challenge ("Go for It!"). These heads are effective if done well, but they usually violate the basic rule that a headline should reveal what an ad is about. They can arouse curiosity enough to make the reader read on, but if the copy does not satisfy that curiosity, the head may make an enemy for the sponsor of the ad.

Copy in Advertisements

Copy is the printed words in an advertisement. It is made up of the headline and the body information. These, together with the illustrations and *logo* (the name of the firm in a distinctive design) plus any other typographic devices, make up the components of an advertising layout.

The person who makes the layout should understand something about what makes an effective piece of body copy. The key word is *words.*

Copy is composed of words, and the best copy is composed of the best words that can be found. Words that stop you in your tracks, words that sell you, words that get you going, words that reflect the consumer's point of view rather than the seller's. "You" words help draw the audience into the message. "Selling" words help the copy get action.

> Those who write advertising copy . . . should, I believe, constantly bear in mind that, if advertising copy is to be at all effective in contributing to the eventual sale, it should not venture beyond its limited province of informing favorably; of inciting curiosity; of building belief; of creating understanding; of developing the urge to investigate and see for oneself.[1]

The copywriter must understand the product thoroughly, know what it can do, how it does it, and its assets and shortcomings when compared with similar products. The writer must also have a clear idea of the target audience. The key to successful communication here, as it is in all other areas, is to be able to create common understanding and believability. An understanding of the target audience's characteristics and behavior is vital.

▼ The Advertising Campaign

All the skills of the designer working with personnel in the advertising agency or its client are brought into play when an advertising campaign is created and executed. A look at a successful campaign can provide insights valuable to anyone involved in graphic communications.

An advertising campaign in its most fundamental form is a series of advertisements that repeat one basic message. The first step in developing

Fig. 12-10 Tremendous time savings can be realized by composing complicated advertisements on a computer with a program created for the job. This computer is ready to go with the AdWorks program.

Fig. 12-11 This grocery advertisement was created on a computer using the AdWorks program. It was produced in 2 hours and 28 minutes from start to finish. Layout required 14 minutes, text input 90 minutes, product art 30 minutes, and printing 14 minutes.

a campaign is intensive research to determine the "position" of the product, service, or idea to be promoted, that is, where the advertiser desires to place his or her client or product in the minds of the target audience. For instance, Avis, the car rental firm, developed a campaign around its position by advertising "we're number two." The campaign emphasized that because of this position Avis tried harder than its competition.

A plan of action is developed that includes the objectives of the campaign, major selling points, goals, and a budget to carry out the plan. Although millions of dollars have been spent on campaigns by large corporations, campaigns do not necessarily require a large expenditure of money. Many successful campaigns have been short term and relatively inexpensive.

Such a campaign was the "Electric Drought" campaign created for the Sierra Pacific Power Company by Doyle-McKenna and Associates, Inc., both of Reno, Nevada, a city in a semidesert region of the mountain West. The goal of the campaign was to tell target audiences the situation faced as a result of drought conditions and to secure their understanding of measures to be taken.

Here was the problem. A nationwide drought greatly reduced the availability of low-cost hydroelectric power. Sierra Pacific normally purchases this power from sources in the Pacific Northwest. Because of the shortage of power the utility had to pay more for electricity. This resulted in its seeking a substantial rate increase from consumers.

The state public utility commission permits fuel and power purchase costs to be passed on to customers once a year. But the commission must approve the amount of increase. Sierra Pacific found it necessary to seek a 10 percent rate increase in this situation.

Research indicated that six out of ten Sierra Pacific customers had no idea of the effects of the drought on the power supply. Much of this research was obtained through focus groups, in which a cross section of the population met with advertising agency and utility personnel who elicited their opinions and perceptions.

The agency and the client, after studying the findings, set the following objectives for the campaign:

1. Build awareness quickly with employees, the media, and the public that a nationwide drought was causing unusual conditions that would increase power costs and electric rates.
2. Create strong recall for this message in the brief five-week period before the rate increase was sought.
3. Establish an effective base of believability that the drought was the "villain" and Sierra Pacific and its customers were both victims.
4. Show steps the utility was taking to help mitigate the increased costs and hold down rates as much as possible.

The strategy devised and the action taken included:

1. Informing Sierra Pacific employees of the drought impact on rates and preparing them to answer customer inquiries.
2. Conducting news conferences which included an aerial tour of drought-stricken reservoirs.
3. Placing advertisements on television and in newspapers. Large space advertisements were used to demonstrate how the company relied on

other areas for hydroelectric energy. The advertisements also emphasized that the drought was creating unusual problems.

The campaign targeted the total market area of the utility, but placed extra emphasis on reaching and developing understanding among the influential persons in the community.

No campaign, whether an advertising, public relations, or editorial effort, is completed until it is evaluated to determine if the goals have been reached. In this case, at the end of just five weeks, research showed

Fig. 12-12 Newspaper advertisements created for the Sierra Pacific campaign explaining the drought situation. (Courtesy Sierra Pacific Power Co., Stan Berdrow, vice president, communications and public affairs; Bob Alessandrelli, manager, creative and production services; Doyle-McKenna & Associates, Inc.)

1. About half the electricity you use is produced here in Nevada...

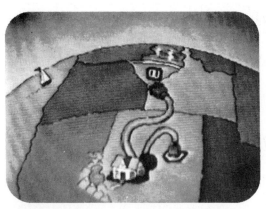

2. But, to keep rates low, Sierra Pacific Power Company <u>plugs into</u> other low cost sources...

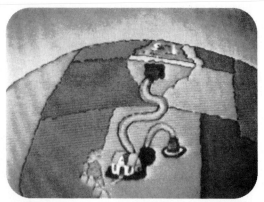

3. Like surplus hydroelectric power from the Pacific Northwest.

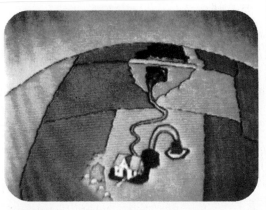

4. But this year's drought has cut back this supply of **inexpensive** electricity and that means rates may go up.

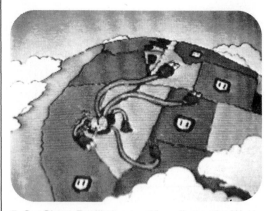

5. So, Sierra Pacific is searching all over the West for other electric bargains...

6. ...to keep your rates as low as possible.

Fig. 12-13   (Above and on facing page) Television storyboards for the Sierra Pacific campaign. These were 30-second spots.

1. To keep rates low Sierra Pacific Power Company goes shopping for about half the electricity you use...

2. And like you, we shop for the best bargains.

3. That's why we buy a lot of hydroelectricity from the Pacific Northwest...It's usually the best buy.

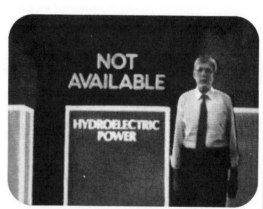

4. But inexpensive hydroelectric power is gone because of this year's drought...

5. ...Purchased energy costs more...and rates will have to go up in the future.

6. In the meantime, we're buying any electric bargains we can find, to keep your rates as low as possible.

77 percent of the public could recall the drought messages. This, agency and utility officials felt, was an unusually high awareness score.

All told, 57 percent of the public believed the drought was the cause for a rate increase. Five weeks previously 60 percent had no idea that the drought would cause electric bills to go up. Instead of the utility being blamed for the rate increase, a majority of the public was convinced the drought was the real culprit.

▼ Creativity and Strategy

In developing a campaign or a single advertisement, two words can be especially helpful for the designer—*creativity* and *strategy*. Creativity in advertising design means, in part, finding a new and unique way to present an idea graphically. Strategy means developing a plan to present this message in a way that will achieve the desired goals.

Creativity and strategy are not just helpful to the designer of advertisements. They can be applied to the creation of any communication.

▼ Effective Design Checklist

- Generally, the most effective layout will *lead* the reader through the advertisement by the design and placement of the elements.
- In advertising layout, art can be used to stir to consciousness a need or desire that was not felt before.
- In an advertisement, art that contains people should show people with whom the intended readers can identify.
- Even though the Z pattern takes advantage of the way the eye often scans a page, if all ads followed this route none would stand out.
- The design, as well as the copy, should emphasize benefits.
- White space can be used to unify a layout if it is kept to the outside of the space in which the elements are placed.
- One item usually should stand out in an advertisement layout, or one item should be emphasized. But the amount of emphasis is a creative decision and most likely will vary with each situation.

▼ Graphics in Action

1. Select two or more of the headlines used as examples on page 275. Visualize how you would design art to illustrate the messages of the headlines. Make rough sketches of the art.

2. Use the idea in 1, but find art in magazines or in clip art books to use as illustrations.

3. Find an advertisement in a newspaper or other publication that you believe is poorly executed. Redesign the advertisement and prepare an explanation of what you did and why.

4. Prepare a rough layout for a grocery store advertisement featuring one item from any full-page newspaper grocery store advertisement. Make a rough 4 by 6 inches. Use art clipped, traced, or sketched (black-and-white art only), maybe from the full-page advertisement itself. Use the store's logo.

5. Mark up the rough created above.

6. Select either the A or B copy following and create a layout. The layout should reflect the type of business and the product or service advertised. Create art or find it as in 2. Make the advertisement 5 by 7 inches and do the markup.

A.

At Your Service
Come and experience the classic
cuisine of Northern and Southern
Italy . . . tonight.
The Grotto
35 Elmwood Avenue
Phone 786-6531

B.

Time to Redo
A new coat of paint can give your
home a brand new look . . . outside
and in! Let our team of painting
perfectionists do a beautiful and
complete job. Satisfaction
guaranteed. Estimates cheerfully
given at no obligation.
A-1 Painters
156 Rutgers Street
Phone 853-3907

Notes

1 Walter Weir, *On the Writing of Advertising* (New York: McGraw-Hill, 1960), p. 7.

CHAPTER THIRTEEN

Designing the Magazine

Courtesy *Smart Magazine*

[SMART SET]

NATIVE DANCE

Into the mainstream.

THE AMERICAN INDIAN DANCE Theatre has given nearly sixty performances in Paris and has brought down the house in Yemen and Algeria, but the company comes to New York for the first time this fall when it opens at the Joyce Theater on September 19. After a two-week run there, the troupe will tour the United States and Canada for a month, and it will be featured on PBS's *Dance in America* in October.

Though its dances are steeped in tradition, this twenty-four-member company, which represents sixteen different North American tribes, is only two years old. Until it was founded in 1987, there was no permanent professional showcase for indigenous American folk dancing. Director Hanay Geiogamah (of the Oklahoma Kiowa/Delaware tribes) has abbreviated and reshaped diverse ritual dances into an intertribal repertoire whose dramatic effect never undermines authenticity. In costumes made from feathers, colored beads, rawhide strips, and furs, dancers invoke, praise, appease spirits of the earth and the air; others impersonate birds, buffalo, deer. In one piece, the celebrants stand erect, trembling and swaying until they resemble a field of grass ruffled by a voluptuous wind, while in the "Fancy Dance" men in an impossible array of feathers prance and preen, transformed into pheasants in a mating competition.

—DAVID DANIEL

Inset: A Fancy Dancer. Right: Dancers SoldierWolf, Tailfeathers, Roach, DeCora.

OLLIE DAY (INSET); TAYLOR KING

30

▼ Big and Small, Magazines Are Basically the Same

"I need help!" The voice on the other end of the line was a familiar one to the journalism instructor. Deborah had been one of his advisees when she was a student. She had opted for the radio/television sequence of courses. She had no interest whatsoever in the print media and avoided the graphics courses, while taking the bare minimum of writing and editing courses.

After she spent two years in an entry-level position at a television station, opportunity came knocking at Deborah's door in the form of a job offer as information officer for the local real estate board. The increase in salary was substantial and the future opportunities appeared bright.

The board members explained Deborah's duties.

"Of course you will continue publication of our monthly magazine," one member remarked.

Deborah didn't get the job.

This chapter and the one that follows are for Deborah and all the other people in the world of communications who find that their jobs will involve producing or helping to produce a magazine.

Most people visualize one of the popular publications when magazines are mentioned. *Reader's Digest* or *TV Guide*, with their more than 17 million copies an issue; *Time, Newsweek, Sports Illustrated, Good Housekeeping, Better Homes and Gardens. Glamour, Seventeen,* and *Rolling Stone; National Geographic,* with its stunning photography, or *Scientific American,* with its precise, detailed text and charts. These are formidable publications. Planning and producing a magazine on such a scale would seem to require technical knowledge that is beyond the abilities of many communicators.

But the same principles of selecting type and illustrations and placing them on a page in the most effective way possible apply to small eight-page association publications as they do to these giants. And the small association monthly can be just as attractive in its way and do just as effective a job of communicating with its target audience. It does take planning and the application of basic principles of good design. We begin by discussing general planning and then move on to the specifics of magazine design.

First of all, the design and production of a magazine can be fun. Of all the areas of printed communications, none is more interesting and challenging. Also, the satisfaction of accomplishment when you hold the printed copy of your magazine is rarely exceeded in any other communications activity.

Where do you begin? Although our concerns are layout and design, we cannot separate this from editorial planning.

In the ideal situation, the staff will include a design specialist—the *art director.* This person has the primary responsibility for the creative aspects of producing the magazine.

The art director should be versed in all the visual aspects of a printed communication, especially type selection and arrangement for effective readability, legibility, and suitability. He or she should be experienced in the production end of the business as well.

There are many examples of outstanding design in magazines, but there are also examples of designs that are beautiful as works of art and still

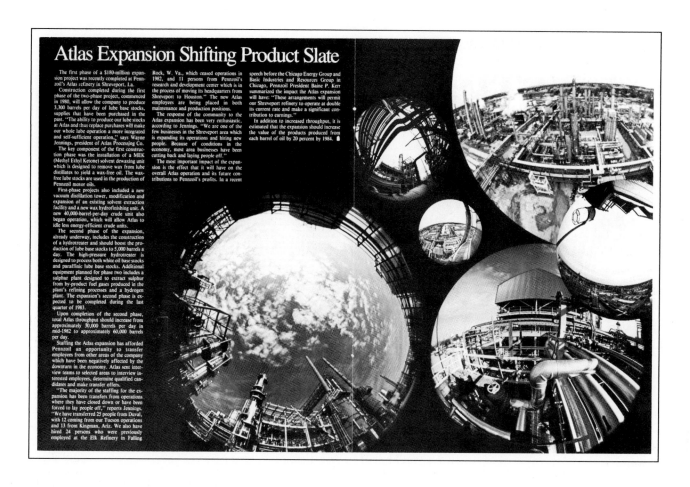

Fig. 13-1 A stunning design for a magazine spread. (As mentioned, though, reverse type should be used with caution since extensive use may negatively affect readability.) (Courtesy Pennzoil *Perspectives*)

hamper the effectiveness of communication. These can include using decorative rules and borders that create barriers rather than unity or advancing hot colors that dominate the visual area.

Too often the tip-off that a publication is in trouble is its use of "screaming" graphics—which figuratively shout for help as the publication goes down for the third time. Too often a publication will suddenly take graphics seriously for the first time when it hits turbulent waters. Examples in recent years include the old *Saturday Evening Post*, a giant in its day, in the months before its demise in 1969, and the *Chicago Daily News* in the final weeks of its life in 1978.

▼ Good Design Aids Content

As in all printed communications, good magazine design must aid and illuminate the content. The combination of good design and poor content can fail, whereas what appears to be poor design and good content can sometimes survive and even prosper. Obviously, however, we should aim for good design and good content.

This partnership concept—good content and good design—extends to the editor and art director. The editor should attempt to keep design

Fig. 13-2 A two-page spread from *Smart,* a national magazine produced on desktop computer equipment. This spread uses the A-T-S-I approach with bold art to attract attention (note that the art crosses the gutter but the fold does not detract from its effectiveness). Note, too, the handling of the short article, which is boxed. (Reprinted from *Smart* magazine, Vol. 1, No. 5)

within the limits of its function while encouraging the art director to provide creative advice that results in the bright and unique. Both should strive to produce a publication that will attract and hold readers.

The editor who has a good art director as a partner is fortunate indeed. However, in many of the more than 10,000 periodicals of various sizes and shapes that appear each month in America, the editor and art director are one and the same. This is especially true with small house publications for business and industry and organizations like the local Girl Scout council, real estate board, and so on. Many of the entry jobs in communications involve producing these small magazines.

Let us assume, then, that we are going to design a magazine from the ground up, or that we are going to give an existing magazine an overhaul to make it more effective. The first step is planning. Time spent on planning and writing the plans down on paper will pay for itself over and over again.

▼ Long-Range Decisions

In designing or redesigning a magazine there are several long-range decisions that should be made immediately. These include:

- *The function, or purpose, of the magazine:* We should determine exactly why it is being published, what we hope to accomplish by sending it out into the world.
- *The personality of the magazine:* Printed communications, like people, project images. What sort of image do we want our magazine to project? Is it dignified and reserved or is it informal and aggressive?
- *The audience we wish to attract:* What sort of person do we want to read our magazine? What are our readers or potential readers like, where are they located, what are their interests?
- *The formula for our magazine:* This means the kinds of information, articles, and features we will include in each issue and how this material will be presented.
- *Will our magazine contain advertising?* If so, how will our editorial formula affect the potential for advertising? Will our target editorial audience be a target market for certain goods and services? If so, the advertising people take over here and determine the markets, appeals, and potential advertisers who will be interested in what we have to offer.
- *When and how often our magazine will appear:* If the publication is a weekly, timeliness is usually an important factor in the formula, and the design format should reflect this timeliness.
- *Design and typographic decisions:* These will include the basic format— the page size, margins, number of columns per page, and the typefaces to be used for standing heads, article titles, captions, and body matter.
- *Editorial style decisions:* Standard practice for spelling, grammar, punctuation, capitalization, and so on, as well as decisions concerning all the physical aspects of the magazine can be included in a *stylebook* or manual. All staff members should use this stylebook as a guide while doing their jobs.

Title and Cover Policies

The title or headline policy has to be determined. Two philosophies are usually found here. One is to adopt one or two families of type for all titles and the other is to select specific types for each article.

Another decision concerns the magazine cover policy. For example, what will be the logo (or nameplate) design, what sort of art will be used on the cover, how will the various typographic elements be used from issue to issue? Other typographic decisions to be made in the planning stage include the arrangement of heads, subheads, masthead, regular feature titles, table of contents, column widths and placement, and/or placement of advertising.

All of this planning should be done with the basic tenet of effective magazine design in mind: *The physical appearance of the publication should reflect the editorial content and appeal to the audience for which it is intended.*

Much of a magazine producer's success is related to the ability to isolate the target audience and build a product that will appeal to this audience. Proper design can help us achieve this goal.

For instance, a magazine devoted to religious concerns should look the part. It must by its appearance say "I'm concerned with religion," just as it must say "Let's go hunting and fishing" in its look if it is to appeal to the hunting and fishing enthusiast. A magazine aimed at buiness- women should wear a different typographic "suit" than one appealing

Fig. 13-3 This cover from *Field & Stream* accomplishes the four functions of a cover: It identifies the magazine in a way that sets it apart from others, it attracts the attention of the intended target audience, it sets the tone or mood of the contents, and it lures the reader inside. (Reprinted with permission of *Field & Stream,* a publication of Times Mirror Magazines, Inc.)

to structural engineers. These stylistic differences need to be considered in the initial planning stages.

Visualization in Magazine Design

The next step for the editor or designer is *visualization*—the construction of mental pictures—of the magazine's eventual appearance.

One word of caution. Although our magazine's appearance may say "hunting and fishing," it is important that a new and unique way be found to say it. And it is important that we say it without resorting to screaming graphics. Our publication should stand out from others that are trying to say the same thing. Thus our planning needs to include a study of the competition and how it is saying what we want to say.

In planning the initial design, then, we must always keep the potential reader in mind. A scientific magazine, for instance, appeals to a precise, orderly mind. It should use precise, orderly types—medium Sans Serifs; clean, sharp modern Romans, such as the Bodonis; straightforward solid line rules and thick-and-thin straight-line combinations of the Oxford rules. It should have accurate and clear-cut placement of elements. Color should be used for emphasis where it is needed, but color should never come on so strong that it shouts.

Scientific American is considered one of the best-designed magazines, and it applies these typographic principles with great success even though it might appear rather forbidding to the reader who is not especially interested in the world of science.

Whatever basic design approach is chosen, it should not be changed in the future without a great deal of study and replanning. The basic design plan should remain constant, as it can help achieve identity and continuity from one issue to the next. It should help the reader recognize the magazine instantly, just as *Time's* distinctive red color, all-cap nameplate in Roman type, and cover art have for years.

As the general planning progresses, we should also keep in mind the admonition of one magazine designer, "If you're dull, you're dead."

The Four Fs of Magazine Design

One way to approach the physical design of a magazine, is to consider the "four Fs" of magazine design—function, formula, format, and frames.

Function   The *function* part of this approach is obvious, but it will be helpful to jot down the things the magazine should accomplish. Will it be an internal magazine, meant for members of an organization or employees of a company? Or will its appeal be external, aimed toward people outside the organization? Or will it have a combined goal of dual appeal, internal and external? Will it be aimed at recruiting new members or new support? Will it be a "how to" publication, or will it be a publication that relies on layouts with lots of photos? All these functions will affect the physical form of the publication.

Formula   The *formula* is the unique and relatively stable combination of elements—articles, departments, and so on—that make up each issue. The elements that should be considered in devising the formula for a

Fig. 13-4 The precise, careful graphics and type arrangements of *Scientific American* might not appeal to everyone, but they are just right for a publication aimed at people interested in the sciences. (Copyright 1989 by Scientific American, Inc. All rights reserved. Page 55 photographs by S. Varnedoe.)

magazine include the sort of articles to be included such as fiction, uplifting essays, interpretative pieces, or whatever.

The formula also includes the sort of illustrations, drawings, and/or photographs to use. It includes the departments that will be a regular part of each issue as well as editorial or special interest columns, poetry, cartoons and jokes, fillers, and other miscellaneous material. Efforts should be made to produce content that has balance and consistency—and variety to prevent dullness.

Once we have defined the function of the magazine and developed the formula to achieve this function, we need to consider our special typographic concerns—the format and frames of the publication.

Format The *format* includes the basic size and shape of the magazine plus the typographic constants or physical features that stay the same from one issue to the next. These constants include the cover design, masthead, break of the book (space allocation), placement of regular features, folio line techniques, and techniques for handling *jumps* or the continuation of articles from one part of the publication to the other.

Magazines select a format, or basic design pattern, incorporate it in their stylebook, and stick to it. There are a number of things we must consider when deciding the format of our publication. These include:

Fig. 13-5 *National Geographic,* an American tradition, has always used the *standard* format. The cover has gained recognition through the years with the yellow frame. Note the excellent legibility of the reverse type and the red on yellow teasers used in this dramatic layout. (Photo by Stephen M. Dowell © 1989 National Geographic Society)

1. The press capacity of the printer who produces the magazine plus the most efficient way to use paper with a minimum of waste.
2. Ease of handling and mailing. How will the publication be delivered to the reader? If it is to be mailed in an envelope, that could be a factor in determining its size. How will the publication be handled by the reader? If it will be the type of publication that is kept and filed, it might be wise to select a size that will fit standard file cabinets or binders. If it is the sort of publication that will be carried around in a pocket or purse, that is another factor to consider in determining size.
3. The content. Large picture layouts require elbow room to be effective. If the publication will use many pictures or diagrams, a large page size will probably be best.

There are, however, some common format dimensions that have evolved. One set of dimensions determined by designers sorts out magazines by size according to the sizes of the type areas. For example,

Pocket:	2 columns wide by 85 lines deep
Standard:	2 columns wide by 119 lines deep
Flat:	3 columns wide by 140 lines deep
Large:	4 columns wide by 170 lines deep

(*Note:* The lines used to measure depth are agate lines, based on agate type, which is 5½ points in height. There are fourteen agate lines in 1 inch.)

Another common method of classifying magazines is by page size:

Miniature	4½ by 6 inches
Pocket	6 by 9 inches
Basic	8½ by 11 inches
Picture	10½ by 13 inches
Sunday supplement	11 by 13 inches

There are variations to these standard sizes, however, and not all magazines will fall into these categories.

The most common magazine page size these days is the basic 8½ by 11 inches, or slight variations thereof. This size represents the most efficient use of paper, is easy to address and mail, and fits nicely in binders and file cabinets. From the graphics point of view, the basic page is large enough to permit good use of art and to give considerable latitude for interesting layouts.

Frames   Magazine *frames* are the outer page margins, the white space between columns of types and pages, and the use of white space to "frame" the various elements such as headings, titles, subheads, bylines, and art.

A basic decision must be made regarding page margins. There are two possibilities. One is to use *progressive margins.* In this type of margination, as explained in Chapter 4, margins are designed to increase in size as they progress around the page. The gutter margin, or inside margin, is the smallest. On the right-hand pages the margins increase clockwise, and on the left-hand pages they increase counterclockwise. Progressive margins may be chosen if the designer wants to give a visual impression of high quality, or "class."

Direct government expenditures for research and development are on the upswing, and while the budget outcome is muddied, Congress is likely to give the Reagan Administration its nearly $48 billion in requested spending authority, if not more.

As analysts are quick to note, the bulk of the spending increase is earmarked for defense programs. The Pentagon would get about $30 billion, or nearly two-thirds of the total R&D funds.

In trying to sort out potential winners and losers, can the Federal Government out-guess the marketplace?

The Congressional Research Service says that, while the United States spends considerably more on R&D as a percentage of Gross National Product than do the governments of its major competitors, it drops below both Japan and West Germany when military programs are excluded. The American Association for the Advancement of Science (AAAS) estimates that spending for non-defense R&D in fiscal 1984 will actually decline when measured in constant 1972 dollars.

Within the non-defense category, the Reagan Administration projects increases in basic research in the so-called "hard" physical sciences and in engineering, while basic research in the "soft" fields—such as biomedical, social and behavioral sciences—as well as applied research would hold steady or decline. In seeking to boost some basic research now, the President "is merely undoing some of the damage he inflicted earlier," wrote Daniel S. Greenberg, editor of *Science & Government Report*.

Louis Schorsch, science specialist at the Congressional Budget Office, said of the shift in R&D priorities at the White House: "There's been a really dramatic reorientation toward defense, and a tendency on non-defense to pull the government back

and rely on the market. That's a pattern you see across the government as a whole, and it shows up in the R&D budget as well. The category that has been cut the most is applied research—the category that links the laboratory and the factory."

Some of the cuts in applied research and demonstration projects involved energy projects that the Administration said should be funded privately, if at all. Nevertheless, the idea of pushing more laboratory advancements into the marketplace appeals to many. This is something the Japanese do quite well, through government-industry-university partnerships in projects with high commercial potential.

Active Role For Government?

House and Senate members, in their ongoing quest for a national industrial policy, have considered a range of suggestions through which the government would play a more active role in the commercialization of research work. U.S. companies in both the smokestack and high-technology fields may soon be outgunned by government-subsidized competitors overseas unless the Federal Government offers help, according to this line of thought.

These suggestions smack of centralized government planning, and run counter to traditional notions of free-enterprise economics. But in truth, the lines between government-funded research and business-sponsored application have long been blurred.

Agriculture, the biggest U.S. export business, is nurtured by federal research programs as well as price supports. Commercial aviation, which accounts for the biggest chunk of manufactured exports, likewise benefits from federal R&D. The government-financed space program has generated so many marketplace products that the space agency publishes a roundup called "Spinoffs." And in the highest high-technology realm, the Federal Govern-

22

ment was the first sponsor, and is still the biggest buyer of American-made supercomputers. Currently the Pentagon's Defense Advanced Projects Research Agency is funding leading-edge research into artificial-intelligence computing, which has unlimited commercial possibilities.

But just because the Federal Government already provides this kind of support, it does not necessarily follow that it should do more of the same. Ideology aside, there is always the practical question of whether the government, in trying to sort out potential winners and losers for special government assistance, could out-guess the marketplace.

Spurring Business

The Reagan Administration is taking a different tack. It supports various initiatives designed to spur business into doing more on its own rather than injecting the government into private-sector R&D choices.

These initiatives include proposed changes in antitrust laws. The Administration, along with several congressional Democrats, has proposed legislation designed to encourage more joint venture research by competing companies. The new Microelectronics and Computer Technology Corp., established by a dozen semiconductor and computer firms, stands as a model for this type of venture. Business representatives say they still fear government antitrust challenges or private triple-damage suits, but if legislation achieves its stated goal, it would result in more research at no direct cost to the government.

Tax policy also can be used to promote R&D. The R&D tax credit, inserted into the big Economic Recovery Tax Act of 1981, was scheduled to last only two years but is likely to be extended. The Administration meanwhile is broadening the definition of eligible expenditures to include computer software as well as hardware.

Patent policy similarly is linked to R&D activity, and the Administration, through a presidential memorandum, has instructed federal agencies to allow contractors, to the extent allowed by law, to retain title to inventions that are developed under federal sponsorship.

Legislation, introduced by Sen. Charles Mathias (R-Md.) and 13 bipartisan co-sponsors, is also pending to restore up to 7 years of patent life lost because of government testing and review requirements. Companies and inventors, who develop new products and then have to wait as long as half the 17-year patent term lifetime to receive government clearance, claim this stymies innovation. "This degradation undermines the basic rationale of the patent system," said Sen. Mathias, "at a time when [industries] face rising research and development costs and stiffening competition from overseas, where full patent protection is more dependable."

The National Laboratories

Presidential Science Advisor George A. Keyworth II also instigated a year-long review by the White House Science Council of the Federal Government's own 755 laboratories, which use about one-third of the federal R&D budget, or $15 billion. The study, directed by Hewlett-Packard Chairman David Packard, concluded that the missions of the national labs should be more tightly defined and their work consolidated. It said greater collaboration with industry and academia should be achieved through, for example, more personnel exchanges and joint projects.

State governments are hoping to foster more R&D within their borders as a way of spurring economic growth.

Finally, the Administration and Congress have moved toward agreement on expanded federal aid to the nation's

23

Regular margins, or those with identical dimensions, are most common. They should take up about half of the total area of the page. Quite often the *folio lines* (page number, name of magazine, date to appear on each page) are placed in the margins of magazine pages. This is another typographic decision to make.

Another white space or frame decision concerns the space between columns. A minimum of 1 pica of white space should be allotted here. Less will make the page look too crowded. A pica and a half is not too much. On a basic 8½ by 11 inch page or larger, 2 picas of white space between columns on a two-column format is about right. Too much white space between columns will destroy unity and make the page appear fragmented.

There is a tendency for those new at the game to crowd elements and jam them together. Don't be afraid to use plenty of white space; make a minimum of 1 pica between art and body copy, between head and subhead, and between all other elements.

Once concrete decisions are made concerning the function and formula of a magazine and the basic format, including handling of the frames, we can turn to the specific elements that will be combined to make our typographic package. But before we do, there are two areas that we should address: basic terminology and redesigning an existing publication.

Fig. 13-6 Frames or margins are important design elements. Most of the white space allotted to margins should be in the outside dimensions of the page to help provide unity and enhance the framing effect. Note the placement of subheads on these inside pages to add balance and break up long copy. (Reprinted with permission from DuPont *Context*)

▼ Basic Magazine Terminology

Following are some basic terms that anyone engaged in magazine editing, design, or production should understand:

- *Bleed:* The extension of an illustration beyond the type area to the edge of a page.
- *Break of the book:* The allocation of space for articles, features, and all material printed in the magazine.
- *Contents page:* The page that lists all the articles and features and their locations.
- *Cover:* Includes not only the front page but the other three pages making up the outside wrap of the magazine as well.
- *Folio:* The page number, date, and name of periodical on each page or spread.
- *Gutter:* The margin of the page at the point of binding, or the inside page margin.
- *Logo:* The magazine's nameplate, appearing on the cover, masthead, and so on.
- *Masthead:* The area, often boxed or given special typographic treatment, where the logo, staff listings, date of publication, and other information regarding the publication is listed.
- *Perfect binding:* A binding method that uses a flexible adhesive to hold the backs of folded signatures together while they are ground to size and more adhesive is applied. The cover is put in place while the adhesive is still wet.
- *Saddle stitch* (also called *saddle wire*): The kind of binding in which staples are driven through the middle fold of the pages.
- *Self-cover:* A magazine cover printed on the same paper stock as the rest of the magazine.
- *Sidestitch* (also called *sidewire*): A method of binding in which staples are driven through stacked printed sections (signatures) of the magazine.

Fig. 13-7 Binding styles for magazines.

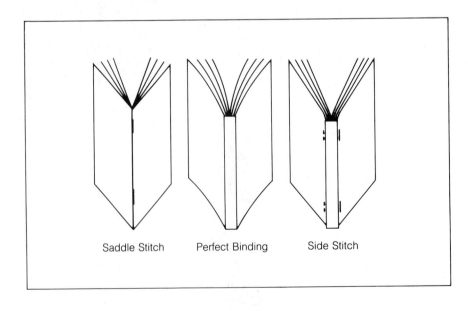

Saddle Stitch   Perfect Binding   Side Stitch

- *Signature:* A large sheet of paper printed on both sides and folded to make up a section of a publication. For instance, four pages might be printed on each side of a sheet and then the sheet folded and cut to make an eight-page signature.

▼ Redesigning a Magazine

The general approach to designing a new publication can be used to redesign an existing magazine as well. But there are a few additional points we need to consider before creating a new look for an existing publication.

The first step in redesigning should be a complete study of the present format. Every item from the smallest typographic device to the logo should be examined. The primary examination should center on the audience and its needs. This might require contact with the audience through research and surveys to find answers to questions such as:

1. Does the present format do a good job of bringing the editorial content and the audience together?
2. Does the present design give the audience what it wants?
3. Will a different format serve the audience better?
4. Does the audience prefer different kinds of illustrations (photos instead of diagrams, for example)?

Changes must take into account factors that may not be obvious. Talking with the printer and compositor is important. It might be impossible to produce the changes sought with the equipment available. Much time and effort could be wasted planning changes it would be impossible for the equipment to duplicate.

If the publication is produced in house, on equipment we possess, the equipment manufacturer can be a valuable source of graphic and redesign ideas and equipment capabilities.

We should solicit ideas from as many sources as possible but not abdicate the job of decision-making. A publication designed by a committee usually turns out to be a hodgepodge collection of graphic ideas that is seldom effective.

We must also avoid the tendency to design by "shotgun." This is the temptation to try to outdo graphic design on each succeeding issue by packing each issue with more attention-getting graphics than the one before. Shotgun design is seen more and more, often to the detriment of the publications on which it is practiced.

If the publication carries advertising, our study should include the types of advertising, the most common sizes, and their present arrangements on the pages. For example, a proliferation of small ads will present problems in page layout that could be avoided if the magazine runs fewer but larger ads.

While it might not be possible to control the size and design of the advertisements, which usually come to the publication from a variety of sources, it is possible to control the arrangement of the ads on the pages.

Creative Communication

Is experience the best teacher?

The idea battleground in design is free-thinking, free-flowing, open-ended, no-rules creativity versus knowledge, discipline, experience. The struggle goes on.

Some psychologists have found that experience can restrict creativity. Some experienced editors or designers might not view a problem with a fresh outlook. They might classify it with past problems and select solutions from past experience. Only if the problem type cannot be recognized is it studied in depth. Otherwise, old solutions are fitted to new problems.

What is the answer? Creative problem solvers suggest the answer might be brainstorming—using group interaction to stimulate individual thought. Participants are encouraged to generate as many ideas as possible without evaluating them. After a period of time, the ideas are grouped, duplicates and obviously impractical suggestions eliminated, and the others rank ordered.

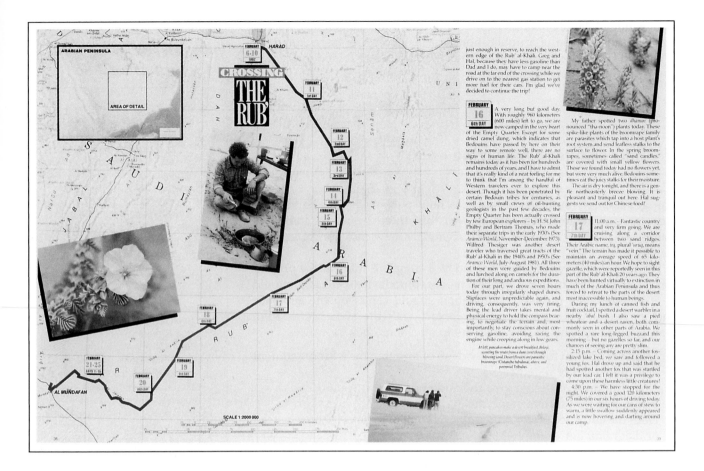

Fig. 13-8 A number of creative approaches to magazine design are illustrated in this two-page spread. Note how the large map is broken out of an overview smaller map to orient the reader. Date squares from the large map travel route are used in place of initial letters in the text. The smaller photos are placed on angles to add variety and interest to the spread, and they are accented with shadow boxes. The map is bled on three sides, and it jumps the gutter at a spot where the saddle-stitch staples do not interfere with any of its details. The title in reverse is not lost in the layout as it might have been if surprinted in regular type. (Reproduced by permission of *Aramco World*)

After considering the various patterns for placement of ads on a page, we should adopt a standard placement arrangement. This will help provide orderly design throughout and add to the unity of the whole magazine package.

In addition, changing column widths will affect advertising. If the format is changed from three columns to two, for instance, this could lead to all sorts of complications regarding space sales and advertising rates.

Finally, sometimes there are so-called typographic sacred cows (such as standing heads on regular columns, locating certain features in specific spots, and the type styles used for titles and logos) that cannot be changed, or can be changed only with great internal organizational difficulty. These must be taken into account.

Steps in Redesign

Once a thorough study of the present design has been made, the redesigning process can begin. Below is a useful step-by-step outline for overhauling a publication.

1. *Start at the back.* Magazines are designed from the back to the front. Why this backward approach? First of all, a new design usually means a new logo or new cover design. This indicates that there is something

new inside—and there had better be! A new label for the same old can of beans does not fool anyone very long. Another reason for starting with the back is that's where most of the design trouble usually can be found. Those little mail-order ads, fillers, and jumped conclusions of articles get dumped in the back. So, clean it up first and then go on to the features. Plan a basic uninterrupted reading pattern, but start with the back.

2. *Group, consolidate, organize.* Analyze the small items and rearrange them. This will often lead to an organizational format that will improve the appearance of the whole publication. How can that be? Well, the secret of a well-designed publication is *departmentalization*. Grouping related items together creates an orderly plan that makes sense to the reader. It helps sort out the contents of the publication.

   If small ads are carried, consider them first. What is their typographical tone? Are they crammed with lots of copy? Do they try to scream at the reader with large, bold type, heavy borders, lots of reverses? Separate them into groups and arrange the groups by sizes. If they are scattered helter-skelter throughout the magazine, group them together in some way—perhaps by subject category. Arrange them on a layout sheet in an orderly pattern (see Figure 13-9). Ads are more effective if grouped by subject matter—schools, camps, travel, trading post—rather than being scattered randomly throughout a publication.

3. *Analyze departments and their heads.* Too many departments—household hints, sports shorts, and so on—can clutter a publication in two ways. It is difficult to design effective layouts if many short departments must be accommodated. Also, the constant repetition of many similar department heads can lead to a choppy format. Check to see if departments can be combined or eliminated if there are too many in the publication.

   When redesigning department or section heads, keep in mind the criteria for good heads: The head should blend well with the copy, help say what the department is all about, and be uncluttered with illustrations and ornaments. If illustrations are used, they should be simple. Avoid art that could become dated in heads—art that is in fashion now but probably will be out soon. The head should be clean, brief, and to the point. It should be easy to adapt it to one-, two-, or three-column widths.

4. *Consider the constants.* There are other constants to review in addition to department heads. (Constants are the typographic devices that do not change from issue to issue.) These include the table of contents, masthead, editorial section, letters section, fillers, next issue's previews, and special advertising sections.

   The entire sequence of constants must be handled individually and then collectively to present a harmonious whole. In designing the constants as well as in all the other parts of the total format, all the principles of design and type selection we have discussed should be put into practice.

5. *Finally, the cover.* Most communicators are tempted to redesign the cover right off, but we leave it to last for a very good reason. A thorough study of all the factors that must be considered in designing a magazine will provide a bonus. Designing an effective cover now should be easy.

Even though the cover is considered last it probably is the most important element in the entire format. It is the display window—what people see first. It can cause the reader to pick up a periodical or pass it by. The ideas about cover design in Chapter 14 are useful in redesigning as well as in creating a new publication.

Fig. 13-9 An effective way to reduce advertisement clutter in magazine design is to group the small ads under appropriate headings. This aids the advertiser, too, as it helps flag the target audience. (Courtesy *Nevada* magazine)

▼ Effective Design Checklist

- Provide for ample margins and white space. If the margins and white space between the columns are inadequate, the entire magazine will have a cramped, jammed-up appearance.
- Define editorial and advertising content clearly. If the reader cannot tell easily where one begins and the other ends, the whole publication will appear disorganized and confused.
- Make sure type styles relate to content. The typefaces and illustration designs must harmonize with the editorial content. Some designers call this "type and content marriage." Good type marriages are a sign of professionalism.
- Pictures should be selected and cropped for content first. Do the pictures help eye movement through the page? If art causes eye movement out of the page or through the page in an uncontrolled way, the page patterns are those of a beginner, and we want to be pros.

Fig. 13-10 This magazine page has many design problems. Most are violations of basic principles of design. How many can you spot? Poor spacing between words and lines, uneven leading in the subhead, mixture of display types, improper use of Oxford rules, and art that does not communicate are a few of the shortcomings.

Fig. 13-11 A student used computer-created type, rules, screened block, and art to make this mechanical of a magazine cover. It was done as a project for the Graphics in Action assignment at the end of this chapter. (Courtesy Kim Rusche)

- Special departments should relate to the basic editorial theme of the magazine. Unrelated departments, especially if there is a proliferation of them, tend to break down the basic editorial impact of the magazine.
- Check the last pages as well as the first. The back of the book should not be a dumping ground. The final pages should be well organized if the entire package is to be as attractive as possible.
- The table of contents should be neat, clean, and functional. If the table of contents and the masthead are cluttered, crowded, or disorganized, redesign them.
- Everything should blend together. Check to see that the stories and articles are laid out to present a unified magazine package. If they seem to be individually designed with no thought given to what else is contained in the issue, they need work.
- Check the body type—the basic reading matter type. It should be a clean, readable design—preferably a Roman face that is big on body. Check the leading and see if the type size is near the optimum for the width of the columns.
- Consider the design from the reader's standpoint. Some magazines are designers' dream productions but nightmares for readers. Don't let layouts become ego trips. Make them functional. See that each element in each layout is there for a purpose. That purpose should be to attract the reader, hold interest, and make reading a pleasant experience.

▼ Graphics in Action

1. Select a magazine that appeals to a special interest audience and analyze its design and typography. Try to determine how the design, selection of type styles, and arrangement of design elements were made to appeal to the target audience. If there are shortcomings in the design, explain the changes you would recommend.
2. Develop a formula for a small magazine. Have its target audience be a group or organization, special-interest or hobby, to which you belong or to which you have a special affinity. Make a written plan to present to possible financial backers of the proposed magazine.
3. Develop a format for the small magazine to carry out the philosophy of the formula developed in 2. (Both 2 and 3 can be combined with Graphics in Action 4 and 5 in Chapter 14 for a major project. It might be well, then, to restrict the size of the magazine to, say, eight or twelve pages, depending on the amount of time that can be devoted to this project.)
4. Design a basic cover layout for a magazine for an organization of which you are a member or for the magazine proposed in 2 and 3. Use an 8½ by 11 page with a type area 44 by 57 picas.
5. Examine the design flaws of the magazine page reproduced in figure 13-10. Redesign the page to eliminate the shortcomings.

Inside and Outside the Magazine

Now we are ready to put our plan into action. The designer faces the task of creating prototype page layouts that will be examined and evaluated by everyone involved in the project.

▼ Designing the Magazine

How does the designer proceed? What are the first steps in the actual creative process?

We are living in a visual age, a world of visual excitement. It takes the combined talent of the artist, the knowledge of the psychologist, and the skills of the designer and printer to produce pages that will encourage the reader to anticipate each issue and reach for one particular magazine among all the others seeking attention.

The Break of the Book

The first decision to be made in crossing the bridge from the planning stage to the layout stage is what is known as *the break of the book*. This is the term editors and art directors use to designate what is going to be placed where.

There are several basic philosophies concerning how to break the book. One is called the *traditional plan*. Under this arrangement, the constants—the regular columns and features, the masthead, the table of contents, and any editorial page material plus the advertisements—are grouped in the first few and the last few pages. The middle pages are reserved for the current issue's main articles and stories. The strongest article in the issue is used as the lead article for this middle section. Several major magazines that use this method of breaking the book include *Reader's Digest, Smithsonian,* and *National Geographic.*

Another approach is called the *front to back system.* In this arrangement, features and articles are spread throughout the magazine. A great many company and association publications follow this approach with perhaps only a contents page at the front of the book.

Of course, the editor and designer can strive to be different and create an entirely new plan.

Laying Out the Pages

Once basic decisions are made, it is time to go ahead and place the graphic elements on paper. This is done with a layout or *grid sheet* for each page. These grid sheets can be obtained for most standard magazine page sizes and column widths from graphic supply concerns or they can be made and reproduced on a duplicator or office copier.

Grids are available in quarter scale, usually four to an 8½ by 11 inch sheet, for making thumbnails. Some editors start their layout work by folding sheets of blank paper into proportionate small sizes, as in making thumbnails for advertisements, but slitting them so they are actually blank

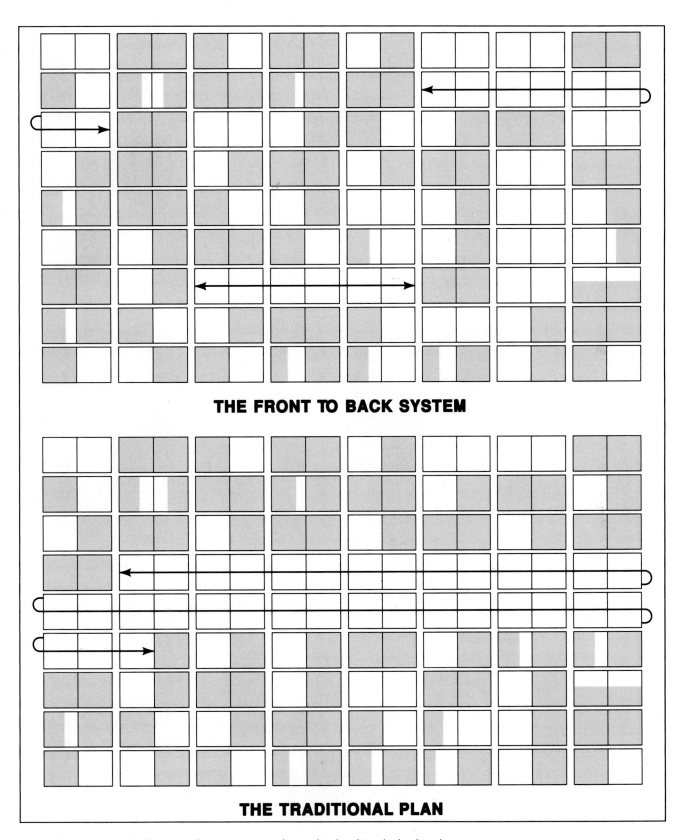

THE FRONT TO BACK SYSTEM

THE TRADITIONAL PLAN

Fig. 14-1 Examples of grid systems for a magazine to be used in breaking the book and keeping track of the progress of the magazine's contents. Shaded areas indicate the placement of advertisements.

Fig. 14-2 Examples of magazine pages designed with the use of a three-column grid. (Courtesy S. D. Warren Company, a subsidiary of Scott Paper Company)

miniatures of the full size. It is possible, too, to make a miniature ruler with a strip of cardboard and label it with an inch scale to the proportion of the full-size sheet. For instance, on the miniature ruler for making thumbnails, one-fourth of an inch might equal 1 inch on a full-scale ruler. Another approach is to cut a cardboard grid that has pica squares ruled on it in nonreproducing blue and have 1 pica equal 1 inch.

The designer uses these tools to rough in all the material the full-size layout will contain and then uses the resulting thumbnail as a guide in keeping things straight when full-size layouts are made.

An easy way to keep account of progress in laying out a magazine is to put the full-size layouts for each page in numerical sequence up on the office wall. Using a small piece of masking tape, the pages can be taken down, worked on, and replaced.

This method of planning an issue has two advantages. The status of the whole issue can be seen at a glance and facing pages can be seen side by side. It should be kept in mind that the magazine reader sees two facing pages at a time, and good planning should include making sure that these pages are compatible.

Screens Will Replace Drawing Boards

The use of pencil, pens, and paper and layout and grid sheets is rapidly diminishing. *Time* now uses a device in its layout department that replaced layout sheets, art knives, and waxers. The new device is called the Vista, and when it went into use it was believed to be the only one of its kind in the world.

The machine has two adjoining video screens. The story, headline, art, and other graphic elements that the designer wants to place on a page are displayed on one screen. The other screen is a monitor that shows what the completed page will look like in color. The designer can move the elements around, alter their sizes, and arrange them on the screen just as most art directors now arrange them on paper. As a *Time* staffer noted, the machine "gives you the freedom to revise and adapt quickly. It can turn the work of hours into as little as twenty minutes."

Whether a magazine is laid out with a ruler and pencil or felt pen, or whether it is composed on state-of-the-art equipment, the basic principles of good design in general and magazine design in particular still apply.

▼ Planning the Cover

All pages of printed communication are important, but if one page must be singled out as the most important of all in magazine design, it has to be the front cover. The cover creates the all-important first impression. It not only identifies the publication but also says something about its personality.

A good front cover should accomplish four things:

- It should identify the magazine in a way that sets it apart from the others.

- It should attract attention, especially from the target audience.
- It should lure the reader inside the magazine.
- It should set the tone or mood of the magazine.

In addition, when the magazine is sold in supermarkets or on newsstands, the cover plays an important role in the selling process.

Every cover should automatically carry certain information. In addition to the logo, the date of issue, volume, and number should appear—especially if the magazine is the type that will be filed and indexed for future reference. The price, if the magazine is sold, should be included. This seems self-evident, but it is surprising how many small magazines fail to include this pertinent information on their covers.

Self-Cover or Separate Cover?

One of the first decisions that has to be made is whether the magazine will be self-covered or have a separate cover. A *self-cover* is a cover printed on the same paper stock as the body of the magazine. Many magazines invest in a more expensive and more substantial paper stock for the cover.

However, the self-cover has its assets. It is, of course, less expensive. In addition, for the small publication on a limited budget, the self-cover can provide color on inside pages at virtually no extra cost. This is possible because most magazines are printed in *signatures*, or sections of pages in units of four, with the number of units in a signature decided by the press capacity. Thus, when color is purchased for the cover, it can also be used on the other pages of the signature containing the cover. This could be four, eight, sixteen, or even thirty-two pages, depending on the size of the magazine and of each signature in its manufacture.

Another cover decision concerns advertising. A very limited number of publications sell the front cover page at a premium as a source of income. These include a few business and professional publications. One is *Editor & Publisher*. However, most editors find front-cover advertising objectionable. The question about cover advertising might be whether the interest-arousing and selling asset of a strong cover should be sacrificed for the additional money obtained by selling the front cover.

Many editors have found that a cover that sets the theme for an issue, identifies what the issue is all about, or ties in with a strong lead article works best. Others report that covers with closely cropped human interest pictures are good attention-getters.

Covers That Sell Contents

A trend today is toward greater use of the cover to sell the contents. More and more *blurbs* are being used on covers. These teasers are designed to lure readers into picking up the magazine for the contents.

In designing blurbs it is important to realize that the cover is the magazine's store window. It is the poster or billboard that will advertise and, hopefully, sell what the publication has to offer. Blurbs should be written and designed much like the copy on posters and billboards—short and to the point. The message should be one that can be taken in at a glance. Advertisers say a billboard message should be read in five seconds. Blurbs on magazine covers should be equally short, never more than three lines of type for each blurb.

Fig. 14-3 A magazine cover created on the computer. The Pagemaker program was used, as was Arts & Letters graphics by Computer Support Corporation, Dallas, Texas. The schematic drawing of the man was made in thirty seconds.

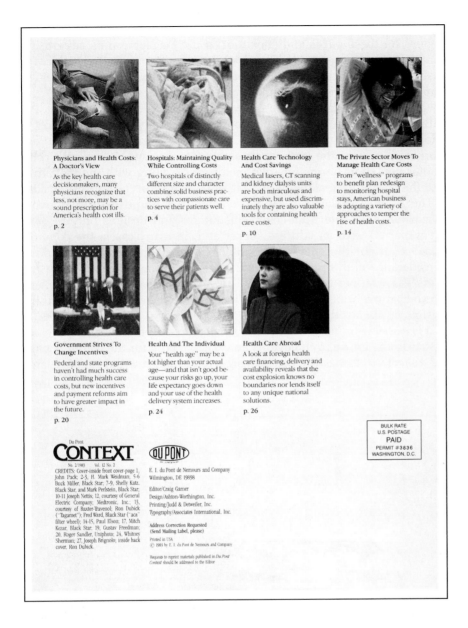

Physicians and Health Costs: A Doctor's View

As the key health care decisionmakers, many physicians recognize that less, not more, may be a sound prescription for America's health cost ills.
p. 2

Hospitals: Maintaining Quality While Controlling Costs

Two hospitals of distinctly different size and character combine solid business practices with compassionate care to serve their patients well.
p. 4

Health Care Technology And Cost Savings

Medical lasers, CT scanning and kidney dialysis units are both miraculous and expensive, but used discriminately they are also valuable tools for containing health care costs.
p. 10

The Private Sector Moves To Manage Health Care Costs

From "wellness" programs to benefit plan redesign to monitoring hospital stays, American business is adopting a variety of approaches to temper the rise of health costs.
p. 14

Government Strives To Change Incentives

Federal and state programs haven't had much success in controlling health care costs, but new incentives and payment reforms aim to have greater impact in the future.
p. 20

Health And The Individual

Your "health age" may be a lot higher than your actual age—and that isn't good because your risks go up, your life expectancy goes down and your use of the health delivery system increases.
p. 24

Health Care Abroad

A look at foreign health care financing, delivery and availability reveals that the cost explosion knows no boundaries nor lends itself to any unique national solutions.
p. 26

BULK RATE
U.S. POSTAGE
PAID
PERMIT #3836
WASHINGTON, D.C.

CONTEXT

Du Pont

No. 2/1983   Vol. 12 No. 2
CREDITS: Cover-inside front cover-page 1, John Pack; 2-3, H. Mark Weidman; 5-6 Buck Miller, Black Star; 7-9, Shelly Katz, Black Star, and Mark Perlstein, Black Star; 10-11 Joseph Nettis; 12, courtesy of General Electric Company; Medtronic, Inc.; 13, courtesy of Baxter-Travenol; Ron Dubick ("Tagamet"); Fred Ward, Black Star ("aca" filter wheel); 14-15, Paul Elson; 17, Mitch Kezar, Black Star; 19, Gustav Freedman; 20, Roger Sandler, Uniphoto; 24, Whitney Sherman; 27, Joseph Brignolo; inside back cover, Ron Dubick.

DUPONT

E. I. du Pont de Nemours and Company
Wilmington, DE 19898

Editor/Craig Garner
Design/Ashton-Worthington, Inc.
Printing/Judd & Detweiler, Inc.
Typography/Associates International, Inc.

Address Correction Requested
(Send Mailing Label, please)

Printed in USA
© 1983 by E. I. du Pont de Nemours and Company

Requests to reprint materials published in *Du Pont Context* should be addressed to the Editor

Fig. 14-4 *Context* solves several problems—what to do with the back cover, how to handle the contents page, how to keep the postage bug and address label from marring the front cover—by putting all this information on the back cover. It brightens the page by using full-color illustrations with teasers for each article. (Reprinted with permission from DuPont *Context*)

Blurbs should also relate closely to the article titles. In one magazine recently, the blurb on the cover announced "Solar Pioneers." The title of the article referred to was "To Catch the Sun." Readers might have trouble relating the two. Make sure the blurbs can be identified with the articles or features quickly and without confusion.

Traditionally, the blurb was placed along the left side of the cover. This was because of the way many magazines were stacked on newsstands so that they overlapped and only a portion of the left side of each cover was visible to the passerby. However, it isn't necessary that this pattern be followed for most magazines today.

Consistency Is Important

The inside cover pages and the back cover can cause some design problems. The key to designing these pages is consistency. Select a graphic pattern for these pages and stick to it. If the magazine runs advertising on the inside front cover and on both back-cover pages, the problem is easily solved. Usually the cover pages sell well because of their high impact. The back cover is seen almost as much as the front cover, and the inside cover pages receive high viewing as well.

But if advertising is not a part of the covers, the designer needs to adopt a cover design philosophy. *Context*, the magazine of the DuPont Corporation, uses the back cover as a contents page. It includes color photos and short blurbs for each major article along with the masthead and the post office mailing permit tag plus an area for the address sticker.

The address sticker can be a problem. It is unfortunate that many magazines have fine front-cover illustrations and designs only to have them marred with an address sticker and, in many cases, the computerized code for the checkout counter. Cover planning should include consideration of this problem. Some magazine designers solve it by using a mailing wrapper or envelope. If, however, costs or other factors rule this out, it might be worthwhile to incorporate the address area into the design, as *Context* does.

It is possible to include some regular features on the inside covers; for instance, the inside front cover is an excellent spot for the contents page for a small magazine. The better-designed magazines avoid running large

Fig. 14-5 The *Reader's Digest* uses the front cover as both a table of contents and a display area for teasers. Art is used on the outside back cover. Both techniques can be used for company or organization publications.

"CAT IN THE DELPHINIUMS"—A PASTEL BY BRITISH ARTIST ANNE ROBINSON

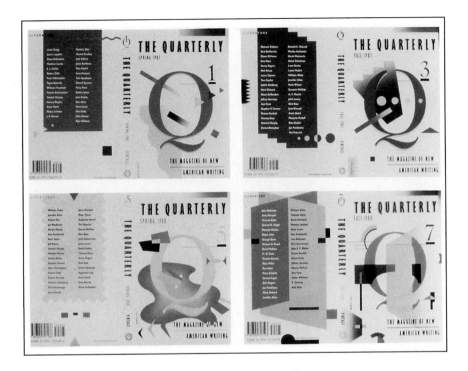

Fig. 14-6 These four covers from *The Quarterly* were designed by Lorraine Louie. Note the creative changes made in the format from issue to issue. At the same time enough consistency in design was maintained to insure continuity and recognition for the publication.

amounts of reading matter on the inside covers and solve some outside-back-cover problems by extending the front-cover illustrations to cover the back as well.

The *Reader's Digest* solves the problem nicely by using the front cover as a blurb and contents page and filling the back outside cover with a full-page illustration.

Cover Design Preparation

A good preparation for designing the cover of a magazine is to browse through an assortment of magazines and see how well the designers have used the attributes of good typography to help accomplish the job of the cover. Ask yourself the following questions to guide your evaluations.

1. *Does the cover identify the magazine and reflect its personality?* The logo should be large enough to be recognized immediately and the style should be compatible with the magazine's personality. (An artist can be commissioned to make a distinctive logo for the publication at a reasonable cost.)

   A strong, aggressive type such as a Sans Serif, Square Serif, or Roman in boldface would be good for a bold, aggressive magazine. A medium or light old style Roman is a good choice for a dignified publication. Scripts or light or medium Romans with swash first letters project an image of gracefulness. Square Serifs or strong modern Romans are good for scientific or technical publications, but they should not be too bold. There are enough type styles available to project almost any image desired.

   Cover illustrations should reflect the personality of the magazine, too. Most magazines use a combination of type and illustrations to project their identity in the cover design.

Fig. 14-7 The art-title-subhead-byline-initial letter technique is applied here. The art is bled on three margins for added impact. Note the title and initial letter are in modern Roman for unity and the subhead and byline in Sans Serif for contrast. Cartoon art has high readership and attention value. (Reprinted by permission from *Sierra*, May/June 1989. Illustration by Elwood Smith. Design by Martha Geering.)

2. *Does the cover attract the target audience and get its attention?* The type styles and illustrations used should relate to the audience and reflect its characteristics and interests. If a magazine is appealing to an aggressive target audience, using a lot of elements on the cover will project the image of a magazine that is full of vitality and action.

3. *Does the cover lure the reader inside?* The basic cover design consists of a strong logo at the top and an illustration relating to the tone of the magazine or the theme or, possibly, the lead article. Designers have found, however, that type is the most effective means of luring the reader inside. If blurbs are used, check to see if they appeal to the target audience. Check, too, to see if there is an element of timeliness. Blurbs should stress the benefits the reader will receive by turning to the articles being touted.

4. *Does the cover create identity from issue to issue?* Once a type style for the logo is adopted it should continue from issue to issue. The placement of the logo plus the date of issue and so on should also be basically the same from issue to issue. The general approach to handling blurbs should be consistent. There should be enough consistency in the cover design to create continuity and identity from one issue to the next.

5. *Does the cover contain the essential information?* Check to see that the date, price, and volume and issue numbers are present. It is surprising how often the lack of these essentials isn't noticed until after page proofs are made.

▼ Essentials of Page Design

There are two points to keep constantly in mind during the designing of a magazine. One is that the reader sees two pages at a time. The other is that all the basic principles of design—balance, proportion, unity, contrast, rhythm, harmony—should come into play.

The A-T-S-I Approach

With this in mind, we can plan the actual design of the pages in the magazine. A good way to get things started is the art-title-subhead and/or byline, initial-letter approach, or the *A-T-S-I formula*. It is a simple and safe way to make layouts attractive and functional. Once this basic concept of magazine page arrangement is mastered, the designer can experiment with more daring use of white space and placement of elements.

In building a page using the A-T-S-I approach, a large attention-getting photograph or other art is used to attract the reader. This is followed by a well-conceived title line. The subhead or byline, or both, are designed to move the reader toward the start of the article. Then the initial letter signals the beginning of the article and serves as a bridge to the reading matter from the other elements in the design.

The Axis Approach

Another method of arranging the elements on a magazine page in an orderly manner is called the *axis approach*. Here the title, subhead, and byline, if used, follow one of the basic rules of good magazine design—*line up the elements*. But here they are lined up on an axis. The axis usually is one of the between-columns alleys.

In placing elements on a magazine page there are a number of points to keep in mind:

1. *First and most important, square up the elements.* This means to line things up and keep things even. Square up the elements where the eye tends to square them. For example, the top of a small illustration above another should line up with the top edge of a large illustration nearby. Type should be lined up along the margins of the type area, or if the lines are indented they should be lined up on the designated indentation. Cutlines for art that is bled to the outer edge of the page should be lined up with the type form and should not be placed in the white area, or margins, of the page.
2. *Distribute the elements throughout the layout.* Elements should not be bunched up all in one place. If the art, title, blurbs, and byline are placed at the start of the story, the page can be thrown out of balance. In addition, it will leave the page with columns of dull, gray type. Place the elements around the page to create better balance and to make the page more interesting.
3. *Keep the elements from fighting each other.* Illustrations that are next to each other but are unrelated will fight, especially if one is not large enough to dominate. Articles in side-by-side single columns will fight

each other for the reader's attention. Elements should be placed so they will harmonize and create a unified whole rather than causing dissonance and disunity.

4. *Be consistent in making layouts.* Those just starting in magazine graphic work sometimes attempt to gain variety and interest by changing the types and styles of subheads and initial letters within one article. They end up with the layout counterpart to the Victorian house cluttered with gingerbread. Consistent use of a carefully chosen style on such items will help layouts achieve simplicity, harmony, and attractiveness.

One school of typographical thought contends that all heads should be in the same type family to preserve consistency and harmony. Many magazines follow this philosophy and use the same types and basic arrangements of elements for all articles throughout every issue.

Other designers say that type should be used to establish an individual mood for each article. Begin by studying the tone and feel of an article, as that will help in selecting and arranging graphic elements to ensure attractiveness and compatibility of graphics and content.

5. *Avoid monotony.* Although the layout should be consistent, an effort should be made to make layouts fresh and unusual. Ask these design questions constantly: Would this article work better with a wider column? Could I use a larger type size for this head? Would a rule placed

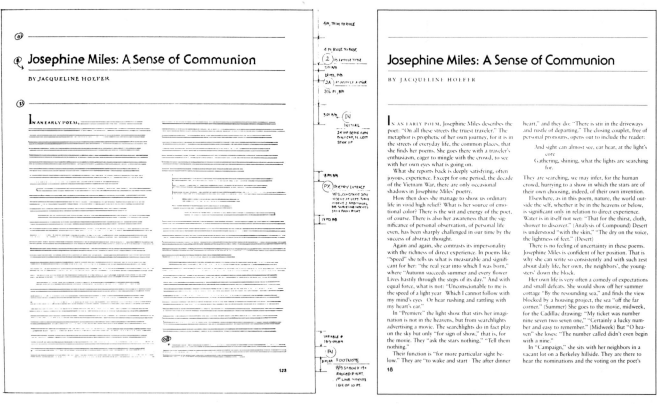

Fig. 14-8 A layout prepared by a professional designer. Note the precise placement of elements and instructions. Designers are architects of the printed word and their "blueprints" are devised with great attention to detail.

Fig. 14-9 The page as it appeared in print after being composed in accordance with the instructions on the layout.

By Robert Wernick

The Greatest Show on Earth didn't compare to home

John Ringling ran the world's most famous circus, and when he built a house and museum, he made them into a three-ring extravaganza

A bronze version of Michelangelo's *David* commands ordered parterres and colonnades of the sculpture garden courtyard, 350 feet long, of The John and Mable Ringling Museum of Art.

John Ringling, the Circus King—some six-foot-four in his silk stockings, 270 pounds, an estimated $200 million in capital assets—was a formidable figure to encounter at any time; seldom more so than on the day in the early 1920s when he summoned an architect to his office and announced that Mable wanted them to build a house.

It was to be a house worthy of a man who was the youngest, and last survivor, of the five Ringling brothers who had built one of the greatest circuses of their time, which they called, with pardonable pride, the "Greatest Show on Earth." The house would be called Ca' d'Zan, Venetian dialect for the "house of John." It would rise from the shores of Florida's Sarasota Bay, beginning with a grand marble staircase with a large boat dock where Mable could board her gondola, and rise to a tower that would dominate the palm-flecked coast. It would have 30 lofty and luxurious rooms—banquet room, ballroom, game room—and its various facades were to contain features of two of the build-

Photographs by Marvin E. Newman

63

Fig. 14-10 This is a well-organized page in which the title can be read at a glance. Notice how the various elements are lined up and that a consistent ragged right type setting is carried through in the title, subhead, and caption lines. (Courtesy *Smithsonian*)

here be effective? How would this look with an oversized initial letter? Is this page too gray, and do I need to use a pulled quote in single or double column width or some other device to give it punch?

6. *Don't overdo bleeds.* Bleeds are a powerful tool for the magazine designer and there is a temptation to overdo them. Don't. And don't bleed art in a helter-skelter manner. Too many bleeds or too great a variety of bleeds creates a haphazard look that destroys simplicity and beauty in a layout.

7. *Avoid placing large pictures on top of small ones.* When small pictures appear immediately under large ones, the smaller ones look crushed. Normal placement of a group of pictures should build them up like toy blocks so they don't seem to topple over.

8. *Keep flashy arrows and fancy artistic devices to a minimum or, better yet, avoid them altogether.* These devices now look old-fashioned. Today the rule is for clean and simple designs that look neat and modern. Studies have shown that readers do not like pictures with rules around them (but in some cases rules can help frame a picture and define the area it occupies), type printed over art (surprinting), oval-shaped art, and similar effects.

9. *Keep the pages dynamic.* Today the rule is simple, uncluttered, dynamic designs. In printed communication the message must be alive to be effective.

Traditionally, printed pages have had a strong vertical thrust. Pages that consist basically of reading matter usually are designed with the vertical column dictating where the copy is placed. As a result, most magazine pages consist of two or three columns of vertical, gray reading matter separated by a thin strip of white space.

This vertical flow is monotonous and if pages are to look alive this flow should be broken up by initial letters, subheads, illustrations, and/or white space.

Making Pages Interesting

Here are a few points to keep in mind when arranging elements on magazine pages to change direction and make the pages more interesting:

- *A vertical plus a vertical can equal monotony:* A page of all single-column copy and single-column illustrations will tend to reinforce the vertical thrust and continue the monotony. Change the direction somewhat to make the page more interesting. This means adding a horizontal thrust. White space can be used to do this. Titles can be spread over several columns. A strong horizontal picture can be used. All will help create a visual change of direction.

- *A change of form can make a page interesting:* Generally we think in terms of squares and rectangles. A change of form from the single vertical or horizontal unit can help create interest. For example, an L-shaped arrangement of illustrations provides a change of direction and adds variety to the page.

- *There is movement in the content of art:* Quite often the center of interest in an illustration can help create a change of pace. The direction the subjects are looking or the direction a moving object is going can create motion on a page. But we must be sure the direction is leading where we want the reader to go. Such illustrations must be placed in a position that ensures the direction is *into* the article, or the page, rather than away from it.

- *White space is a good directional tool:* Breaking out of the standard margination, with more white space at the top or side of the page, with uneven column endings, can help break up vertical thrust. Color used in charts and graphs or in multicolumn heads can draw attention away from the vertical movement of the columns.

Although we have emphasized some rather fixed typographic rules for making magazine designs, the challenge always is to strive for the fresh and unusual. But the fresh and unusual should still be in balance and help to create a harmony among the elements as well as presenting a unified whole.

Exploring the North 40

Hunting for fossils or mushrooms, caring for the animals, and downing hearty home-cooked meals are a few of the pleasures of a Nova Scotia farm.

● ● ● ● ●

Many early settlers of Haywood County rumbled down into the region via the "Great Philadelphia Wagon Road."

at day's end they pile into their truck, drive to the levee, boat back to Bloody Bayou and jump in. "We just strip

"Out here I just sits back and lets it go."
— Alcide Verret

down and leave our clothes on the dock," said Gwen. "They smell awful."

"I need to be able to turn around at the end of the day and see the results of my labor — to see a hillside that I yarded."
— AL REINHART, 32, LUMBERJACK

▼ ▼ ▼

To be stuck in the Sahara with a dead car is never a pleasant prospect.

▲ ▲ ▲

nstrator in the Square terms of governnot even lections. iousness arty met nen with s of the ed in the s a bitter people's" people's" nanmen or China Deng and eful and lchemizents into

Prescott **Bush recently completed a multimillion-dollar deal for construction of a hotel and golf course in Shanghai.**

United States made no comr by Chinese security police for Bush to which he had b before the president left Bei more than express "regret.' ing to the by tens of ing. In spit dom fight the Soviet resolutely democrac to wonde hundreds fully marc mocracy t God forbi were firec Sandinista

toward the birds. It is mid-May on Grimsey, an arctic isle off Iceland's north coast, time for the annual harvest of seabird eggs.

OVER THE CLIFFS FOR EGGS

"In the old days, mainlanders called our island the 'country's breadbasket,'" says Bjarni Magnusson, 51, as he takes a break between rope descents. "Even

Fig. 14-11 Cut-in heads can break up the gray of the printed page and add life and brightness. The variety of possibilities is endless but the rules of good typography should be applied in selecting types, spacing, and lines and in placing rules.

Creative Communication

Know when to stop.

The creative process can be endless. On the other hand, a puzzle has an answer. A mathematical problem is either right or wrong and that is the end of it.

But the creative design process can go on and on. There is no one final, absolutely correct answer. In creative design one of the skills worth developing is knowing when to stop.

Knowing when to stop involves knowing the capabilities of the equipment and the materials available, the characteristics of the target audience, and the attributes and limitations of the medium used to transmit the completed layout in its final visual form.

Allen Hurlburt, magazine art director, explains adherence to the basic principles of good design combined with creativity in describing his philosophy on balance:

> The balance in modern layout is more like that of a tightrope walker and her parasol than that of a seesaw or measuring scales. A tightrope walker in continuous and perfect balance is not much more interesting than someone walking on a concrete sidewalk.
>
> It is only through threatened imbalance, tension, and movement that the performance achieves interest and excitement. For the modern magazine designer and the tightrope walker, balance is a matter of feeling rather than formula.[1]

▼ Useful Design Elements

A number of design elements lend themselves especially to magazine design. These elements can also be used for brochures and similar printed communications.

Initial Letters

Initial letters are in. We seem to go through periods when they are popular and other times when they aren't. One of the advantages of modern computer typesetting is the ease with which devices such as initial letters can be used. In the days of hot type composition, it was expensive to use initial letters. Slugs had to be sawed and fitted.

Initial letters can be effective typographic devices. They can aid the reader in bridging the gap between art, title, and article. They can have a unifying influence. They can open space on the page, breaking the monotony of columns of type. They also can help provide balance if placed properly.

In addition, initial letters can be used in the same type style as the title of the article to give unity and consistency throughout. (A different type style tends to destroy harmony. In some cases, though, magazines have used different type initials effectively by selecting the same typeface for the initials as is used in the magazine logo. But care must be taken to ensure type harmony among head and subhead, initial letter, and body type.)

Care is also needed in placing initial letters. An initial letter should never be placed at the top of a column except at the beginning of an article. It should be placed far enough from the top of a column so it will not confuse the reader. Nor should it be placed too close to the bottom of a column.

The distances between initial letters should be varied throughout an article. They should be placed in unequal spots in the copy and they should never line up horizontally in two adjacent columns.

Adopt a standard method of handling initial letters. There are a number of possibilities. Two common types are rising initial letters and dropped initial letters. Rising initial letters extend above the first line of the body copy. The baseline of rising initials should line up with the baseline of

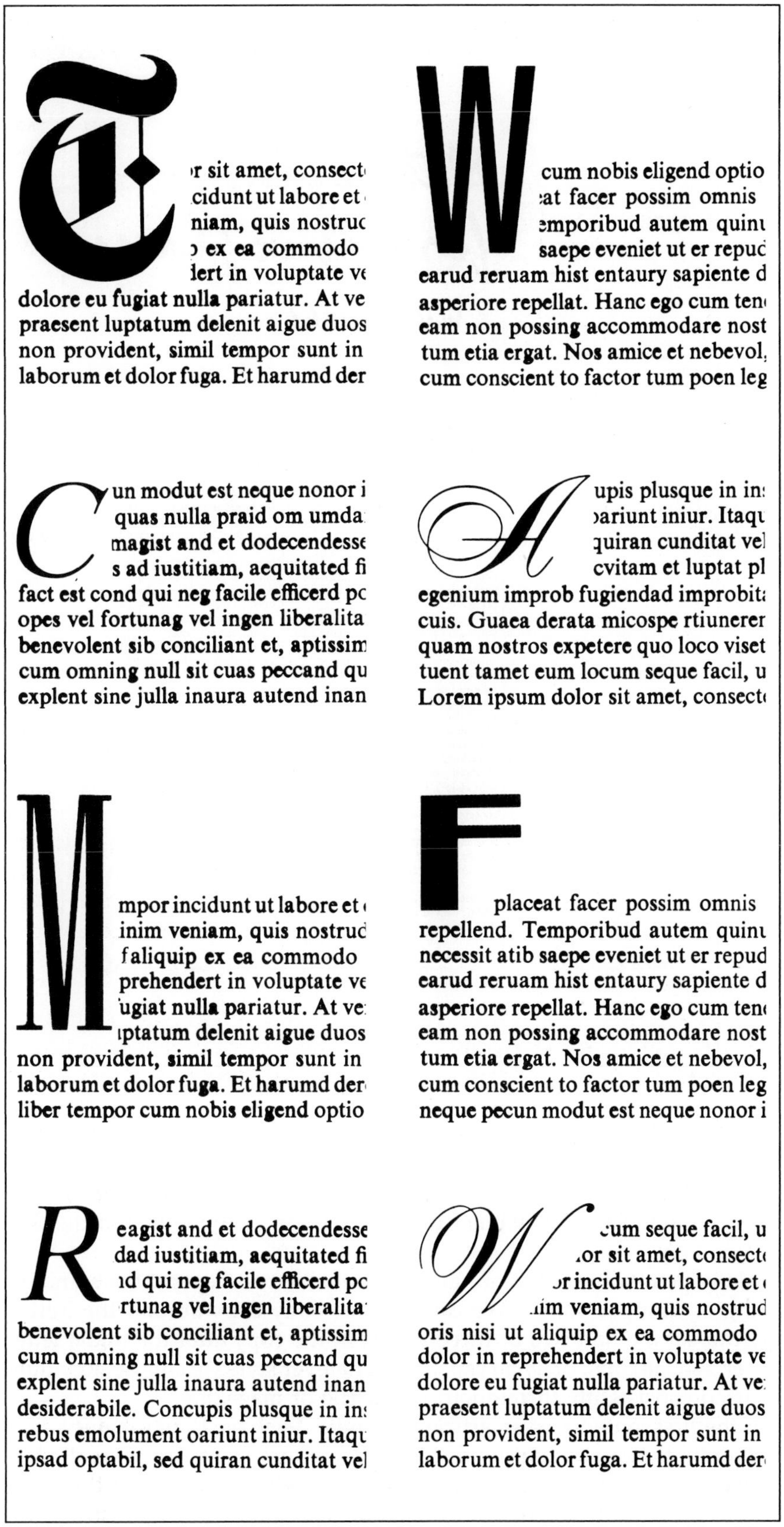

Tr sit amet, consect cidunt ut labore et niam, quis nostruc) ex ea commodo lert in voluptate ve dolore eu fugiat nulla pariatur. At ve praesent luptatum delenit aigue duos non provident, simil tempor sunt in laborum et dolor fuga. Et harumd der

Wcum nobis eligend optio at facer possim omnis emporibud autem quint saepe eveniet ut er repud earud reruam hist entaury sapiente d asperiore repellat. Hanc ego cum ten eam non possing accommodare nost tum etia ergat. Nos amice et nebevol, cum conscient to factor tum poen leg

Cun modut est neque nonor i quas nulla praid om umda magist and et dodecendesse s ad iustitiam, aequitated fi fact est cond qui neg facile efficerd pc opes vel fortunag vel ingen liberalita benevolent sib conciliant et, aptissim cum omning null sit cuas peccand qu explent sine julla inaura autend inan

Aupis plusque in ins oariunt iniur. Itaqu quiran cunditat vel cvitam et luptat pl egenium improb fugiendad improbit cuis. Guaea derata micospe rtiunerer quam nostros expetere quo loco viset tuent tamet eum locum seque facil, u Lorem ipsum dolor sit amet, consect

Mmpor incidunt ut labore et inim veniam, quis nostruc f aliquip ex ea commodo prehendert in voluptate ve ugiat nulla pariatur. At ve iptatum delenit aigue duos non provident, simil tempor sunt in laborum et dolor fuga. Et harumd der liber tempor cum nobis eligend optio

Fplaceat facer possim omnis repellend. Temporibud autem quint necessit atib saepe eveniet ut er repud earud reruam hist entaury sapiente d asperiore repellat. Hanc ego cum ten eam non possing accommodare nost tum etia ergat. Nos amice et nebevol, cum conscient to factor tum poen leg neque pecun modut est neque nonor i

Reagist and et dodecendesse dad iustitiam, aequitated fi nd qui neg facile efficerd pc rtunag vel ingen liberalita benevolent sib conciliant et, aptissim cum omning null sit cuas peccand qu explent sine julla inaura autend inan desiderabile. Concupis plusque in ins rebus emoluent oariunt iniur. Itaqu ipsad optabil, sed quiran cunditat vel

Wcum seque facil, u or sit amet, consect or incidunt ut labore et nim veniam, quis nostruc oris nisi ut aliquip ex ea commodo dolor in reprehendert in voluptate ve dolore eu fugiat nulla pariatur. At ve praesent luptatum delenit aigue duos non provident, simil tempor sunt in laborum et dolor fuga. Et harumd der

Fig. 14-12 Initial letters are receiving increased attention because of the ease of using them now compared with the days of hot metal type when sawing and adjusting materials was necessary. Initial letters add variety to a page and break up the gray reading matter, but they must be used carefully.

Fig. 14-13 Good unity, harmony, and balance are illustrated by this page from *Tierra Grande*. The type style renders the feel of native plants. The initial letter in the same typeface indicates the start of the article. The art is distributed on the page for good balance and harmony. What other typeface might have served equally well for the title of this article? (Courtesy Texas Real Estate Research Center, Texas A & M)

Native plants for native places
by William C. Welch

Anyone who has seen the dogwoods bloom in East Texas or gazed upward at the towering trunk of an old bald cypress tree growing along a Central Texas river can sense the tremendous possibilities for landscaping offered by Texas' native plants. Although still a long way from maximizing the potential of native plants, Texans in recent years have seen an increase in the preserving and planting of these native species and in the numbers of well planned communities where native plants have been retained during construction.

Developers and property owners are coming to realize that a setting of native pines, dogwood and wild azaleas has more visual appeal than the asphalt-paved parking lot of a shopping development. There is a growing awareness that carefully selected and positioned native species can beautify a landscape and help relate it to its natural environment.

Certain advantages accrue to landscapes created with these native plants rather than with their imported relatives. Often the native plants are more resistant to drought, insects and disease. In addition individual characteristics make native plants ideally suited for a particular terrain, be it swampy, rocky,

wild olive

bald cypress

arid or otherwise. A native plant can be found which will thrive where imported plants will die.

Unusual challenges for landscaping with native plants can be found in some of our more densely populated areas. Subdivisions around Austin and San Antonio, for instance, often are located in hilly areas to take advantage of views. The ecology of these sites is very delicate, with a thin soil layer over rock supporting a few small trees and shrubs, such as Texas persimmon (*Diospyros texana*), live oak (*Quercus virginiana*), agarita (*Mahonia trifoliolata*) and sumac (*Rhus lanceolata*).

If the property owner has a stereotyped concept that landscaping should consist of planting various broadleafed evergreen trees and flowering shrubs, he may clear the site of all the "brush." Then, after spending considerable time and money trying to provide topsoil, adequate irrigation and reduced soil ph, as well as dealing with insect and disease problems, the property owner wonders why maintaining the landscape is such a big and expensive job.

With some careful thinning, pruning, transplanting and a few well placed

groundcover areas, the property owner may have had the makings of an attractive and functional landscape development had she or he not been so hasty in removing the existing native plants on the site. Mature landscape material is expensive to buy and difficult to find. Builders, developers and homeowners need to evaluate very carefully what is growing naturally on the site before destroying it.

The diversity of temperature, topography, soil and rainfall in Texas results in a tremendous variety of plant life (more than 4,500 species). Every part of the state has its own unique character derived from plants as well as topography, soils and rock formations. A few of our distinctive plants useful for landscaping include the following:

Texas mountain laurel (*Sophora secundiflora*) — This evergreen large

possum haw holly

the first line of the body copy. A dropped initial is cut into the body copy with the top of the letter lined up with the top of the letters in the first line of copy.

Many possibilities exist of handling initial letters beyond the standard rising or dropped initial. We can contour body copy around the initial. We can set the first two or three words in the copy after the initial letter lowercase, all caps, or small caps. But whatever style we choose, it should be used consistently throughout the article or magazine.

If the first sentence of an article is a quote, we have to figure out a way to handle the quote mark with the initial letter. There are several alternatives. We can use quote marks from the body type, or from the same type as the initial letter, or leave them out. The latter alternative, however, can cause confusion for the reader.

Always be aware when using initial letters that, as with many typographic devices, poorly planned ones look a lot worse than none at all.

Titles

The trend is toward short titles. The reader should be able to read the title as a unit and not have to read each word alone. The space between words in titles should never be wider than the lowercase *x* of the type being used. And a title should never be extended to fill an area if it means putting so much space between words that the unity of the line is destroyed.

Letterspacing, the placing of more space than normal between each letter in a word, should be used very carefully if at all. If words in a title are letterspaced, the space between the letters should be minuscule. And the

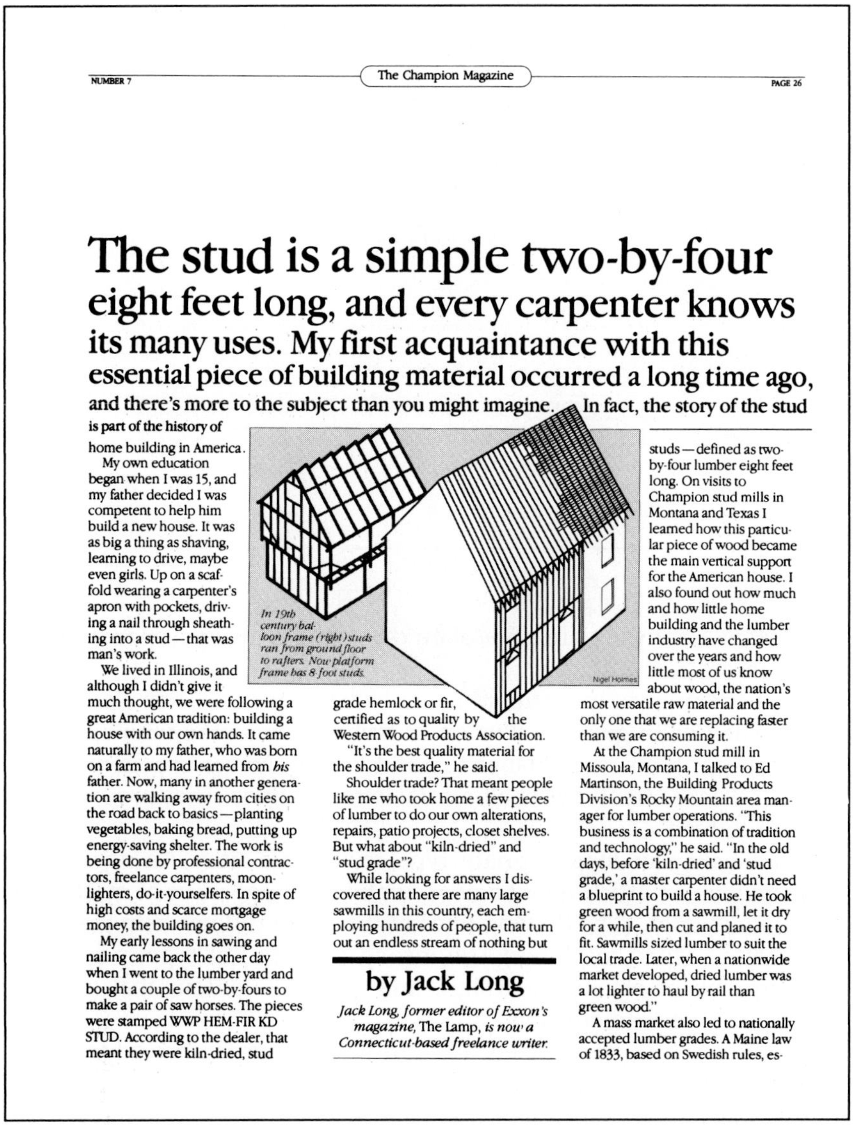

NUMBER 7 — The Champion Magazine — PAGE 26

The stud is a simple two-by-four eight feet long, and every carpenter knows its many uses. My first acquaintance with this essential piece of building material occurred a long time ago, and there's more to the subject than you might imagine. In fact, the story of the stud is part of the history of home building in America.

My own education began when I was 15, and my father decided I was competent to help him build a new house. It was as big a thing as shaving, learning to drive, maybe even girls. Up on a scaffold wearing a carpenter's apron with pockets, driving a nail through sheathing into a stud—that was man's work.

We lived in Illinois, and although I didn't give it much thought, we were following a great American tradition: building a house with our own hands. It came naturally to my father, who was born on a farm and had learned from *his* father. Now, many in another generation are walking away from cities on the road back to basics—planting vegetables, baking bread, putting up energy-saving shelter. The work is being done by professional contractors, freelance carpenters, moonlighters, do-it-yourselfers. In spite of high costs and scarce mortgage money, the building goes on.

My early lessons in sawing and nailing came back the other day when I went to the lumber yard and bought a couple of two-by-fours to make a pair of saw horses. The pieces were stamped WWP HEM-FIR KD STUD. According to the dealer, that meant they were kiln-dried, stud grade hemlock or fir, certified as to quality by the Western Wood Products Association.

"It's the best quality material for the shoulder trade," he said.

Shoulder trade? That meant people like me who took home a few pieces of lumber to do our own alterations, repairs, patio projects, closet shelves. But what about "kiln-dried" and "stud grade"?

While looking for answers I discovered that there are many large sawmills in this country, each employing hundreds of people, that turn out an endless stream of nothing but studs—defined as two-by-four lumber eight feet long. On visits to Champion stud mills in Montana and Texas I learned how this particular piece of wood became the main vertical support for the American house. I also found out how much and how little home building and the lumber industry have changed over the years and how little most of us know about wood, the nation's most versatile raw material and the only one that we are replacing faster than we are consuming it.

At the Champion stud mill in Missoula, Montana, I talked to Ed Martinson, the Building Products Division's Rocky Mountain area manager for lumber operations. "This business is a combination of tradition and technology," he said. "In the old days, before 'kiln-dried' and 'stud grade,' a master carpenter didn't need a blueprint to build a house. He took green wood from a sawmill, let it dry for a while, then cut and planed it to fit. Sawmills sized lumber to suit the local trade. Later, when a nationwide market developed, dried lumber was a lot lighter to haul by rail than green wood."

A mass market also led to nationally accepted lumber grades. A Maine law of 1833, based on Swedish rules, es-

In 19th century balloon frame (right) studs ran from ground floor to rafters. Now platform frame has 8-foot studs.

Nigel Holmes

by Jack Long

Jack Long, former editor of Exxon's magazine, The Lamp, *is now a Connecticut-based freelance writer.*

Fig. 14-14 The *Champion Magazine* uses the same typeface for the title (The stud . . .) or intro line, which is printed in red, and the lead-in sentences as is used for the reading matter. (Courtesy Champion International Corporation)

Fig. 14-15 Several unobtrusive design techniques give this magazine page interest and life. Note that both the subhead and title are centered for consistency. The large initial letter in the same typeface as the title adds impact. The page is given life by the placement of the art on an angle. Quite often not enough white space is allowed for type wrapped around art. Here there is adequate white space and it is consistent. Details like first words in paragraphs in boldface can pep up an otherwise uncomplicated page. (Courtesy *J. D. Journal*, Deere & Company)

space between any lines in a title should be tight. Too much space between lines can destroy unity.

Script or Cursive types should never be letterspaced. They were meant to be joined or to give the illusion of being joined—the whole point of these types is their resemblance to handwriting. Never use Scripts, Cursives, or Black Letter types in all-capital letters in titles. They are extremely difficult to read.

Avoid using an inappropriate type style for a title. An agricultural magazine ran an article about a new combine, a large piece of machinery, in Coronet Script. (The flowing lines of the script did not help create the image of a rugged machine.) The incongruous use of type can destroy the harmony of elements, design, and editorial content that makes a completed layout so effective.

Also be careful with stylized titles, in which an artist has added flourishes or sketches to the letters. Sometimes this can be effective because it is unusual; other times it is just plain amateurish, and it can destroy the proper horizontal direction of the line as well.

Bleeds

Bleeding art, that is, extending it beyond the normal type area into the margin to the edge of the page, can add variety to the page and impact to the art. It can help create a change of direction to break up monotony, and it can create the impression that the illustration goes on and on.

Bleeds are good design devices, but they should be used with thought. If bleeding will give greater impact to the layout, use it. If the photo content is such that an image of a vast expanse can be enhanced by bleeding, do it. If a change of direction or a breakout from the monotonous can be accomplished by bleeding art, bleed it. But give it some thought first. Also, be sure to check with the production staff or the printer about any possible mechanical limitations for handling bleeds.

Captions

Editors and art directors agree that, in theory at least, a good piece of art should not need a caption. It should tell the story all by itself. Well, that would be great if the art could talk. But often art needs assistance—a caption—in telling the story. A good caption should be as brief as possible and it should add information, not simply restate what is obvious in the art.

How should captions be handled?

One rule of thumb in magazine design is always include a caption unless there is a compelling reason to leave it out. Have you ever been frustrated by a lovely scene used in a magazine to set a mood or present a pleasing

Fig. 14-16 A well-designed arrangement of pictures. The small pictures are not crushed by the larger ones, the space between the art is equal all the way around, and the captions are placed in an orderly and convenient position. All elements are lined up. (Courtesy *J D Journal*, Deere & Company)

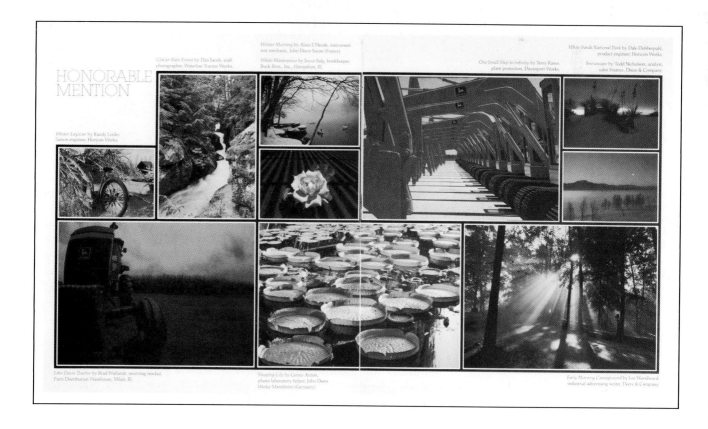

MISSION TO
MARS

By MICHAEL COLLINS TRUSTEE, NATIONAL GEOGRAPHIC SOCIETY

Photographs by ROGER H. RESSMEYER STARLIGHT

Illustrations by
PIERRE MION and ROY ANDERSEN

THE BLUE-AND-WHITE EARTH is gone now, its remnant a mere pinpoint of light in their window, and dimmer than Venus. Their destination, out the opposite porthole, is a plump red beacon, larger and more inviting as the weeks go by.

The crew enjoy discussing these impressions with one another and with Mission Control, although with a ten-minute time delay conversation with Earth is becoming increasingly stilted. Everyone is proud of the fact that—at the halfway point—they are all strong and well, physically and mentally. Even their plants seem to be thriving.

The decision to launch them had been long and tortuous, its roots as old as the telescope. For centuries astronomers had marveled at the fuzzy red ball in their eyepieces and lavished on it their most fanciful dreams.

Intelligent beings lived there, they theorized, digging long straight canals to bring water to cities hidden in desert oases. Later, when the likelihood of a lifeless planet crushed these hopes, there remained a special fascination with our planetary neighbor, in many ways the closest thing to an Earth twin in this solar system.

Unlike the other planets, such as the gaseous giants Jupiter and Saturn, Mars seems friendly, accessible, even habitable. As author James Michener put it: "Mars has played a special role in our lives, because of the literary and philosophical speculations that have centered upon it. I have always known Mars."

I grew up not only knowing the place but also wanting badly to

Blazing the trail for a piloted mission to Mars, a Soviet probe will reach 17-mile-long Phobos (left), larger of the planet's two moons, in the spring of 1989. Former astronaut Michael Collins describes a possible scenario for a U. S.-Soviet manned voyage in the year 2004.

PAINTING BY PIERRE MION

733

Fig. 14-17 This two-page spread from *National Geographic* illustrates several features of a carefully designed magazine layout. The overall approach is art-title-byline-initial letter. The high-impact art is bled on three sides to expand it and give the feel of the infinity of the universe. The art crosses the gutter to help unify the two pages. Unity is also achieved by using the same Roman for all display and text type and an italic version for the caption. Elements are lined up along an axis formed by the gutter between the art and the type. (Copyright 1989 National Geographic Society)

visual experience and then searched through every page to try to find out where the scene was located? Never frustrate your readers.

There are two basic approaches to placing captions. One is to place them adjacent to the art they describe. The other is to cluster captions to refer to several pictures in a spread or an article. In either approach, the designer should make it easy for the reader to match caption and art.

Avoid all-capital-letter captions. Settle on a caption width that is never wider than the width of the art. A caption that is narrower than the art lets air into the page, while captions wider than the art look awkward. Select a type style that harmonizes with the body type. Often an italic of the body type works well. Avoid caption widths that violate the optimum line length formula (one and a half times the lowercase alphabet). If captions are wider than the formula, divide them into two or more columns under the illustration and leave at least a pica of white space between the columns.

Crossing the Gutter

When a layout extends through the *gutter*—the margin between two facing pages—onto the adjacent page, be very careful in crossing that gutter. In fact, avoid crossing the gutter except in the center fold of a saddle-wire-bound publication. On a two-page spread, crossing the gutter may be needed for unity and maximum impact. But care should be taken in running a photo across the gutter because the fold may detract from its effectiveness. If a photo that contains people is used, the center of interest may be destroyed or someone's face creased.

Also be very careful in crossing the gutter with type. In fact, this should be avoided if there is a chance problems will arise. A head can be placed across the gutter if it is planned carefully so that the crease comes between the words. Care must be taken, too, to see that the pages line up in printing and the head is in a straight line from one page to the other.

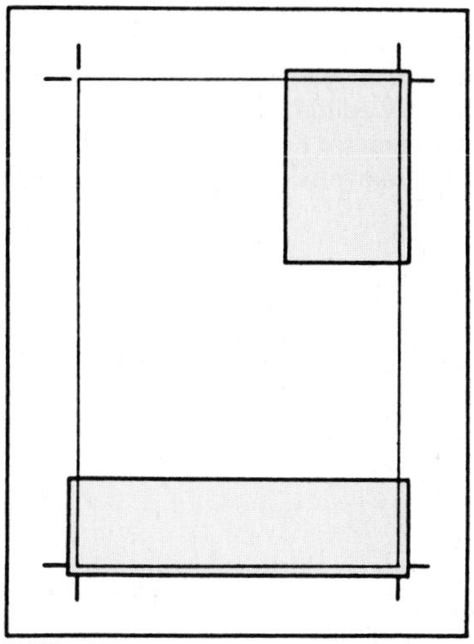

⅛″ Bleed Areas

Fig. 14-18 Bleeding art to the extremities of the page can give it greater impact and added weight as a design element. In planning bleeds the layout should indicate the bleed extending about 1/8-inch beyond the margin for trimming after the page is printed.

Placing Advertisements

There are a number of layout patterns for placing advertisements on the page. One plan is especially suitable for magazine design. This is to fill full columns with advertisements and leave full columns open for editorial matter.

Some magazines, especially the shelter publications (as the "homemaker" magazines are called), use an "island" makeup plan for inside pages. Advertisements are isolated, usually in the center of the page, and surrounded with reading matter.

In planning the graphics for your magazine, remember that in today's world people are constantly being exposed to improved graphics and new graphic techniques. A publication must be concise, complete, and attractive. The trend is toward simpler layouts with bigger and fewer pictures, shorter but bolder heads and titles, larger and bolder initial letters. More tightly edited stories, concisely written, can provide the space needed for better display.

▼ Effective Design Checklist

- Square up elements. Square items with the margins. Keep even (that is, line up) the top edges of illustrations that are similarly placed. Square up elements where the eye tends to square them. For example, the top of a small illustration above another should line up with the top edge of a nearby large illustration.
- Distribute elements throughout the layout. Bunching illustrations, title, subhead, and byline at the start of the story creates imbalance and can leave columns of gray type.
- Keep elements from fighting each other. An illustration or a head placed alongside another can fight it. Keep peace in the family by isolating unrelated heads and illustrations.

Fig. 14-19 The "magazine" plan for placement of advertisements. Full columns of advertisements leave full columns for editorial matter. This is an arrangement for two-column advertisements. (Courtesy *Smithsonian*)

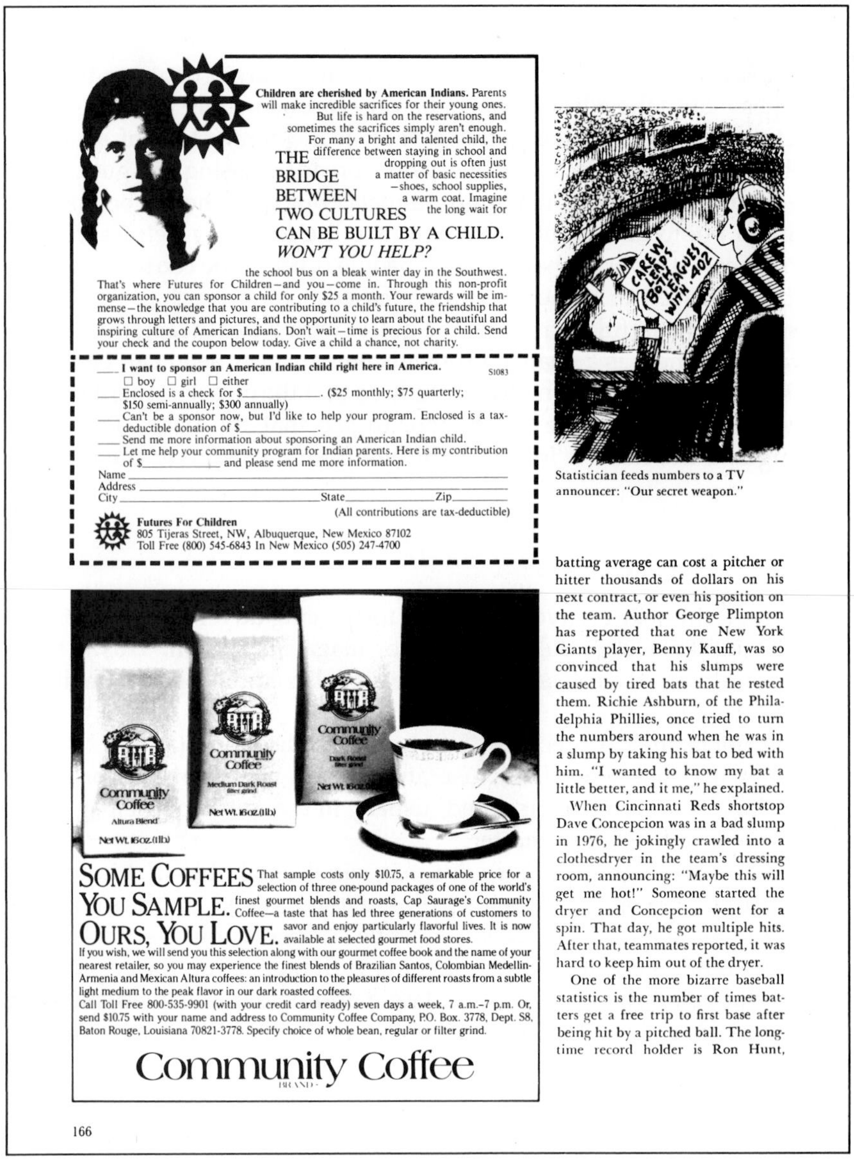

Children are cherished by American Indians. Parents will make incredible sacrifices for their young ones.

But life is hard on the reservations, and sometimes the sacrifices simply aren't enough.

THE BRIDGE BETWEEN TWO CULTURES CAN BE BUILT BY A CHILD. *WON'T YOU HELP?*

For many a bright and talented child, the difference between staying in school and dropping out is often just a matter of basic necessities—shoes, school supplies, a warm coat. Imagine the long wait for the school bus on a bleak winter day in the Southwest.

That's where Futures for Children—and you—come in. Through this non-profit organization, you can sponsor a child for only $25 a month. Your rewards will be immense—the knowledge that you are contributing to a child's future, the friendship that grows through letters and pictures, and the opportunity to learn about the beautiful and inspiring culture of American Indians. Don't wait—time is precious for a child. Send your check and the coupon below today. Give a child a chance, not charity.

_____ **I want to sponsor an American Indian child right here in America.** S1083
  ☐ boy ☐ girl ☐ either
_____ Enclosed is a check for $_____. ($25 monthly; $75 quarterly; $150 semi-annually; $300 annually)
_____ Can't be a sponsor now, but I'd like to help your program. Enclosed is a tax-deductible donation of $_____.
_____ Send me more information about sponsoring an American Indian child.
_____ Let me help your community program for Indian parents. Here is my contribution of $_____ and please send me more information.
Name _____
Address _____
City _____ State _____ Zip _____
(All contributions are tax-deductible)

Futures For Children
805 Tijeras Street, NW, Albuquerque, New Mexico 87102
Toll Free (800) 545-6843 In New Mexico (505) 247-4700

Statistician feeds numbers to a TV announcer: "Our secret weapon."

batting average can cost a pitcher or hitter thousands of dollars on his next contract, or even his position on the team. Author George Plimpton has reported that one New York Giants player, Benny Kauff, was so convinced that his slumps were caused by tired bats that he rested them. Richie Ashburn, of the Philadelphia Phillies, once tried to turn the numbers around when he was in a slump by taking his bat to bed with him. "I wanted to know my bat a little better, and it me," he explained.

When Cincinnati Reds shortstop Dave Concepcion was in a bad slump in 1976, he jokingly crawled into a clothesdryer in the team's dressing room, announcing: "Maybe this will get me hot!" Someone started the dryer and Concepcion went for a spin. That day, he got multiple hits. After that, teammates reported, it was hard to keep him out of the dryer.

One of the more bizarre baseball statistics is the number of times batters get a free trip to first base after being hit by a pitched ball. The long-time record holder is Ron Hunt,

SOME COFFEES YOU SAMPLE. OURS, YOU LOVE.

That sample costs only $10.75, a remarkable price for a selection of three one-pound packages of one of the world's finest gourmet blends and roasts, Cap Saurage's Community Coffee—a taste that has led three generations of customers to savor and enjoy particularly flavorful lives. It is now available at selected gourmet food stores.

If you wish, we will send you this selection along with our gourmet coffee book and the name of your nearest retailer, so you may experience the finest blends of Brazilian Santos, Colombian Medellin-Armenia and Mexican Altura coffees: an introduction to the pleasures of different roasts from a subtle light medium to the peak flavor in our dark roasted coffees.

Call Toll Free 800-535-9901 (with your credit card ready) seven days a week, 7 a.m.–7 p.m. Or, send $10.75 with your name and address to Community Coffee Company, P.O. Box 3778, Dept. S8, Baton Rouge, Louisiana 70821-3778. Specify choice of whole bean, regular or filter grind.

Community Coffee
BRAND

166

- Be consistent in making layouts. Select a pattern for handling heads, captions, and all other graphic elements and stick to it for continuity, identity, and consistency.
- Seek the fresh and unusual. Seek new ways of doing things, study the work of others, and don't be afraid to experiment. But keep the new and unusual within the bounds of sound principles of design.
- Use good judgment with bleeds. There are so many ways to bleed photos that there is hardly a wrong way. However, remember that varying bleeds too much will bring a haphazard look to pages. "Mini-bleeds," or small photos that are bled, usually are not effective. Some designers say the full-page bleed is the only acceptable one.
- Don't place initial letters at the tops of columns, except at the start of the article. Adopt a consistent typographic plan for handling initial

Fig. 14-20 Arrangement for single-column advertisements in a magazine. The advertisements fill full columns and leave full columns open for editorial material. (Courtesy *Smithsonian*)

letters. Also, avoid placing subheads at the top of columns as this can confuse the reader.

- Don't place cutlines in the margins. If you design a bleed for a page do not place the cutline so it extends into the margin. Stop the cutline at the margin.
- Don't place big pictures on top of little pictures. The little pictures will look crushed. The same principle applies to advertisements. In placing ads on the page it is better to put the bigger ads at the bottom of the columns and build them up like building blocks.
- Don't use arrows, pointing fingers, and other fancy ornamentation unless there is a good reason, such as creating an atmosphere or mood. Dynamic designs are usually neat, clean, and open.

▼ Graphics in Action

1. Find a one-page magazine layout that includes the title, byline, a sub-head, art, and initial letter, etc., and design a two-page spread based on this page. Increase the size of the body type area by one-third and add additional art as you see fit. Prepare a grid for the layout on paper or light white posterboard at least 11 by 17 inches. Each page will have three 14-pica columns with 1½ picas of space between 10-inch-deep columns (you may have or can purchase prepared grids to fit these dimensions).

2. Find an article in a newspaper and design a cover for a "flat" magazine (see page 294) based on the article.

3. Find an article in a pocket-size magazine such as *Reader's Digest*, *TV Guide*, or *Ford Times* and redesign it for a flat or basic-size page.

4. Use the formula you developed for a small magazine in Chapter 13 "Graphics in Action" number 2 to develop a plan for breaking the book for the publication.

5. Make a complete dummy (a rough simulation of a completed magazine; it contains the layouts for all the pages in the order in which they will appear in print) for the small magazine you developed in Chapter 13 "Graphics in Action." The dummy should carry out the magazine's formula and format philosophy in its design.

6. Make prototype page pasteups for selected pages of the dummy created in number 5. Clip from publications the body type, titles, sub-heads, art, and all other elements that closely resemble those you specified for your layouts. (The elements do not have to make sense to give a visual idea of how the pages would look if actually produced.)

Notes

1 Allen Hurlburt, *Publication Design* (New York: Van Nostrand Reinhold, 1976), p. 28, 31.

CHAPTER FIFTEEN

The Newspaper and the Designer

Photograph: Tony Kelly

. . . immediate opening, head writing, layout, pasteup part of the job . . . copy editor wanted with flair for design . . . we are looking for an illustrator/designer who has a flair for informational graphics . . . seeking highly creative graphic artist for newly created position of editorial artist . . . must demonstrate excellent writing and editing skills and a solid background in creative layout . . . we're seeking editors who know how to lay out pages and use color . . .[1]

Fig. 15-1 The modern newspaper. The *Washington Times* is a consistent winner in the annual competitions of the Society of Newspaper Design. Its design is considered one of the very best in the United States.

The age of graphics has arrived in newsrooms across the country. More and more advertisements are appearing in the help-wanted columns of newspaper trade publications seeking journalists with graphic skills or artists with an understanding of the newspaper profession. Both are expected to have computer skills, as well.

Design has come to the fore as an important—even vital—part of producing the more than 1,700 daily and nearly 9,000 weekly newspapers in the United States and Canada. In 1979 a group of newspaper designers organized the Society of Newspaper Design with 22 members. Today its membership has grown to more than 2,200.

New opportunities are opening for young people interested in journalism and graphics. Old hands in the newsroom have decided it is time to take graphics seriously. How do they start the study of newspaper layout and typography? What is a good way to become acquainted with the world of newspaper design?

One approach to the understanding of any creative skill is to examine what has gone before and consider how that history affects the art today. Will Durant wrote in the preface of his monumental work, *The Story of Civilization,* that the study of history will help people "to see things whole, to pursue perspective, unity and understanding."

A study of the past is especially helpful in newspaper design. Many of the practices, page arrangements, design forms, and even terms used today were developed during the evolution of the physical form of newspapers. Many of the design changes being initiated today are efforts to break out of this tradition to make newspapers more attractive and more readable and to avoid repeating the mistakes of the past.

In this chapter we trace the development of newspaper design and see where it seems to be going in the computer age. In addition, we examine designing or redesigning a newspaper in the light of present trends.

▼ The Colonial Era Newspaper Format

The first newspaper designer in America was a renegade Englishman who fled his country a jump ahead of the sheriff. Only one issue of his newspaper was printed and it was promptly suppressed.

Benjamin Harris arrived in Boston sometime in 1690. The single issue of his newspaper, *Public Occurrences, Both Foreign and Domestic,* appeared on September 25, 1690. Fourteen years passed before another newspaper was produced in what is now the United States. The *Boston News-Letter* was issued by John Campbell, the postmaster. It continued publication for seventy-two years.

Both newspapers were produced in a format similar to that of the early newspapers in England. They were the first of what might be called the Colonial era of American newspaper design. They were small and made little or no effort to display the news. However, they did have some distinctive typographic characteristics that some designers today would endorse.

Public Occurrences was four pages about 7½ by 11⅜ inches. It was set in type about the size of 12 point in columns about 17 picas wide. There

Fig. 15-2 John Peter Zenger's *New-York Weekly Journal* in typical Colonial era "bookish" newspaper format.

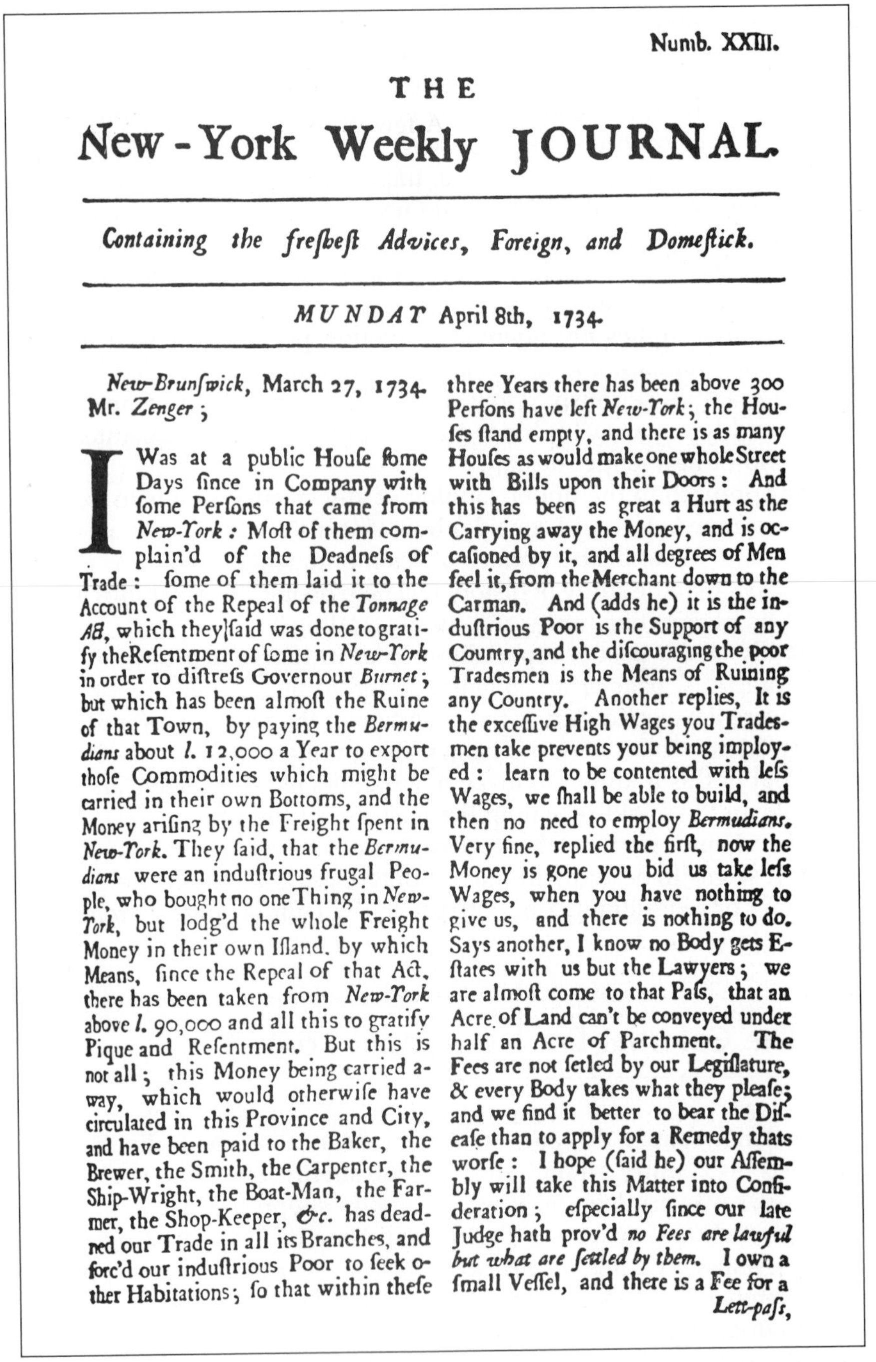

Numb. XXIII.

THE

New - York Weekly JOURNAL.

Containing the freſheſt Advices, Foreign, and Domeſtick.

MUNDAT April 8th, 1734.

New-Brunſwick, March 27, 1734. Mr. *Zenger* ;

I Was at a public Houſe ſome Days ſince in Company with ſome Perſons that came from *New-York* : Moſt of them complain'd of the Deadneſs of Trade : ſome of them laid it to the Account of the Repeal of the *Tonnage Act*, which they ſaid was done to gratify theReſentment of ſome in *New-York* in order to diſtreſs Governour *Burnet* ; but which has been almoſt the Ruine of that Town, by paying the *Bermudians* about *l.* 12,000 a Year to export thoſe Commodities which might be carried in their own Bottoms, and the Money ariſing by the Freight ſpent in *New-York*. They ſaid, that the *Bermudians* were an induſtrious frugal People, who bought no one Thing in *New-York*, but lodg'd the whole Freight Money in their own Iſland, by which Means, ſince the Repeal of that Act, there has been taken from *New-York* above *l.* 90,000 and all this to gratify Pique and Reſentment. But this is not all ; this Money being carried away, which would otherwiſe have circulated in this Province and City, and have been paid to the Baker, the Brewer, the Smith, the Carpenter, the Ship-Wright, the Boat-Man, the Farmer, the Shop-Keeper, *&c.* has deadned our Trade in all its Branches, and forc'd our induſtrious Poor to ſeek other Habitations ; ſo that within theſe

three Years there has been above 300 Perſons have left *New-York* ; the Houſes ſtand empty, and there is as many Houſes as would make one whole Street with Bills upon their Doors : And this has been as great a Hurt as the Carrying away the Money, and is occaſioned by it, and all degrees of Men feel it, from the Merchant down to the Carman. And (adds he) it is the induſtrious Poor is the Support of any Country, and the diſcouraging the poor Tradesmen is the Means of Ruining any Country. Another replies, It is the exceſſive High Wages you Tradesmen take prevents your being imployed : learn to be contented with leſs Wages, we ſhall be able to build, and then no need to employ *Bermudians*. Very fine, replied the firſt, now the Money is gone you bid us take leſs Wages, when you have nothing to give us, and there is nothing to do. Says another, I know no Body gets Eſtates with us but the Lawyers ; we are almoſt come to that Paſs, that an Acre of Land can't be conveyed under half an Acre of Parchment. The Fees are not ſetled by our Legiſlature, & every Body takes what they pleaſe; and we find it better to bear the Diſeaſe than to apply for a Remedy thats worſe : I hope (ſaid he) our Aſſembly will take this Matter into Conſideration ; eſpecially ſince our late Judge hath prov'd *no Fees are lawful but what are ſettled by them.* I own a ſmall Veſſel, and there is a Fee for a *Lett-paſs*,

were two columns to a page, and the columns were separated by white space rather than column rules. Two three-line initial letters appeared on page 1. They were the only typographic efforts to add variety to the body matter.

The *Boston News-Letter* followed a similar pattern.

The Colonial era newspapers were produced by printers rather than journalists or publishers. Many of these printers were book and general commercial printers first and newspaper producers second. They used the same typefaces for their newspaper printing and their book work. As a result, these newspapers resembled early-day books in page format. They were set in large types on wide columns, and the columns were usually separated by white space. A few printers used vertical rules between columns.

Many of the design changes in recent years have included a return to some of the characteristics of the Colonial newspaper. These include the larger body type, wider columns, and white space rather than rules to separate the columns.

During the more than 200 years that elapsed between the Colonial format and the format of today, newspapers went through some wrenching changes in appearance. The designer of newspapers—whether the newspaper is a metropolitan daily, a company employee paper, or a university or school weekly—will find it worthwhile to trace these changes and see how newspapers evolved.

▼ Traditional Newspaper Format

The traditional format dominated newspaper design for nearly a hundred years. But as newspapers proliferated, competition began to affect the business. There was a increasing effort to be first with the news, to obtain the largest circulation, to get the most advertising, to make the most money.

The large margins of the Colonial newspapers were reduced to get more news and advertisements on the pages. Body matter was set in

Fig. 15-3 The traditional format emerged in the middle 1880s and dominated newspaper design in the United States for a century. This is Joseph Pulitzer's famous *World* which was published in New York City.

smaller type so more material could be fitted on the page. In the 1800s, eye-fatiguing 6-point type for reading matter was common.

Column widths were reduced until the 13-pica column became standard. Instead of ample white space, vertical rules were placed between columns so they could be crowded more closely together. The hairline rule on a 6-point base became standard.

Increasing interest in the news, especially during the Mexican War, 1846–1848, led to greater display of titles on news stories. There had been an occasional head on stories before the 1840s. Most, however, were of one to four lines, and all were restricted to a single column.

For instance, the *Hartford Courant* printed stories in 1812 with single-line heads in bold and italic types a few points larger than the body type. They were *label heads* that declared such things as "A Double Murder," "Interesting Debate," and "Overwhelming Calamity." In 1815 the *National Intelligencer* headlined the defeat of the British at New Orleans with a single-column line, "A Most Incredible Victory!"

The Mexican War seems to have been the event that triggered the expansion of headlines. Additional lines were added with short dashes

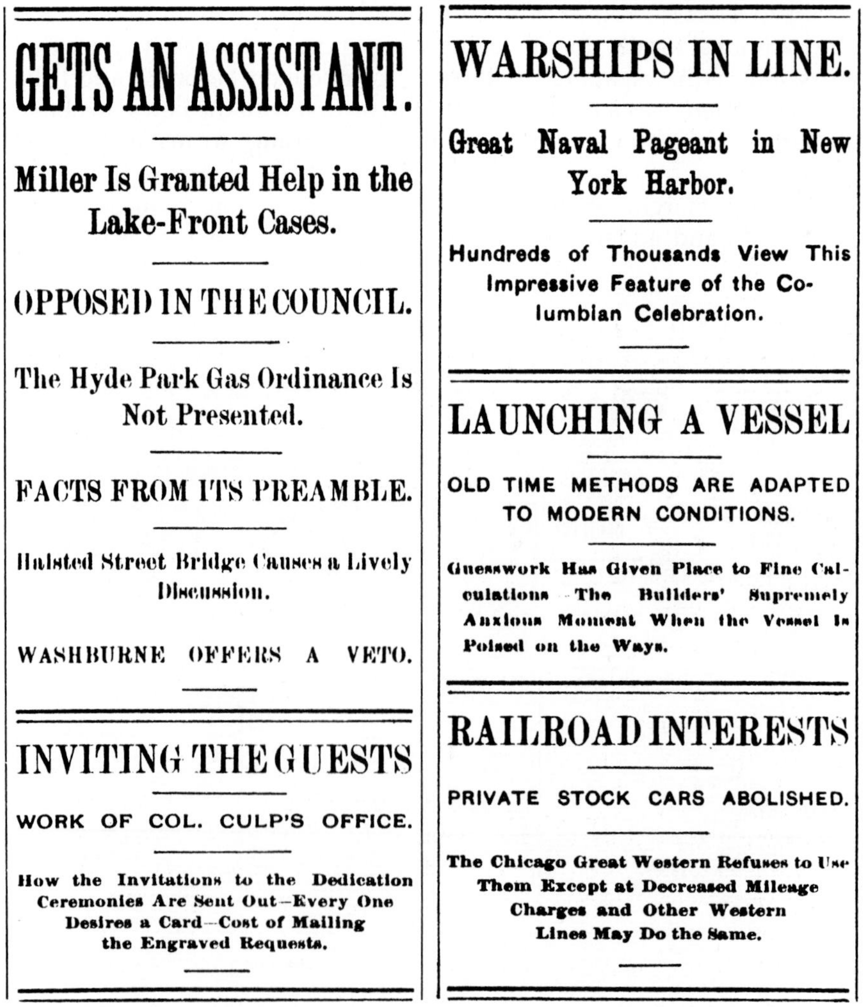

Fig. 15-4 Typical traditional headline forms. This style was popular from the middle 1880s through the early 1900s and is still found in some newspapers. The top all-cap lines are bar lines followed by two-, three-, and four-line inverted pyramids except for the third deck in "Launching A Vessel," which is a hanging indent. The decks are separated by jim dashes, and the heads are separated by Oxford cutoff rules.

GETS AN ASSISTANT.

Miller Is Granted Help in the Lake-Front Cases.

OPPOSED IN THE COUNCIL.

The Hyde Park Gas Ordinance Is Not Presented.

FACTS FROM ITS PREAMBLE.

Halsted Street Bridge Causes a Lively Discussion.

WASHBURNE OFFERS A VETO.

INVITING THE GUESTS

WORK OF COL. CULP'S OFFICE.

How the Invitations to the Dedication Ceremonies Are Sent Out—Every One Desires a Card—Cost of Mailing the Engraved Requests.

WARSHIPS IN LINE.

Great Naval Pageant in New York Harbor.

Hundreds of Thousands View This Impressive Feature of the Columbian Celebration.

LAUNCHING A VESSEL

OLD TIME METHODS ARE ADAPTED TO MODERN CONDITIONS.

Guesswork Has Given Place to Fine Calculations—The Builders' Supremely Anxious Moment When the Vessel Is Poised on the Ways.

RAILROAD INTERESTS

PRIVATE STOCK CARS ABOLISHED.

The Chicago Great Western Refuses to Use Them Except at Decreased Mileage Charges and Other Western Lines May Do the Same.

between each unit. These units became known as *decks*. The short dashes were and are called *jim dashes*.

The gold rush in California followed hard on the heels of the Mexican War with more big news, and the single column, multideck headline became a standard design form in American newspapers. During the Civil War it was not uncommon for a newspaper to print a headline with up to twelve decks employing as many as six different type styles.

There was good reason for restricting headlines to single columns rather than using multicolumn heads, which we would do if a big news story broke today. Some of the larger newspapers were printed on the Hoe type-revolving cylinder press. The type was held in place on the big rotating cylinders with the help of wedge-shaped column rules anchored in the curved bed of the cylinder. Most column rules were made of brass, and printers were reluctant to cut them.

There were the design aesthetics of the era as well. Printers believed that to "break the column rule"—to spread a layout over two or more columns—disfigured a page. The *New York Herald* ran two-column headlines in 1887, but it left the rule between the columns in place and divided the headline on either side of the rule. The rule ran right through the headline!

When decks of more than one line were composed, the practice was to center each line. This led to the headline pattern called the *inverted pyramid*, in which each succeeding line is smaller than its predecessor and all are centered to give the appearance of an upside-down pyramid.

The single-line head or one-line deck became known as a *crossline* or *bar line*. If the line filled the column width, it was called a *full line*. These terms are still used today.

Other traditional headline patterns were developed. A head in which the first line is a full line and each succeeding line is indented, usually an em, and justified became known as a *hanging indent*. The *Wall Street Journal* uses a hanging indent in its head schedule.

A head in which the top line was set flush left, the middle line centered, and the third line flush right with all lines as nearly equal in length as possible was called a *step head*. Step heads could be two, three, or even more lines deep but all had a step-down pattern.

During the "yellow journalism" era, which started in the 1890s, more and more multicolumn headlines appeared. Headlines became larger and bolder. The single-line *banner head* that stretched across the width of the page made its appearance.

All these typographic innovations became part of the traditional pattern of newspaper design.

The Tabloid Format Arrives

In the 1920s two cousins, Joseph Medill Patterson and Robert McCormick, who were members of the family that owned the *Chicago Tribune*, started a half-sheet newspaper. The *tabloid* was born. Tabloid newspapers—newspapers with small pages usually half the size of the broadsheet—had been tried before but no one had been successful in this country.

But the time was right for a small newspaper, tightly written, full of pictures and snappy headlines, and aimed at the big-city subway rider,

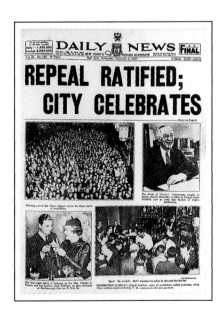

Fig. 15-5 The *Daily News*, New York was the first successful tabloid newspaper, and it continues today as one of our circulation leaders. The front page uses the *poster* format which billboards stories found inside the newspaper.

Fig. 15-6 *Newsday,* the superbly designed New York and Long Island tabloid, uses the poster format with a definite magazine-type layout to emphasize stories found inside. *Newsday's* inside pages (left) have a strong magazine design, and they are good sources of ideas for students and editors or designers of "magapapers." This index page illustrates the trend toward sectionalism in newspapers, in which the newspaper contains special interest sections or "magazines." (Copyright 1989, Newsday, Inc.)

to be a success. And a success it was. Soon tabloids were springing up in most major American cities.

As a result of its flashy design, the tabloid was tagged a "sensational" journal. However, the tabloid page size has many assets as a design form and deserves a solid place in the communications spectrum.

Characteristics of the Traditional Format

Narrow columns, along with rigid and precise headline patterns, became trademarks of the traditional newspaper format. Some newspapers continue to use this format. Some designers are adopting traditional characteristics today where they are appropriate to the overall design philosophy. The traditional format is characterized by:

- Column rules separating narrow columns. The rules are cast on a 4-point base for many newspapers, further cramming the type together.
- Headlines with a number of decks, all separated by jim dashes.

- Nameplates often embellished with "ears" or type material on either side at the top of the page. These contain weather, edition logo, promotional material, slogans, and so on.
- Cutoff rules separating unrelated units such as stories, photos, and cutlines.
- Rules above and below the folio lines, the full-width lines under the nameplate giving the volume and issue number, date, city of publication, and similar information.

Fig. 15-7 *The Litchfield County Times* maintains the traditional format for its front page. It uses the optimum line width, however, creating a six-column page. This newspaper has won awards in the Society of Newspaper Design competitions.

- Banner heads, sometimes used every day regardless of the importance of the news. These banners are followed by readouts, or decks.
- Boxes, bullets, ornaments, and embellishments used liberally.
- Many headlines set in all-capital letters, particularly the top decks in the head.
- Types from several families often used in the head schedule.

In addition, the design plans of the front and inside pages of the traditional newspaper usually follow definite preconceived patterns. (These patterns are discussed in Chapter 16.)

As with all design, though, it is difficult to categorically classify the patterns of all newspapers within clear-cut time and design periods. Some publications changed slowly and some never changed at all. But the traditional approach to newspaper design began to come under serious challenge in the late 1930s and early 1940s with the emergence of what might be called the "functional" design philosophy.

▼ Functional Newspaper Design

The *functional design* philosophy is based on the concept that if an element does not perform a function it should be eliminated, and if another element does the job better it should be used.

John E. Allen led the revolt against the traditional, highly formalized style of newspaper design. His editorship of the *Linotype News*, regarded as the nation's typographic laboratory, and his authorship of three books on newspaper design gave authority to his recommendations. His campaign started in 1929 with what he called "streamlined" headlines. These were heads set flush left and ragged right.[2]

In arguing for the new flush left and for abandoning the complex head designs, Allen made these points:

1. The traditional headline form is difficult to write and often it is necessary to use inaccurate or inappropriate words because of the rigid unit count.
2. All-cap lines are hard to read compared with lines set in lowercase.
3. Flush left heads allow more white space into the page and give heads more breathing room.
4. Traditional head forms are difficult and take more time to set in type.

These points made sense to many designers, and the new style was adopted by more and more newspapers. It was based on the idea that the purpose of typography and graphic design is to make the contents understandable and inviting to read.

Designers examined each element of the newspaper page and evaluated its worth in terms of effective communication. They proceeded from the thesis that if a functional newspaper were to be designed, the first step was to define its function—this is a good starting point for any design project, incidentally.

The functions of most newspapers can be summarized to include *informing, interpreting, persuading,* and *entertaining*. The design and layout of any newspaper should help it achieve four specific goals:

1. Increase readability and attract the reader.
2. Sort the contents so the reader knows at a glance which information is the most important and what each part of the newspaper contains.
3. Create an attractive and interesting package of pages.
4. Create recognition so the paper can be readily identified.

A number of innovations in design were adopted in the late 1930s and early 1940s to help accomplish these goals. After the adoption of the flush left headline form with fewer, if any, decks, other efforts were made to let light into the pages and brighten them. White space was used wherever

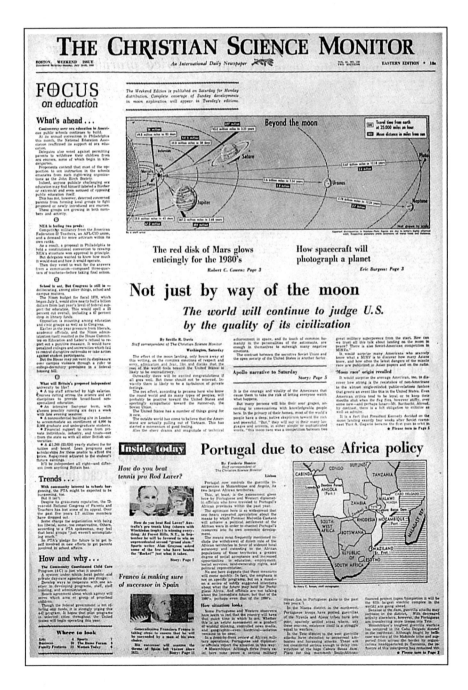

Fig. 15-8 *The Christian Science Monitor* was an early convert to functional modern design. Heads were set in capital and lowercase, white space was used in place of column and cutoff rules, multicolumn heads helped break up the traditional vertical thrust, and elements that did not perform a function were modified or eliminated.

possible. Nameplates were simplified. Often, ears were dropped or cleaned up and typographic embellishments in the nameplate area were eliminated in favor of white space.

The top, and in some instances the bottom, rule on the folio line was eliminated. Vertical column rules were dropped in favor of white space between columns and this white space was increased. A pica of white space between columns was considered minimum for effective separation.

Cutoff rules and jim dashes were scrapped in favor of white space, although some newspapers continued to use cutoff rules if they were thought more effective in designating story and art unit limits.

The new design movement favored fewer banner heads but more variety in layout. The optical attraction of the upper left corner of the page was utilized by placing important stories or photos there rather than subordinating them to the traditional lead story in the upper right corner.

Shorter nameplates were another design innovation. Designers frequently used skyline heads and stories, placed above the nameplate and extending across the width of the page. The traditional nameplate, which extended across the top of the page, was reset in varying widths so it could be "floated" or shifted around on the page and set in two, three, or more column widths for variety and change of pace.

Other functional innovations included good display of a page 1 index and highlighting of inside features to attract the readers to the inside of the newspaper. More photos were used and in larger sizes to give them increased impact. Photos were cropped closely and enlarged. More attention was paid to the bottom half of the page—the area below the fold—to get a better balance and to present a livelier look from top to bottom.

Jumped stories were eliminated as much as possible since readership studies revealed that a story loses about 80 percent of its audience when it is continued to another page.

A horizontal thrust was introduced with the use of more multicolumn heads and photos. This helped break up the dullness of column after column of vertical makeup. Captions were shortened and rules, boxes, and ornaments simplified.

Not all newspapers adopted all of the functional design devices. But more and more newspapers did appear in a format that reflected what had become accepted as a basic tenet of good newspaper design:

The appearance of the newspaper should reflect its editorial philosophy and appeal to its particular audience.

The *New York Times*, traditional yet in its basic design, continues to win awards for its adherence to that philosophy.

▼ The Optimum Format Arrives

In 1937 the *Los Angeles Times* restyled its format to functional design, and that year it won the coveted Ayer award for outstanding newspaper design. Twenty-eight years later, the *Courier-Journal*, Louisville, Kentucky, and its companion newspaper, the *Louisville Times*, became the first metropolitan newspapers to usher in the *optimum design era.* The *Courier-Journal* cut its columns per page from eight to six and widened them from 11 picas to 15.

In the 1960s some newspapers began to appear with a "downstyle" head dress. The *downstyle head* is composed in all lowercase letters except for the first word and proper nouns. It further simplifies newspaper design and eases reading.

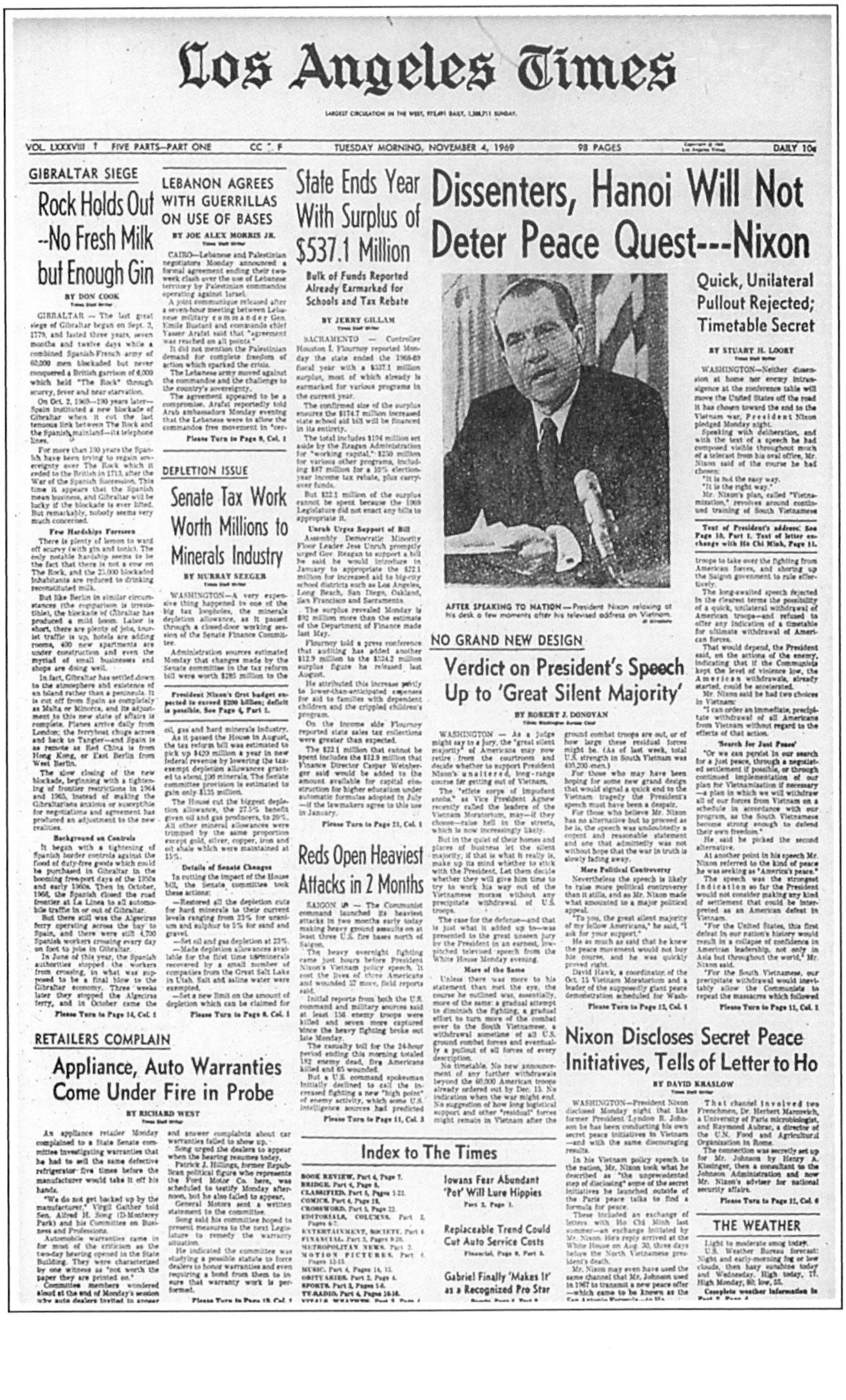

Los Angeles Times

VOL. LXXXVIII FIVE PARTS—PART ONE TUESDAY MORNING, NOVEMBER 4, 1969 98 PAGES DAILY 10¢

GIBRALTAR SIEGE

Rock Holds Out --No Fresh Milk but Enough Gin

BY DON COOK

LEBANON AGREES WITH GUERRILLAS ON USE OF BASES

BY JOE ALEX MORRIS JR.

State Ends Year With Surplus of $537.1 Million

Bulk of Funds Reported Already Earmarked for Schools and Tax Rebate

BY JERRY GILLAM

Dissenters, Hanoi Will Not Deter Peace Quest---Nixon

Quick, Unilateral Pullout Rejected; Timetable Secret

BY STUART H. LOORY

DEPLETION ISSUE

Senate Tax Work Worth Millions to Minerals Industry

BY MURRAY SEEGER

AFTER SPEAKING TO NATION—President Nixon relaxing at his desk a few moments after his televised address on Vietnam.

NO GRAND NEW DESIGN

Verdict on President's Speech Up to 'Great Silent Majority'

BY ROBERT J. DONOVAN

Reds Open Heaviest Attacks in 2 Months

RETAILERS COMPLAIN

Appliance, Auto Warranties Come Under Fire in Probe

BY RICHARD WEST

Nixon Discloses Secret Peace Initiatives, Tells of Letter to Ho

BY DAVID KRASLOW

Index to The Times

Iowans Fear Abundant 'Pot' Will Lure Hippies

Replaceable Trend Could Cut Auto Service Costs

Gabriel Finally 'Makes It' as a Recognized Pro Star

THE WEATHER

Fig. 15-9 The *Los Angeles Times* was one of the first metropolitan newspapers to adopt the optimum format, replacing the traditional eight-column page with a six-column page which allows wider columns to make reading easier.

Wethersfield Actor Lands Choice Soap-Opera Role Page C1

The Hartford Courant

ESTABLISHED 1764, DAILY EDITION, VOL. CXLVII NO. 53   WEDNESDAY, FEBRUARY 22, 1984 — 5 SECTIONS   COPYRIGHT© 1984 THE HARTFORD COURANT CO.   25¢ PER COPY

Marines Begin Beirut Airport Withdrawal

Associated Press

BEIRUT, Lebanon — U.S. Navy helicopters ferried Marine combat troops from their base at Beirut's airport to warships in the Mediterranean Tuesday as the Marine withdrawal from Beirut officially began.

Israeli jets, meanwhile, bombed and strafed positions in the Syrian-controlled mountains east of the capital.

"Today the support people have gone and we're working on the combat gear," said the Marine spokesman, Maj. Dennis Brooks. "Today is the first day of the relocation of the actual 22nd MAU (Marine Amphibious Unit) personnel."

He was referring to combat troops that have stood by since Feb. 7 when President Reagan announced his intention to withdraw the Marines.

Helicopters soared in from the 6th Fleet shortly after dawn Tuesday and the withdrawal of the estimated 1,300 shore-based Marines began an hour later, Brooks said.

Brooks said he could not say how many Marines were withdrawn Tuesday. He estimated it would take "approximately a week, maybe two" to evacuate the base at Beirut's airport, maintained for 17 months by the Marines acting as part of a multinational peace-keeping force.

Since the base was established in September 1982, 265 U.S. servicemen have died in Lebanon.

The airport has been virtually surrounded by anti-government militias since last Wednesday, when Druse fighters drove from the mountains to the coast south of the base, linking up with their Shiite allies and further undermining the government of Christian President Amin Gemayel.

In Washington, Senate Republican leader Howard H. Baker Jr. said he had been told in a White House meeting with President Reagan that the withdrawal "would be finished by the end of this month, barring unforeseen circumstances."

Seven men of an air-naval gunfire liaison company talked to reporters on their way out. They carried M-16 rifles, light anti-armor weapons and grenade launchers.

"I'm ready to go. I've got women to meet and beers to drink," said Lance Cpl. Samuel Lee, 20, of Miami, Fla.

But when asked about the Marines' mission, Lee added: "We were just trying to restore peace. It doesn't look like it happened. It's a shame the U.S. has been here more than a year, and they (the Lebanese) still can't get their act together."

The battleship New Jersey circled about a mile offshore as Marine CH-53 Sea Stallion helicopters took off about every 15 minutes from a landing zone on the western edge of the base.

See Marines, Page A10

A young Lebanese Shiite Moslem boy gives a U.S. Marine a kiss goodbye at the Marine base near Lebanon's Beirut airport. Marine combat soldiers began pulling out of the base Tuesday.   AP

State Loses Bid To Ban Tandems

By MIRANDA SPIVACK
Courant Staff Writer

WASHINGTON — The U.S. Supreme Court cleared the way Tuesday for double-trailer trucks to roll through Connecticut, turning aside the state's argument that the tandem-trailer vehicles are a safety hazard.

The justices, in a brief statement, affirmed a June 1983 decision by U.S. District Court Judge Jose A. Cabranes of Hartford, who said a federal law allowing widespread travel by the twin trucks supersedes a Connecticut ban imposed in April 1983.

The state passed its statute barring the tandems in response to a 1982 federal law permitting the trucks to travel on interstate highways.

The justices did not address the safety questions attorneys for the state had made the cornerstone of their argument. Cabranes had said that it is up to Congress to deal with safety questions.

The high court's action does not affect another part of the Connecticut law, which established a special permit for operators of tandem-trailers.

See State, Page A10

McKinney Charged in Export Flap

By DAVID LIGHTMAN
Courant Staff Writer

WASHINGTON — Rep. Stewart B. McKinney of Connecticut was charged Tuesday by the Commerce Department with violating a U.S. export regulation, an action that could mean a fine of up to $10,000.

The 4th District Republican is the first public official in the seven-year history of the regulations to face such a charge, a Commerce Department official said.

McKinney said, however, he was fulfilling his duties as a congressman and acting only to help a constituent.

The regulation involved is aimed at preventing U.S. companies from cooperating with Arab nations that try to prevent firms

See McKinney, Page A10

Task Force Opposes Deregulation of Nursing Homes

By MICHELE JACKLIN
Courant Staff Writer

The state should continue setting the rates paid by private patients in nursing homes, a special task force said Tuesday, almost certainly ending any consideration of deregulation of the industry by the General Assembly this year.

Deregulation would not only cost private patients millions of dollars a year, but could cost the state up to $71 million in added

Medicaid expenses, the task force said.

Allowing decontrol of nursing home rates was one of the most volatile issues before state lawmakers, with nursing home owners arguing that they needed to be able to charge private patients more to offset the unrealistically low payments they receive for Medicaid patients.

The recommendations of the task force, released at a State Capitol news conference, were adopted on a 12-5 vote, with the three representatives of the nursing home industry in opposition. Joining them were Health

"We have no plans in the immediate future to reintroduce this legislation. We will let the matter rest," said Louis J. Halpryn, vice president of the Connecticut Association of Health Care Facilities.

The recommendations of the task force, released at a State Capitol news conference, were adopted on a 12-5 vote, with the three representatives of the nursing home industry in opposition.

Services Commissioner Douglas S. Lloyd and Audrey Wasik of the state Commission on Long Term Care.

Halpryn and task force member Harry DiAdamo, a spokesman for three non-profit convalescent homes, said the task force membership was heavily weighted against the industry.

"They would have voted the way they did from the very first day," DiAdamo said.

There were other indications

of the deep rift among members. DiAdamo described the data compiled by the commission as "inconclusive" while A. Cynthia Matthews, a Democratic senator from Wethersfield and task force co-chairwoman, said the numbers were "very convincing."

Alleging that many of the state's 296 nursing home facilities are facing financial difficulties

See Panel, Page A10

Death With Dignity Sought for Daughter

By JOSEPH M. COHEN
Courant Staff Writer

The father of 42-year-old multiple sclerosis victim Sandra Z. Foody told a Superior Court judge Tuesday, "I don't want her tortured to death. I want her hooked off the respirator ... so she can die in dignity."

Kenneth F. Foody and his wife, Ann M. Foody, of South Windsor testified Tuesday they had given the past 24 years to caring for their only child and, as devout Roman Catholics, had wrestled with the moral question of letting their die.

The Foodys have filed a Hartford Superior Court lawsuit asking to have their daughter taken off a respirator at Manchester Memorial Hospital, something the woman's doctor said is likely to lead to her death within a "few minutes."

In the Washington Street courtroom of Judge Mary R. Hennessey, lawyers compared the case to the landmark 1976 New Jersey case of Karen Ann Quinlan, and they said it would set Connecticut precedent in so-called death-with-dignity cases.

After hearing almost six hours of testimony and arguments Tuesday, Hennessey gave attorneys a week to file written briefs.

No date for a ruling was set.

There was no testimony or argument in court in opposition to taking Foody off a respirator, although her doctor, Giao Ngoc Hoang of Manchester, said the woman was not brain-dead.

Hoang described Foody as "awake but unaware. I cannot say with certainty she does not feel." But, he said, except for her brain regulating her body temperature and other very basic functions, she seems to be unaffected by external stimulation.

"We are all in agreement her condition is not reversible," Hoang said of the physicians who have consulted on the case.

Hoang said he considered the Foody case a "moral dilemma." "We always want to do what is best for the patient. ... We are not God. Even though we are physicians, we have to live by the law of the land," he said.

Hoang is named as a defendant in the Foodys' lawsuit, along with Manchester Memorial Hospital, hospital acting Executive Director Michael Gallacher, state Attorney General Joseph I. Lieberman and Hartford State's Attorney John M. Bailey.

The lawsuit is intended to en-

See Parents, Page A10

INDEX

AUTHOR DIES — Mikhail Sholokhov, who wrote "And Quiet Flows the Don," has died at age 78. Sholokhov won the Nobel Prize in literature in 1965, the only officially sanctioned Soviet writer to win the prize. Page A11.

Amusements	C1	Editorials	B10	
Ann Landers	C6	Food	E1	
Bridge	C6	Horoscope	C7	
Business	C8	Legal Notices	D6	
Classified	D4	Obituaries	B8	
Comics	C6	Sports	D1	
Connecticut	B1	Television	C4	
Crossword	C6	Wednesday	C1	

WEATHER: MOSTLY SUNNY 26° to 44°F (-3° to 7°C)
Complete Weather   B12

Mondale, Hart Take Their Battle For Top Spot to New Hampshire

Washington Post

MANCHESTER, N.H. — With his handsome Iowa caucus victory in hand, Walter F. Mondale came to New Hampshire Tuesday seeking a win in next Tuesday's primary that he said could "take us the rest of the way" to the Democratic presidential nomination.

Mondale's erstwhile leading opponent, Ohio Sen. John H. Glenn Jr., also turned to the New Hampshire contest, which for him has been transformed by a shocking fifth-place finish in Iowa into a test of political survival. As the disabled Glenn campaign tried to

that his surprise runner-up finish in Iowa would produce a closer race for Mondale in New Hampshire and would shake the Democratic contest down to a Mondale-Hart battle.

Mondale faces a new challenge from a strengthened Colorado Sen. Gary W. Hart, who declared

keep from expiring, its New Hampshire manager announced that Glenn is scrapping his controversial anti-Mondale ads and is making new, "entirely positive" ads.

The Iowa results changed the dynamics of the New Hampshire race, while strengthening Mondale's status as the favorite for the nomination. Instead of facing his main challenge from Glenn, who

See Boosted, Page A4

Mayor Faults Zoning Plan, but Forgos Veto

By JOSEPH RODRIGUEZ
Courant Staff Writer

Hartford Mayor Thirman L. Milner Tuesday criticized the City Council for adopting new downtown zoning regulations, but said he would let them become law — without his signature.

Milner said he would not veto the regulations because he said they ignore the needs of city residents and favor developers.

"As mayor, I am not satisfied with the ordinances that were passed," Milner wrote. "I cannot and will not affix my signature

approving the ordinance as passed."

Deputy Mayor Francisco L. Borges, a key backer of the package, said, "I'm pleased the mayor did not veto the ordinances." He said he was confident the council could have come up with the six votes needed to override a mayoral veto. The Council approved the package by an 8-1 vote.

The ordinances give city officials more control over downtown development. They require retail space in all new buildings and establish new standards for construction design, building size and traffic management. Devel-

opers also will be allowed to exceed maximum building sizes in exchange for providing amenities such as movie theaters and day-care centers.

Milner late last year vetoed key elements of a similar package because it did not offer incentives for builders to provide jobs and housing for city residents.

Milner did not explain in his statement why he did not veto the new package. He could not be reached Tuesday for comment.

Leaders of neighborhood

See Milner, Page A10

Fig. 15-10 The *Hartford Courant* as it appeared in 1984, a carefully designed example of the optimum format. (Copyright 1984, Hartford Courant)

Fig. 15-11 Five years later the *Courant* had joined the redesign era with skyline teasers, more color, and a new typeface for headlines. (Copyright 1989, Hartford Courant)

In 1965 daily newspaper circulation in the United States was 60,358,000. By 1970 the population stood at 203,302,031 and newspaper circulation was 62,108,000. Ten years later the population had grown by more than 23,000,000 while newspaper circulation remained virtually unchanged. During the decade circulation of all the general circulation newspapers in the country only increased by 115,040.

Editors and publishers recognized that something was wrong. The country had more and more people but not more and more newspapers. They looked at their products and decided steps should be taken to make newspaper reading more attractive.

Following the lead of the *Courier-Journal* and others, many newspapers made two basic changes. The size of the body type was increased from the standard 8 point to 9 and even 9½. Columns were widened to approximate the optimum line length for reading ease and speed. This meant changing from the cramped 10½- or 11-pica columns to more comfortable 14- or 15-pica columns.

By the middle 1970s the majority of newspapers had switched to the six-column format for at least their front and section pages. Some newspapers went to five-column pages and virtually all increased the white space between columns.

A leader in urging more readable newspapers was Edmund C. Arnold, who succeeded Allen as editor of the *Linotype News* and who has been recognized as one of America's authorities on newspaper design.

"There are many advantages to the op format," Arnold wrote in *Modern Newspaper Design*. He pointed out that "a line length at optimum is an asset because it enhances communication. The reader likes the longer measure, too, even if he doesn't understand the technicalities involved." [3]

However, even though the optimum format made newspaper reading easier and more pleasant, circulation continued static. Many newspapers have again narrowed their columns back to 12 picas and reduced the body type back to 8 point. For them, the intent of the optimum format has been lost.

▼ The Redesigning Era

Newspaper design today is in a state of flux. The optimum format is with us in spirit if not in actuality. At the same time editors and designers are probing ways to make newspapers more visually exciting. They are struggling to keep up with the rapid changes in life-styles and reader interests. The new technology is also dictating many aspects of form and format.

There is, however, a growing recognition of the importance of blending design with content to develop the most effective communications package. Dr. Mario R. Garcia, a leading designer, wrote in *Contemporary Newspaper Design*, "Improvement in content and emphasis on clear writing and editing, combined with effective graphic innovation should be present" before circulation declines can be reversed. [4]

At the same time, the new technology is opening doors of opportunity for newspaper editors and designers. The mechanical constraints of the past have been eased, allowing the designer to apply what has been called the *total design concept*. This means that instead of being restricted by a page divided into columns, the designer now has an area that can be treated as a blank rectangle (the dimensions of the page) on which to create the most effective layout possible.

The design question has become not how to fill columns but how to create an effective page within its overall dimensions. Some designers have retained the basic vertical approach which has withstood the test of time for so many years. The answer for many others has been found in the *modular* format. This approach to design evolved from the creations of a Dutch painter, Piet Mondrian, and a school of design that is known as the Swiss approach, or the International Typographic Style.

Modrian did most of his work in the early part of this century. By 1917 he was concentrating his efforts on the use of the primary colors—red, yellow, and blue—combined with black and white, and limiting his art forms to straight lines, squares, and rectangles.

These compositions, employing vertical and horizontal lines at 90 degree angles forming crosses, rectangles, and squares, are typical of Mondrian's art. Mondrian, who lived until 1944, has been an important influence on contemporary art and architecture, as well as on the modular approach to the layout of newspapers and other printed communications.

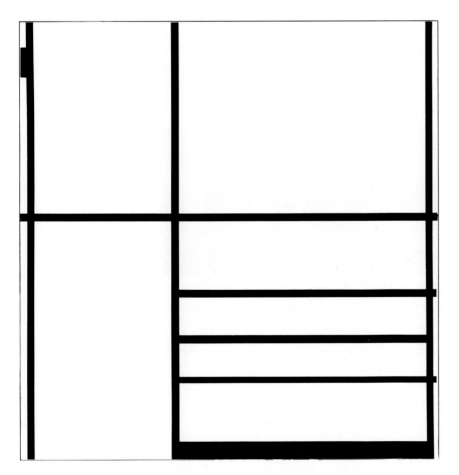

Fig. 15-12 This painting by Piet Mondrian is an example of his style, which inspired the modular makeup of many of today's newspapers. (Courtesy Westvaco Corporation)

As graphic historian Philip B. Meggs notes, "Mondrian used pure line, shape, and color to create a universe of harmoniously ordered, pure relationships. To unify social and human values, technology, and visual form became a goal for those who strived for a new architecture and graphic design."[5]

Another school of design thought emerged from Switzerland in the 1950s. The Swiss approach to design, or the International Typographic Style, had a strong influence on graphic design through the next twenty years and it is being revived today. The objective of the International

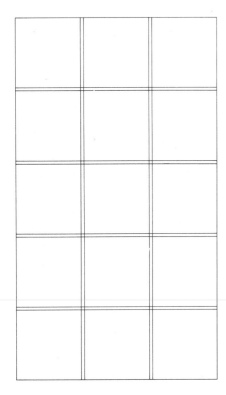

Fig. 15-13   Followers of the International Typographic Style advocate the use of a grid in graphic design. A grid consists of precisely drawn straight lines in rectangles and squares on paper or cardboard printed in nonreproducing ink. It is used for placing elements on a layout. In desktop publishing and computer pagination the grid is created on the monitor. This is a three-column grid.

Fig. 15-14   Here is a rough of a possible arrangement of illustrations and copy on the three-column grid. The modular approach to newspaper design grew out of the work of Piet Mondrian and the philosophy of the International Typographic Style.

Typographic Style is to create visual unity of layouts through using a precisely drawn grid of straight lines in rectangles and squares and placing elements within this frame.

Other characteristics of the International Typographic Style include:

- Use of Sans Serif types with Helvetica being a favorite.
- A tendency to set type flush left and ragged right.
- Combining art and copy to present information in as clear and factual a manner as possible.
- Eliminating exaggerated claims and propaganda in copy and just sticking to the facts.

In addition to its design mechanics, the Swiss movement looks on design as a socially useful activity. The designer defines his or her role as a disseminator of important information to the various segments of society, not just a creator of pleasing designs.

The use of the grid, advocated by this movement, has become a standard practice, especially in laying out pages for publications. In the mod-

Fig. 15-15 The *St. Petersburg Times,* considered one of the best designed newspapers in the United States, illustrates modular makeup. The page contains seven distinct rectangles. Note the information package (A hot rod on water?) which contains the work of a photographer, graphic artist, writer, and editor. On the right the seven rectangles that form the modular makeup of the *St. Petersburg Times* page.

ular approach to newspaper design the page is made up of a series of rectangles, with perhaps a square now and then. Each spatial unit, or module, contains a complete story. It is as if each story with its head and art, if used, is prepackaged in a rectangle or, more rarely, a square so these packages can be arranged in as pleasing a way as possible on the newspaper page.

Fig. 15-16 Consistently recognized for its outstanding design, the *Morning Call* used modular layout for this front page. Note the information package containing the work of a photographer, graphic designer, writer, and editor. (Courtesy *Morning Call*, Allentown, Pennsylvania)

Designers who have adopted the modular plan—and it works well for large-format magazines and newsletters as well as newspapers—have these suggestions for planning a modular page:

- Make numerous thumbnail sketches of newspaper pages made up of various combinations of rectangles. Do not use squares as they are monotonous and uninteresting. Keep in mind the basic design principle of proportion. Select the sketch that is most pleasing and use it as the basic design grid for that particular page.
- One module should dominate the page and presumably contain the lead story. This module should be placed above the fold, usually in the upper left or upper right—the high-interest sections of the page.
- Each module should be self-contained, with a rule to define it. However, the rule should not be too heavy, no more than 4 points wide. Some designers specify color for the rules if available. Fairly generous white space, usually 18 to 24 points, can be used instead of rules.

Fig. 15-17 When the *Oregonian* was redesigned these goals were set: Increase newsstand sales, convert Sunday-only readers to daily readers, reduce median age of readers from 51 to 39, reduce number of people who don't read because of lack of time, gain more public awareness of content and editorial character of the paper. On the left is a front page of the paper after the redesign was completed. One feature of the redesigned Oregonian is the use of a summary deck on headlines. This deck is written in radio news style to give readers the story in a nutshell. Three such decks can be seen with stories on the Metro page on the right. The Oregonian strives to provide brevity for the busy reader while still giving in-depth stories to those who seek more comprehensive coverage. (Courtesy *Oregonian*, Portland, Oregon)

- Modules contain a head and often art. Modules can be art and cutline only, but they should be considered as packages of information. The elements included should transmit the information as clearly and completely as possible.
- The trend is toward the use of more line art in newspaper design. Photographs, when used, should be cropped closely to give impact to the center of interest.
- Plenty of white space should be used within the rules—around the heads and other elements.

Whenever a newspaper is being redesigned or a new publication is being planned, it is important to precede the actual design work by a publication analysis.

▼ How to Analyze a Publication

The designer of a newspaper should plan an analysis of the product on a regular basis. If possible, each issue should be evaluated with the editor. This evaluation could start with a few moments spent in browsing through the entire copy.

The overview of the issue should reveal a general impression of whether the typography maintains the character of the publication. For instance, does the whole issue help identify the publication as a conservative newspaper for a conservative community? Does it help say this newspaper is produced by an organization devoted to academic interests or to the manufacture of heavy equipment? The appearance of the newspaper should reinforce its purpose.

The examination should reveal if the arrangement of elements aids communication. Are there barriers? Are the rules, ornaments, or subheads placed so the flow of copy is interrupted to the point that the reader is confused? Does the layout increase readability? Would readers be attracted to this story or that? Or would they be inclined to pass a story by because the layout appears dull? If you were a reader would you want to spend time with this newspaper?

The entire package should do a good job of "sorting" the contents so the reader can find topical matter easily and without confusion.

Once an overall impression has been formed, each of the typographic elements should be examined and evaluated:

1. *Body type:* Is it legible and readable? Roman types have the highest readability. But the Roman used in newspaper design should have rather soft serifs and not too sharp a differentiation in the widths of the letter strokes. The size of the body type should be checked to see if it is large enough for easy reading. Types that are big on the body (large x-height) are usually preferred. Line length and leading should be examined to ensure their effectiveness.
2. *Headlines:* Is the type selected for the headlines attractive? Does it reflect the tone of the newspaper? Of the feature stories? Even though a bold type is chosen, it should be clear and legible. The same is true for light typefaces. A headline type should have a fairly good unit count. If

more than one family is used in a head arrangement, the types should harmonize but one should dominate. Spacing between words and lines should be examined as well as all the factors of legibility, such as the size of the head for the width of the line.

3. *Typographic color:* Have ornaments, bull's eyes, arrows, and so on been avoided? If typographic devices are used, they should enhance legibility rather than detract from it.

4. *Newspaper constants:* Do the standing heads, department and column heads, and the masthead harmonize with the overall effect of the layout? The constants should be alive and not static. Headings such as "The President's Column," "Washington Week," or similar label heads should be redesigned if they do not have an element that can be changed, preferably with each issue, to illuminate the contents and add life.

5. *Pictures and cutlines:* Are the pictures cropped for proper emphasis? Art should communicate a message and not just ornament the layouts.

Fig. 15-18 The *Sun News,* which serves a resort area, recently underwent a dramatic change in appearance. The plan was to give the paper a contemporary, distinctive look that would stand out on a newsstand where it had to compete with ten other newspapers. On the left is the newspaper before the redesign. After the redesign (right) the *Sun News* carried fewer front page stories to make room for larger art. The nameplate was given a vertical treatment for a distinctive look. This allowed for more room for top-of-the-page teasers. Century Bold was selected for the headlines with Franklin Gothic for accent. Bylines are in Franklin Gothic and Century Oldstyle. (Courtesy *Sun News,* Myrtle Beach, S.C.)

Cutlines should be in a consistent style throughout the newspaper. They should be set in a type style that harmonizes with the other types but still provides some contrast. Cutlines should be set in the proper width, never more than about 18 picas wide. Cutlines under wide art should be broken into columns that approximate the optimum line length. Indented cutlines with ample space between lines brighten a page.

6. *Front page layout:* Can the application of the basic principles of design be seen? Is unity, balance, contrast, and harmony evident in the selection and placement of the elements? Is the optical center used to achieve balance? The front-page design should emphasize the most important story, but usually more than one strong element is needed to make a lively page. However, the page should not be overloaded to the point where it appears to be a conglomeration.

Are there strong elements in the "hot spots"—the four corners of the page? Unless there is a planned vertical thrust to the page, a strong horizontal treatment is desirable. There should be ample but well-planned white space.

7. *Inside page layout:* Is a consistent pattern evident for the placement of advertisements? Advertisements should not be placed haphazardly on a page, and the pattern adopted should be used consistently throughout the newspaper. Advertisements should be placed so they do not destroy reasonable editorial matter display; that is, they should be kept as low as possible on the page. Is there an editorial stopper—a story or art or a combination—on each inside page?

8. *Section pages and departments:* Are these given the same care as page 1? These pages should reflect the purposes and characteristics of the sections in their typographic design. For instance, the sports pages should have headline type that helps to say "This is the sports section," and the family living section should have a typographic dress that helps identify it. The principles of good layout should be evident in these pages as well as in all the others.

Once each part of the newspaper's anatomy has been analyzed and areas of improvement noted, the suggested improvements should be examined to see how they fit in helping to create a unified, attractive publication.

▼ Effective Design Checklist

- Eliminate barriers. Check to see, for instance, if copy broken by subheads, art, or other typographic devices is easy to follow over, around, or under the devices.
- Break up long copy for easy reading. Use extra space between paragraphs, pull out pertinent points and box them, use subheads, or indent and illuminate enumerated points with bull's eyes or other typographic color devices. But don't overdo it.
- Set copy on proper measure for easy reading. Check to see that copy is never set more than two columns wide. Never set it two columns wide if the columns are more than 12 picas. Body matter shouldn't be set more than about 15 picas wide.

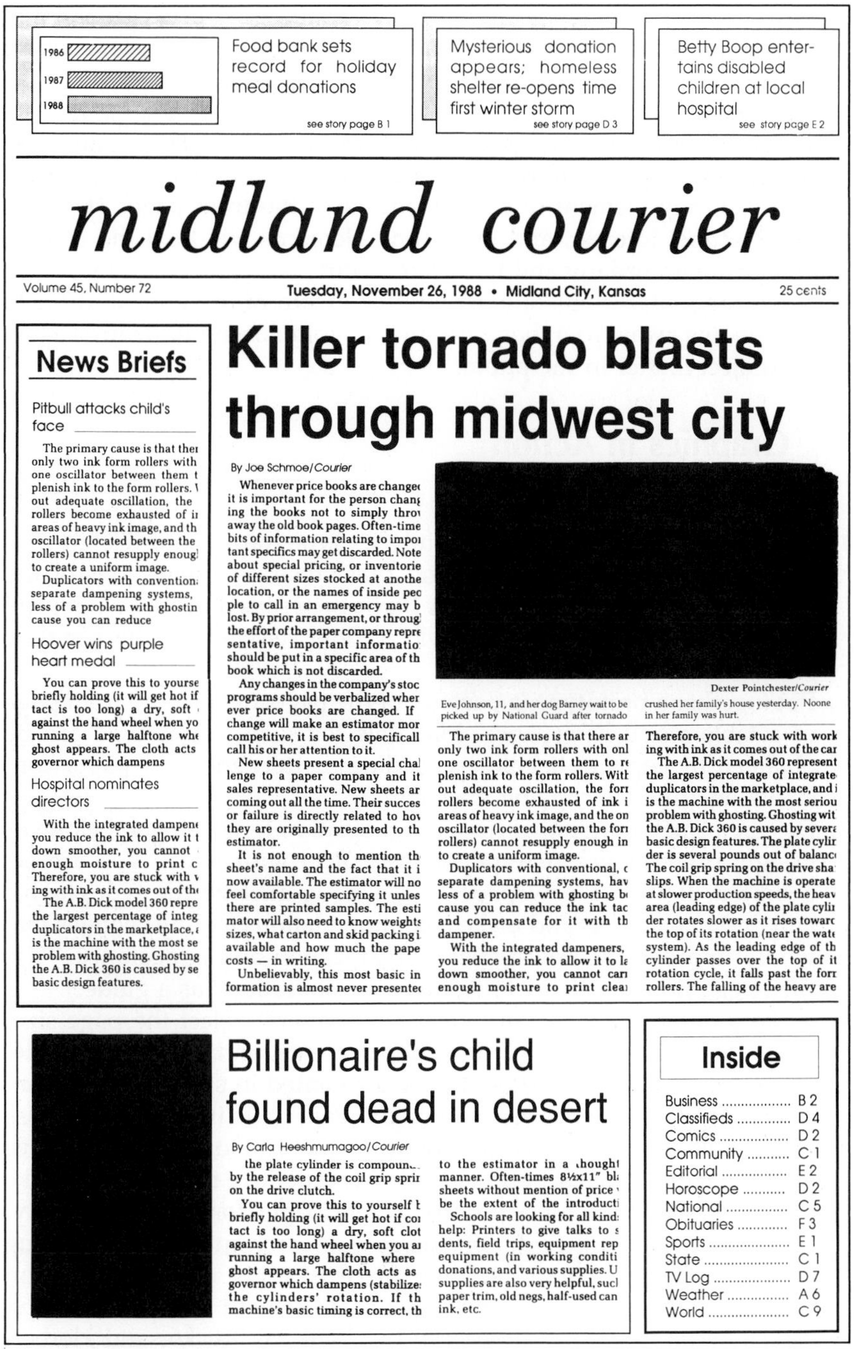

1986
1987
1988

Food bank sets record for holiday meal donations

see story page B 1

Mysterious donation appears; homeless shelter re-opens time first winter storm

see story page D 3

Betty Boop entertains disabled children at local hospital

see story page E 2

midland courier

Volume 45, Number 72 — Tuesday, November 26, 1988 • Midland City, Kansas — 25 cents

News Briefs

Pitbull attacks child's face

The primary cause is that thei only two ink form rollers with one oscillator between them t plenish ink to the form rollers. \ out adequate oscillation, the rollers become exhausted of i areas of heavy ink image, and th oscillator (located between the rollers) cannot resupply enoug to create a uniform image.

Duplicators with convention separate dampening systems, less of a problem with ghostin cause you can reduce

Hoover wins purple heart medal

You can prove this to yourse briefly holding (it will get hot if tact is too long) a dry, soft against the hand wheel when yo running a large halftone wh ghost appears. The cloth acts governor which dampens

Hospital nominates directors

With the integrated dampen you reduce the ink to allow it t down smoother, you cannot enough moisture to print c Therefore, you are stuck with ing with ink as it comes out of th The A.B. Dick model 360 repre the largest percentage of integ duplicators in the marketplace, is the machine with the most se problem with ghosting. Ghosting the A.B. Dick 360 is caused by se basic design features.

Killer tornado blasts through midwest city

By Joe Schmoe/*Courier*

Whenever price books are change it is important for the person chan ing the books not to simply thro away the old book pages. Often-time bits of information relating to impo tant specifics may get discarded. Note about special pricing, or inventorie of different sizes stocked at anothe location, or the names of inside pec ple to call in an emergency may b lost. By prior arrangement, or throug the effort of the paper company repre sentative, important informatio should be put in a specific area of th book which is not discarded.

Any changes in the company's stoc programs should be verbalized wher ever price books are changed. If change will make an estimator mor competitive, it is best to specificall call his or her attention to it.

New sheets present a special cha lenge to a paper company and it sales representative. New sheets ar coming out all the time. Their succes or failure is directly related to ho they are originally presented to th estimator.

It is not enough to mention th sheet's name and the fact that it i now available. The estimator will no feel comfortable specifying it unles there are printed samples. The esti mator will also need to know weight sizes, what carton and skid packing i available and how much the pape costs — in writing.

Unbelievably, this most basic in formation is almost never presente

Dexter Pointchester/*Courier*

Eve Johnson, 11, and her dog Barney wait to be picked up by National Guard after tornado crushed her family's house yesterday. Noone in her family was hurt.

The primary cause is that there ar only two ink form rollers with onl one oscillator between them to r plenish ink to the form rollers. Wit out adequate oscillation, the for rollers become exhausted of ink i areas of heavy ink image, and the on oscillator (located between the for rollers) cannot resupply enough in to create a uniform image.

Duplicators with conventional, separate dampening systems, hav less of a problem with ghosting b cause you can reduce the ink tac and compensate for it with th dampener.

With the integrated dampeners, you reduce the ink to allow it to l down smoother, you cannot ca enough moisture to print clea

Therefore, you are stuck with work ing with ink as it comes out of the ca

The A.B. Dick model 360 represent the largest percentage of integrate duplicators in the marketplace, and i is the machine with the most seriou problem with ghosting. Ghosting wit the A.B. Dick 360 is caused by sever basic design features. The plate cylir der is several pounds out of balanc The coil grip spring on the drive sha slips. When the machine is operate at slower production speeds, the hea area (leading edge) of the plate cyli der rotates slower as it rises towar the top of its rotation (near the wat system). As the leading edge of th cylinder passes over the top of it rotation cycle, it falls past the for rollers. The falling of the heavy are

Billionaire's child found dead in desert

By Carla Heeshmumagoo/*Courier*

the plate cylinder is compoun by the release of the coil grip spri on the drive clutch.

You can prove this to yourself b briefly holding (it will get hot if co tact is too long) a dry, soft clot against the hand wheel when you a running a large halftone where ghost appears. The cloth acts as governor which dampens (stabilize the cylinders' rotation. If th machine's basic timing is correct, th

to the estimator in a though manner. Often-times 8½x11" bl sheets without mention of price be the extent of the introducti

Schools are looking for all kind help: Printers to give talks to dents, field trips, equipment rep equipment (in working conditi donations, and various supplies. U supplies are also very helpful, suc paper trim, old negs, half-used can ink, etc.

Inside

Fig. 15-19 A student created this prototype page on a Macintosh computer. The reading matter is greeking, or simulated type clipped from a publication, but everything else was produced on the computer. Prototypes are mockups that can give the student a good idea of how a page will look. (Courtesy Kim Rusche)

- Don't use overlines with photos. (Overlines are small heads above photos.) Studies show they have no value. If a head is part of the cutline format, it should be immediately adjacent to the cutline.
- Stick to one or two families of type for the head schedule. Plan a basic head schedule with one family of type and use its variations for contrast. If another family is used, use it sparingly and for contrast and accent, rather than basic heads.
- Be cautious with reverses. Be sure reversed type has high legibility. Reverses can be difficult to print well, and they can disfigure a page.

- Crop carefully and enlarge generously. Crop to emphasize the point of interest and eliminate anything that gets in the way of the story-telling quality of the photo. Enlarge to give impact and reader interest and do it generously.
- Never run two unrelated photos next to each other. They will tug at the reader for attention, decreasing the impact of each.
- Have a strong element in each quadrant of the front page. The "hot spots" attract interest, and the strong elements will hold the readers and ensure an orderly and balanced layout.
- Watch the folds. In making layouts, be aware of the folds and try to keep photos away from them. Be sure the fold doesn't destroy or maim the subject matter in the photo.

▼ Graphics in Action

1. Select a daily newspaper and study its typography and design. See how many examples you can find that illustrate applications of the principles of design. (Example: A Black Letter used for the nameplate might contrast well with the style of type selected for the headline schedule.)
2. Put together a prototype (a pasteup of the page) of an entire front page of a tabloid-size newspaper. Materials for such a prototype can be obtained by copying headlines, reading matter, and newspaper constants on an office copying machine.
3. Design a small newspaper for an organization to which you belong or design an employee newspaper for a firm with which you are acquainted. If such a newspaper exists now, analyze the typographic and design elements and make recommendations for any changes that would improve the newspaper's organization and appearance. (Use the suggestions in "How to Analyze a Publication" as a guide.)
4. Make a design plan for an "ideal newspaper." Select the page size, types for headlines and features, departments, and all the design elements that you believe should be incorporated in such a publication.
5. This project is extensive and you might want to do it as a team or class effort. Select a nearby community or neighborhood that doesn't have a newspaper (or that seems ready for a competitive newspaper). Make a basic design plan for a proposed newspaper for that community. This project should include considerable research and actual surveying by the students. A good starting place is study of the chapter on research and surveys in a newspaper management text such as *Newspaper Organization and Management* by Frank W. Rucker and Herbert Lee Williams (Ames: Iowa State University Press, 1969).

Notes

[1] From advertisements in *Editor & Publisher*, January 28, 1984.
[2] John E. Allen, *Newspaper Designing* (New York: Harper, 1947).
[3] Edmund C. Arnold, *Modern Newspaper Design* (New York: Harper, 1969), p. 266.
[4] Mario R. Garcia, *Contemporary Newspaper Design, A Structural Approach* (Englewood Cliffs, N.J.: Prentice-Hall, 1981), p. 23.
[5] Philip B. Meggs, *A History of Graphic Design* (New York: Van Nostrand Reinhold, 1983), p. 23.

Designing and Redesigning Newspapers

Newspapers all across the country—big ones, little ones, metropolitan dailies, community weeklies, and company and campus newspapers—have discovered graphics. There is constant examination and reexamination of the newspaper's appearance as editors and designers seek ways to make their product more appealing and attractive.

Many newspapers have created a new position on their staffs for a person who specializes in the graphic aspects of communication, the *graphic journalist*. The graphic journalist is one who combines the skills of the designer with those of the reporter and editor to devise methods of presenting information in the most effective way possible.

The graphic journalist combines words, type, art, borders, photographs, drawings, and typographic devices to form a unit of information. He or she also works with the editorial staff in designing and laying out newspaper sections and pages. The graphic journalist must understand design principles and graphic and typographic techniques. In addition, the person must have the editor's understanding of news and information and the opportunities for its visual presentation.

The editors of a publication, working with the graphic journalist, strive to apply the following formula: If the rewards received from reading an article were divided by the effort made to accomplish the task, the result would equal the possibilities of the piece being read.

$$\frac{\text{Rewards received}}{\text{Effort to read}} = \text{Chance of being read}$$

Communicators should keep this formula in mind. People will refuse to read material that is full of barriers, choosing instead the many forms of communication that require little effort to absorb.

Another major reason for the growing attention to graphics is the intrusion in the editorial room of the electronic age. Pagination devices are becoming standard tools for putting together newspaper pages, and many journalists are involved in arranging information units to form a page layout on a video screen. These journalists will make graphic as well as editorial decisions.

▼ Developing a Design Philosophy

One of the first steps in sharpening your skills as a graphic journalist, or a communicator who will be involved in the design of printed communications, is to start developing a *design philosophy*. A design philosophy might be defined as your beliefs and attitudes toward all the aspects of graphic design. It would include such things as your preferences for particular type styles for certain situations, how you would project an image through the use of typography and graphics, and how you would apply such subjective ethical terms as taste and judgment to your work.

As Bill Ostendorf, who has helped design more than thirty publications, puts it, "Design is not the implementation of a series of 'correct' or 'right' techniques. Design, like editing, involves taste, judgment, discipline, training, and experience. And it requires some kind of design philosophy."[1]

Design philosophy helps the communicator to bring meaning and direction to the many decisions that have to be made.

How do you develop a design philosophy?

"Unfortunately, no one can really hand you an article or list that provides such a philosophy. You have to develop and nurture your own," Ostendorf notes.[2]

A philosophy of design cannot be acquired by reading books or attending classes, but they can help. It requires a career-long effort to learn everything possible about communications—not just design aspects, but also such things as market research, advertising, public relations, statistics, computers and new technological developments, reporting and editing.

The first step in developing a design philosophy is to make a decision to keep working on it. This should include reading lots of books and professional magazines, attending workshops, and studying the design work of others.

▼ The Approach to Redesign

Let us assume that you have just been appointed editor of a small newspaper. This is your first job and you want to do all you can to make the newspaper as effective as possible. It does not matter if this is a large or small community newspaper or a university or organizational publication, the procedure is the same.

Of course, you examine the content of the newspaper to see if it is supplying what the audience wants and needs. In addition, you consider the method of publication and distribution. Your major concern, however, is with the typographic dress the newspaper is wearing. The decision is made to do a complete overhaul of the design. But how will you start and carry out a redesign project?

One course of action is to follow the procedures professional designers recommend. These procedures offer a guide for designing a new newspaper as well as redesigning an existing publication.

A newspaper to be successful must have three qualities: (1) It must contain the information people want and need. (2) It must attract the audience. (3) It must be interesting.

Design can help make a newspaper attract the desired audience and be attractive. It can also make the newspaper interesting.

But before specific design principles are put to work to create the physical appearance desired, a few general guidelines for designing effective newspapers should be reviewed:

- Typography and graphics can tell the reader what type of publication is being produced. They can say, "This is a hard-hitting, crusading publication." They can say, "This is a dignified publication devoted to accuracy and thoroughness." And they can say, "This newspaper is taking a light, breezy approach to all the activities it is attempting to cover."
- Typography and graphics can provide instantaneous identification for a publication. Readers should recognize immediately that this is a

Fig. 16-1 The trend to redesign for more effective communication is not restricted to the newspaper's front page. Section pages are receiving attention, too. This section page from the *Tampa Tribune* contains an information package as well. The package is the result of work by a writer, editor, photographer, and graphic artist, whose combined skills are used to present the story in the most effective way possible.

certain newspaper, not the *Daily Times* from the neighboring town or the *Employees Gazette* from the plant down the road.

- Typography and graphics can help readers spot the sports, life-style, or other departments or classifications of material the publication contains. They can help the reader sort out the information and indicate which material the editors believe is of special importance and which they consider of minor interest.

Fig. 16-2 Size has nothing to do with good design. The *Swanton Enterprise,* a weekly with a circulation of 2,000, was redesigned in modular format by David W. Richter of the Ohio State University School of Journalism.

Before the principles of good design are put to work, a thorough study of the situation should be made. Everyone in the organization should become involved. The newsroom and advertising, production, circulation, and marketing departments should participate. An analysis of the market and reader demographics should be part of the research.

When the *Oregon Statesman,* a 45,000-circulation daily in Salem, was redesigned, publisher John H. McMillan says the designer spent a week carefully reading copies of the newspaper to become familiar with its form and content. Then decisions were made based on a realistic look at the resources. This included the limitations of the staff and equipment, how changes could be made within the framework of the ongoing work schedule, and what could be done considering the size limitations of the "news hole" (the area that could be devoted to everything but advertisements) and the entire newspaper.

The Redesign Plan

Once a study of the present format and resources is completed, a plan of procedure and a timetable can be organized. Many designers follow a step-by-step outline that looks something like this:

1. Research and set the goals.
2. Survey the readers for their views regarding the newspaper.
3. Devise a realistic timetable.
4. Specify how and by whom decisions will be made.
5. Make the design decisions.
6. Produce a prototype.
7. Evaluate and refine the prototype.
8. Produce a final prototype to be used as a guide in designing actual pages.
9. Put the new design into action, evaluate it, get reader reactions, and make necessary adjustments.
10. Continue evaluation, solicit reader and staff reactions.

Goal Setting for Newspaper Design

There isn't much point in jumping into a design program without first deciding what it is supposed to accomplish. Goals should be set, and these goals might be entirely different from one newspaper or one community to another.

For instance, one of the criticisms leveled at newspapers by the judges in an Inland Daily Press Association makeup contest was that some newspapers were printing sixteen or more stories on the front page. The judges ruled that the effect was too much clutter and decreased readability. The judges also noted that "the most annoying design element on many newspapers is promotions found at the top of the front page in dark, ugly boxes and in color." They pointed out that the devices were so powerful that "they easily pull the reader's eyes away from stories that are less graphically appealing but which should be read nevertheless."

On the other hand, when the *Orlando Sentinel* was redesigned, more stories and visual items were packed into page 1, the nameplate was underlined with a color bar and topped with a promotion in colorful high-tech deco style. And, in the first four months following its redesign, Sunday single-copy sales were up 15 percent. Overall daily single-copy sales were up 13.5 percent from the previous year.

The lesson for the designer is that goals should be set, and the design should help the newspaper achieve these goals. In the case of the *Sentinel*, one of the main goals of the redesign was to increase sales to tourists visiting the city, establish name recognition, and attract attention.

Fig. 16-3 Mark Williams, editorial art director, designed the *Orlando Sentinel*. He used a modular format to reflect the news of the day as well as the character and personality of the publication. Note how the elements fall into a design of rectangles within a rectangle. The elements in the *Orlando Sentinel* page form a pattern of rectangles, which designers of modular pages believe can be more interesting than the vertical thrust of the traditional newspaper. Note that the upper right rectangle below the nameplate area contains a story and its related sidebar. (Courtesy *Orlando Sentinel*)

Each publication, then, should set its goals to fit its individual situation, not necessarily to win awards in contests.

However, there are a number of goals that are valuable for all designers to consider. Study after study has shown that readers like a well-organized newspaper. They like to find information easily, and they like to find it in the same spot issue after issue. A basic objective of any redesign project might be to organize the content to achieve this goal.

Goals should include adherence to the general guidelines for good design such as making the newspaper more visually attractive while building a consistent design theme throughout and designing with simplicity and restraint. That is, the design should never overwhelm the message. Graphics and design elements should not only be colorful but also should convey information to the reader.

A consistent award-winner and a newspaper that is recognized for its effective design is the *New York Times*. Yet this newspaper continues to use a traditional design for its main news section that might be regarded as outdated and old-fashioned. One reason the *New York Times*, regarded as *the* newspaper of record, receives acclaim for its graphics is that it exemplifies the basic premise of effective design. A newspaper's appearance should reflect its editorial philosophy and appeal to the audience it wishes to attract.

After sixteen months of work the *Press Democrat* of Santa Rosa, California, emerged in a completely new format. The design was greatly

Fig. 16-4 The *Santa Rosa Press Democrat* as it appeared before the redesign project. It had a modular look that designer Lou Silverstein found "clean but boring." (Courtesy *Press Democrat*)

influenced by what readers said they wanted in their newspaper. The goal of the *Press Democrat* staff was to produce a newspaper that skillfully delivered the news.

One of the first steps taken was a readership survey. This was made by a Boston marketing firm, Urban and Associates. The firm did a telephone survey of 1,000 households in the *Press Democrat* circulation area. Results of the survey included the discovery that subscribers were reading

Fig. 16-5 After the redesign the *Press Democrat* won the "best front page in California" award of the California Newspaper Publishers Association, and reader reaction was overwhelmingly positive. (Courtesy *Press Democrat*)

the newspaper deeply and critically. They wanted hard news, but they also wanted good reporting that contained more than just the facts—they wanted stories that were interesting and insightful.

As Chris Urban, who conducted the survey, told the *Press Democrat* staff, there was a strong interest in news. At the same time, the lighter and more entertaining stories appealed to a growing readership of younger and more mobile individuals. But the light approach should be in the proper place and with the proper perspective. In other words, the serious news should be handled seriously, and the lighter feature stories could be made to be more fun to read.

Lou Silverstein, chief designer for the *New York Times,* worked with the *Press Democrat* staff on the new design. He is not an admirer of modular design, which was the approach used by the newspaper before the redesign. Silverstein sees modular as a way to cram news into little boxes. It creates the impression that everyone at the newspaper spends time making sure everything is neat. The result is a newspaper that looks clean but boring.

"My criteria for a well-designed news section is that it have a balance between newsiness, surprise, serendipity on the one hand, and on the other a feeling of organization and a demonstrated effort on the part of the editors to make things easy and clear, including the use of graphic tools," says Silverstein.

"Graphics should play an important part in the presentation of news just as every available tool should play an important part for an editor. Ideally, the best visual people in the organization should be involved as early as possible and as deeply as possible in the planning and then the design of important news pages . . . a sophisticated top editor will use his best visual talent in the most integrated way with his own journalistic talents," he adds.[3]

The look of the new *Press Democrat* is altogether different from its previous appearance. It has a more vertical thrust with more one-column heads and stories. The look is more unpredictable than with the modular approach. It is more exciting. At the same time, the whole package is better organized. Each edition is chock-full of material and does not look rigid.

Editor Michael J. Parman says, "Reader response has been overwhelmingly in favor of our new look. We were perceived as being somewhat dull in appearance. Obviously, the wonderful design that Lou Silverstein gave us is anything but dull. We have had exceptional growth the past several years, and I have no doubts that the design has a great deal to do with that."[4]

Press Democrat circulation has increased from 73,000 to 87,000 in the three years since the change, and this has outstripped the growth in population for its circulation area. The newspaper won the "best front page in California" award from the California Newspaper Publishers Association in 1989.

Making Design Decisions

Once goals have been set, the design or redesign of a publication can proceed. The design decision-making process can be organized along general and specific lines. First, let's consider some suggestions for the

(a)

(b)

(c)

Fig. 16-6 A more colorful *Christian Science Monitor* greeted the new year in January 1989. This was the third redesign of the newspaper in thirteen years. Shown is (a) the *Monitor* in 1983 just before the second redesign, (b) the paper when it appeared in a new dress in the fall of 1983, and (c) as it greeted its readers in a more colorful format in January 1989 after the installation of $1.8 million in electronic color prepress and communications equipment. The paper switched from 25-pound newsprint to 40-pound stock to achieve high print quality and produce a crisp and durable newspaper.

overall design and then the specifics, such as the front page, inside pages, section pages, and the editorial page.

Designers generally agree on these rules for good design:

1. The design must communicate clearly and economically with maximum legibility.
2. The design should create identity for the newspaper.
3. The design must communicate with a sense of proportion. That is, the breadth of the design should be controlled by the context of the news of the day.
4. The design must communicate in a style that is easily recognized.
5. The design must communicate with consistency. This consistency should be helpful to the readers in finding content in each edition.
6. The design must accomplish its goal with economy. This means changes in arrangement can be made quickly, and space and materials used with acceptable budgetary restraint.

Effective design can often be achieved if the designer will consider four steps: (1) Square off type masses. (2) Use plenty of white space. (3) Put life in the four corners of the page. (4) Keep it simple.

Although some designers are moving away from the modular approach that was so popular in the 1970s and 1980s, it is still an approach of choice for many, and should be studied. In the modular approach a newspaper page consists of building blocks of pleasing rectangles. Type should be squared off so that each column of type in a rectangle ends at the same depth. The square off at the bottom of columns should be in a straight line to create harmonious rectangles. Type should not be allowed to zigzag across the pages. The rectangles, both vertical and horizontal, should be arranged to create a pleasing combination and an appealing page.

Pages need breathing room. White space used effectively can brighten a page—it is not wasted space but a necessary design element. It can help the reader by isolating and emphasizing elements and by indicating where one item ends and another begins. A crowded, jammed-together page should be avoided.

Judicious editing of reading matter can provide increased white space and increased readability. White space can also be added by adequate separation between columns, indenting heads, using a pica of white consistently between stories, and/or setting captions in a width narrower than the art they identify.

Any element on a page that does not help the reader can be called an ornament. Any type of decoration is an ornament. Ornaments should only be added to layouts if they perform communication functions, such as establishing identity for a publication or specific departments.

When the eye travels through a page it follows a rough Z line from upper left to upper right, to lower left and across to lower right. The four corners of a page are contact or turning points in this Z line. They have been called the *hot spots*. A strong element in each hot spot will help give a page motion and balance.

In making design decisions, the basic principles of design plus the precepts of legibility, readability, and suitability must be applied. In addition, the design should be flexible so changes can be made easily to avoid a rigid, day-after-day sameness.

It should be remembered that nothing lasts forever, and the design should be reevaluated from time to time. Changes should be made as

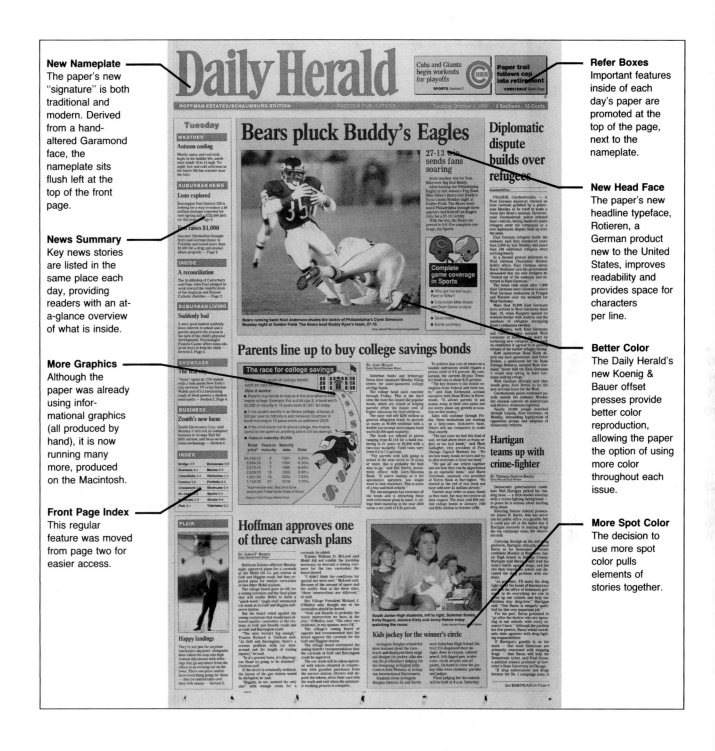

New Nameplate
The paper's new "signature" is both traditional and modern. Derived from a hand-altered Garamond face, the nameplate sits flush left at the top of the front page.

News Summary
Key news stories are listed in the same place each day, providing readers with an at-a-glance overview of what is inside.

More Graphics
Although the paper was already using informational graphics (all produced by hand), it is now running many more, produced on the Macintosh.

Front Page Index
This regular feature was moved from page two for easier access.

Refer Boxes
Important features inside of each day's paper are promoted at the top of the page, next to the nameplate.

New Head Face
The paper's new headline typeface, Rotieren, a German product new to the United States, improves readability and provides space for characters per line.

Better Color
The Daily Herald's new Koenig & Bauer offset presses provide better color reproduction, allowing the paper the option of using more color throughout each issue.

More Spot Color
The decision to use more spot color pulls elements of stories together.

conditions warrant, but they should always be made after careful evaluation and study.

Finally, it is important to communicate what is happening every step of the way with all staff members and solicit feedback. But the final decision should be made only by the person responsible for the design.

Once design changes have been agreed on and everything is in order, it is time to see how these changes will look in a finished page. This can be done by creating a prototype. A *prototype* is a pasteup of a "dummy"

Fig. 16-7 The major design decisions that were made in a complete overhaul of the *Daily Herald*, Arlington, Illinois, are shown in this new front page for the newspaper. (Courtesy *Step-by-Step Graphics* magazine, Peoria, Illinois (309) 688-2300)

(a)   (b)

Fig. 16-8 It isn't just the general circulation newspapers that are redesigning to present a more interesting and readable product. The trend extends to special interest, organizational and association publications. This is the *San Francisco Banner Daily Journal,* which serves the legal community in California, (a) before and (b) after its recent redesign.

page. Headline types are used, illustrations are put in place, and simulated body matter is arranged on the page.

Materials for a prototype can be clipped from other publications as long as they are exactly like the real thing in size and design. They are pasted on a grid sheet so the completed prototype looks just like the designer envisions the page except that the words do not make sense. They just show the form and format.

It is a good idea to photocopy the page. This copy will give a more "printed" look to the prototype and will help it appear realistic.

Prototypes are models of actual pages. During the design or redesign process many prototypes may have to be made before the final design is adopted.

▼ The Body Type

The tendency in newspaper design is to start with the front page and spend most of the time and effort with this small part of the whole package. While the importance of an attractive "display window" cannot be overlooked, the rest of the publication deserves equally serious planning.

Since the basic objective is to get the newspaper read, the first step in creating a design should be an examination of what is read most—the body type. Time spent on good design should not be lost because the reader gives up on a story set in type that inhibits pleasant reading.

The criteria for selecting a proper body type are legibility and readability. Legibility is the visual perception of type, words, and sentences. Readability is the comprehension and understanding of the communication.

Body type should be legible, of course. The variables that make one typeface more legible than another include the serifs, the type size, and the letter design. The size of the typeface on the body, the leading, the set width, and column widths are also factors.

Studies have shown that serif type is preferred for newspaper body type. Serif types have more reader appeal, but this may be changing. Numerals are more legible when set in Sans Serif. Italics, obliques, and boldface types should not be used for body type in newspapers. Reverse type slows reading by almost 15 percent.

Typeset words are perceived not by letter but by shape, and this shape outline is lost when words are set in all caps. Type set in all caps slows reading speed, reduces legibility, and takes up to 30 percent more space.

Quite often the most frequent comment received when a new design is in place concerns the type used for reading matter. When the *Charleston Gazette* was redesigned the body type was increased from 9 to 9.5 points. Don Marsh, the editor of this West Virginia newspaper, reported in *Editor & Publisher*, "Although we've given up about 5 to 7 percent of our news space by using larger type, with our readers the change was the most popular aspect of our redesign."[5]

When the *St. Joseph* (Missouri) *Gazette* was redesigned, one change the

abcdefghijklm
ABCDEFGHI

abcdefghijkm
ABCDEFGHI

Fig. 16-9 Century, a popular reading matter typeface, as it was cast in foundry type (top) and as it was redesigned (bottom) by Tony Stan for International Typeface Corporation. The ITC version was designed for modern printing methods and digital typesetting. Note the more subtle letterfitting, larger x height, and shortened ascenders. The opening in the c has been enlarged, and some serifs have been selectively eliminated.

VENETIAN

Hae abapoa baephi mapobenea nihiopasb taehpro aopah ihpead eahorpheatid basa poihin aenbopana abanpe posi phoreath ahe asboipin maneboa preipha otphaon ihnopo bep aithap abonen aopeha abapoa eha beapiov paoha ahi ihepab eahor pheatbasa poihinaeb opam abpeihp aohapr

FRENCH OLD STYLE

Hae abapoa baephi mapobenea nihiopasb taehpro aopahatoame ihpead eahorpheatid basa poihin aenbopana abanpeposahn bdgl phoreath ahe asboipin maneboa preipha otphaon ihnopoaneihm bep aithap abonen aopeha abapoa eha beapiov paoha ahonthlrk ihepab eahor pheatbasa poihinaeb opam abpeihp aohaphoinbw

ENGLISH OLD STYLE

Hae abapoa baephi mapobenea nihiopasb taehpro aopathatoa ihpead ihpead eahorpheatid basa poihin aenbopana abanpe posahn phoreate phoreath ahe asboipin maneboa preipha otphaon ihnopoane bep ath bep aithap abonen aopeha abapo eha beapiov paoha ahont inepabme ihepab eahor pheatbasa poihinaeb opam abpeihp aohaphoin abapoah

TRANSITIONAL

Hae abapoa baephi mapobenea nihiopasb taehpro aopahatoa ihpead eahorpheatid basa poihin aenbopana abanpe posahn phoreath ahe asboipin maneboa preipha otphaon ihnopoane bep aithap abonen aopeha abapoa eha beapiov paoha ahont ihepab eahor pheatbasa poihinaeb opam abpeihp aohaphoin

MODERN

Hae abapoa baephi mapobenea nihiopasb taehpro aopai ihpead eahorpheatid basa poihin aenbopana abanpe pos phoreath ahe asboipin maneboa preipha atphaon ihnohi bep aithap abonen oapeha abapoa eha beapiov pacha ah ihepab eahor pheatbsa poihineab opam abpeihp aohapho

EGYPTIAN

Hae abapoa baephi mapobenea nihiopasb ta ihpead eahorpheatid basa poihin aenbopan phoreath ahe asboipin maneboa preipha ots bep aithap abonen aopeha abapoa eha beap ihepab eahor pheatbasa poihinaeb opam ab

GROTESK

Hae abapoa baephi mapobenea nihiopasb taehpro aopa ihpead eahorpheatid basa poihin aenbopana abanpe po phoreath ahe asboipin maneboa preipha otphaon ihnop bep aithap abonen aopeha abapoa eha beapoiv paoha t ihepab eahor pheatbasa poihinaeb opam abpeihp aohar

Fig. 16-10 Which letter form would you select for the reading matter in your publication? The variations in these representative types seem slight when viewed individually but when compared they are striking. (The wording doesn't make sense since meaningful words might interfere with study of the letter forms.)

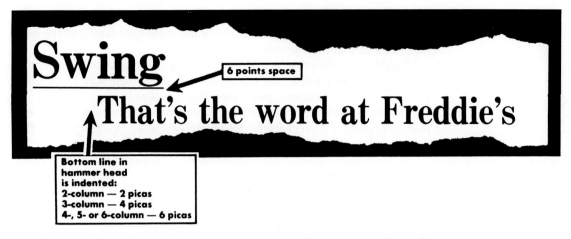

Fig. 16-11 A stylebook and a design manual are important reference sources for a publication's staff. The precise specifications for handling a reverse kicker head are spelled out in this illustration from the typographic manual of the *Philadelphia Inquirer*.

readers seemed to like most was an extra ½-point of leading between the lines. The *Gazette* sets 9-point type on a 9.5-point base.

Body type should be examined for its legibility and its reproduction qualities. It should have a clean and open cut. When type is printed at high speeds with thin ink on absorbent newsprint, the letters tend to spread and distort. The space between the ends of the strokes of the letter *c*, for instance, should have an opening big enough so it won't appear to be an *o* when printed at high speed.

The letter should be strong, or bold, enough to avoid a gray look. It should have sufficient contrast between thick and thin strokes to break monotony, but the thin strokes should not be so thin they tend to fade away.

The typeface chosen for setting the great majority of words in a newspaper should have good proportions. The relationship of height to width should approximate the golden rectangle in proportion. The *x*-height should be ample to give full body but not so large that it interferes with the clear distinction of the ascenders and descenders of the letters. Time taken to select a body type that increases legibility will pay dividends in reader reaction.

▼ The Newspaper "Show Window"

First impressions are hard to change, and the first impression the reader will have of a newspaper is created by its front page and the design elements it contains. These elements include the nameplate (also called the *flag*) and any embellishments in the nameplate area, the headlines and cutlines, any standing heads such as those for the weather and index boxes, plus any other regular features on the page.

The *nameplate* is the newspaper's trademark. It should be legible, distinctive, attractive, and appropriate. As a general rule designers recommend that it should have harmony with the other types on the page. This usually means it should be of the same type family as the basic headlines or it should contrast well with them.

Many newspapers choose a Black Letter type for the nameplate because of the image created of a time-honored, respected institution. Black Letter contrasts well with most types chosen for headlines.

One of the first steps in planning a newspaper is to adopt a *stylebook* that incorporates correct usage regarding grammar, spelling, capitaliza-

Fig. 16-12 This page from the *St. Petersburg Times* illustrates a number of newspaper design terms.

tion, and so on for handling various types of information. It is the newspaper's operating handbook. It is a good idea to create a *typography manual* as well.

The typography manual would include a *head schedule*. This is a listing, with examples, of all the sizes and styles of types to be used in headlines and their unit counts as well as their arrangements. The head schedule becomes a handy reference source and helps create a consistent, well-organized appearance.

Some newspapers have very elaborate typography manuals that are virtual textbooks of design. For the small newspaper, a simple manual

Fig. 16-13   The impact of *USA TODAY* on newspaper design extends to the lexicon of the trade. Here are some design terms coined at the newspaper. (Courtesy *Design*, the Journal of the Society of Newspaper Design)

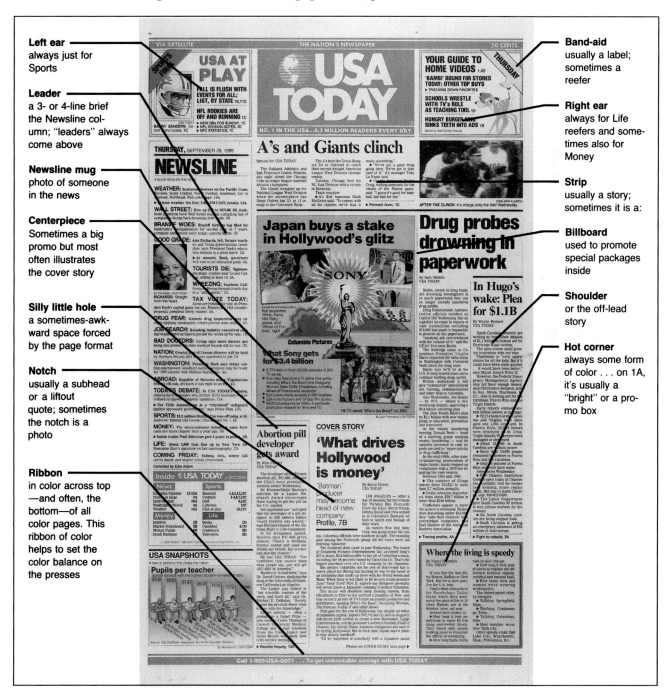

Fig. 16-14 (Left) Many newspapers are adopting a headline form that includes a short paragraph as a subhead rather than the traditional deck. When the *Los Angeles Times* was redesigned in late 1989 it devised a head similar to this one from *The Oregonian*. In its promotion of the redesign *The Times* said the form would give busy readers a quick summary of the stories. (Right) Note that the top deck of this head is in downstyle and the "nut graf" has a solid square instead of an open square as in the head at left. (Courtesy *The Oregonian*)

2 officials 'a-peel' tax distribution

☐ Road improvements are in danger, say Mayor Bud Clark and Portland Commissioner Earl Blumenauer

By GORDON OLIVER
of The Oregonian staff

Council approves housing project despite protests

■ Construction will start in May even though 300 area residents express disapproval. A greenbelt is included in revised plans.

By ANN RUDRUD
NEWS STAFF WRITER

Poles Say Holdouts Quit Mine

Associated Press

Fig. 16-15 Typical flush left head with all words capitalized. Note treatment of source line with rules above and below. (Courtesy *Denver Post*)

can be created that includes samples of the sizes and styles of types and borders available as well as samples of cutline treatments and other typographic features of the publication. Duplicate copies can be made on a copying machine for each member of the staff.

The headline has several functions to perform. It can attract the reader, persuade the reader to consider a story, and help make the page attractive. It can help create identity and personality. The typeface and arrangement should help the headline accomplish its tasks.

Usually the most attractive newspapers stick to one type family for best appearance and effective design in headlines. Another face that harmonizes with the basic family is sometimes chosen for contrast. Headline type is selected for legibility, personality, durability, range of series available in the family, and its unit count.

The guidelines below are helpful when placing elements on the front-page grid:

1. Study other newspapers, especially those that have been recognized for their outstanding design (see "Creative Communication" box in this chapter). Spend some time on visualization. Try to form an idea of how the page should look when completed.

2. Make some thumbnail sketches of possible arrangements. Then select the sketch that most closely resembles the page arrangement you think is best.

3. Decide where the nameplate will be placed. Will it be a permanent fixture at the top of the page? Or will a "skyline" story, headline, or promotion device be placed above it? Will a *floating flag* (a smaller version of the nameplate) be used? Often newspaper designers create several sizes of a nameplate in several column widths. These can be

181 feared dead
in Spain jet crash

UNITED PRESS INTERNATIONAL

Gas 'abuse' found

$100 million refund due

Swing

That's the word at Freddies's

Pilots ignored warning, tape shows

26 passengers died on blazing jetliner June 2

moved around the page. But if the flag floats, care should be taken not to lose it on the page and thus lose the newspaper's identity.

4. Decide on strong elements for each hot spot on the page. Usually the major elements are placed in the upper-left and upper-right quadrants. Tradition dictated that the lead story of the day be placed in the upper-right corner, but this is no longer necessary.

5. Consider placing strong elements in the two lower quadrants of the page as well as in the top ones. This will cover the hot spots and help to create a well-balanced page.

6. Check to see if headlines fight each other, if heads "tombstone" or "butt." Tombstones are heads of identical form placed side by side. If they seem to confuse the reader they should be changed. If heads are side by side they should be distinctly different, in most cases.

7. Check each design element on the page to see that it is performing a function. If not, consider eliminating it.

8. Have the page exhibit the attributes of basic design principles—balance, harmony, proportion, unity, rhythm, and contrast.

▼ **The Inside Pages**

Since the change to optimum format, the design of inside pages has been in a state of turmoil. When the broadsheet (the full-size newspaper page) was changed from eight to six columns and the tabloid from five to four, the placement of advertisements changed from several basic patterns to confusion and often chaos. It was possible to find reading matter set in

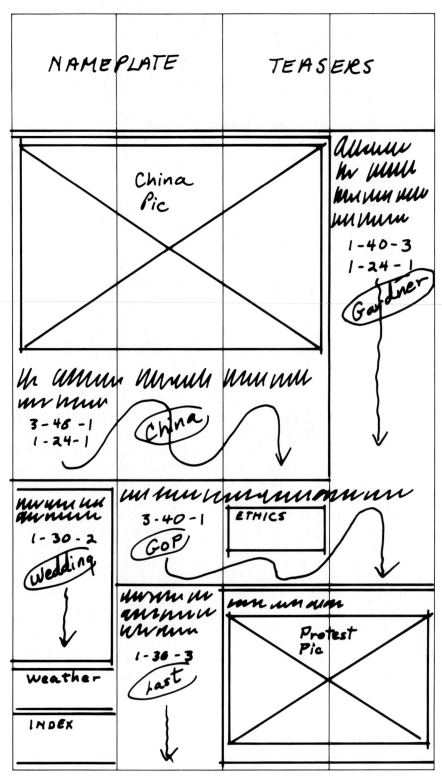

Fig. 16-21 A rough dummy of the *Seattle Times* front page. Newspaper page dummies are usually made to one-half scale on 8½ by 11 sheets. Each story has an identifying slug (a word or two) and a code to indicate headline size. The code 3-40-1 for the GOP story means a headline three columns wide in 40-point type and one line deep. The code 1-40-4 for the Gardner story means a headline one column wide in 40-point type and three lines deep.

Fig. 16-22 The *Seattle Times* is a consistent winner in the Society of Newspaper Design awards competitions. Just three typefaces are used in its basic head schedule: Imperial, Vega, and Cairo. (Reprinted with permission of the *Seattle Times*)

Fig. 16-23 The standard newspaper page sizes and column widths adopted by many newspapers in 1984. (Courtesy American Newspaper Publishers Association)

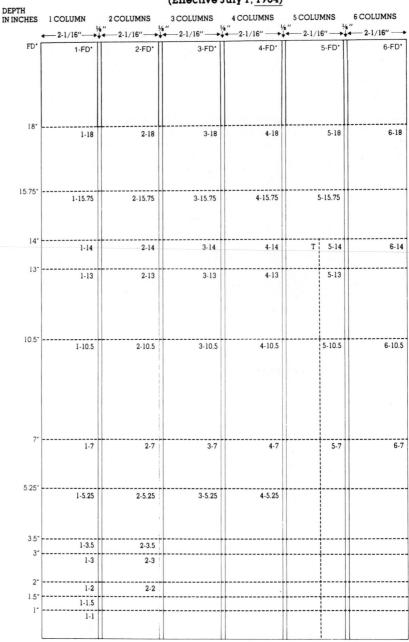

The Expanded
SAU™ Standard Advertising Unit System
(Effective July 1, 1984)

- 13 inch depths are for tabloid sections of broadsheet newspapers.
- T is a full page size for 21½ inch cut-off tabloid newspapers. It measures 14 x 9⅜".

1 COLUMN 2-1/16"	5 COLUMNS 10-13/16"
2 COLUMNS 4¼"	6 COLUMNS 13"
3 COLUMNS 6-7/16"	DOUBLE TRUCK 26¾"
4 COLUMNS 8⅜"	**FD—FULL DEPTH**

*FD = full depth of 21" or longer, as according to individual newspapers' printed depth as indicated in Standard Rate and Data Service listing. Printed depth generally varies in newspapers from 21" to 22½". All newspapers can accept 21" ads and may float the ad if their printed depth is greater than 21".

as many as four or five different widths on one page. The reader was faced with a hodgepodge that defied comfortable reading and the basic rules of legibility.

The situation has improved, however, as most newspapers have adopted the *Standard Advertising Unit.* The SAU was devised by the American Newspaper Publishers Association in an attempt to bring standardization and order out of the chaos caused by the use of various column widths.

The SAU system starts with a 2 1/16-inch column with 1/8 inch between columns. Note the substitution of inches for points, picas, and agate lines. The full-page width contains six columns for advertisements. The ANPA suggests that only inches and fractions of inches be used to measure advertisements. It also recommends fifty-seven modular sizes for national advertisements but leaves size decisions for local advertisements up to the individual newspapers.

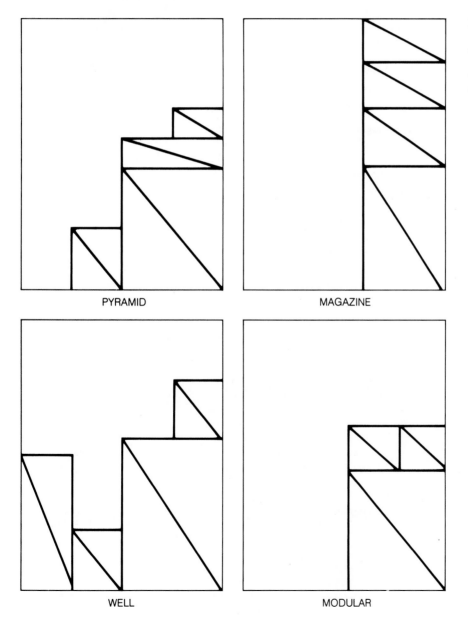

PYRAMID

MAGAZINE

WELL

MODULAR

Fig. 16-24 Some of the more popular inside page arrangements for placement of advertisements. The key to attractive inside pages is to adopt a consistent pattern and keep the top of the page, especially upper left-hand corner, open as much as possible for editorial display.

The new system has helped designers produce more attractive inside pages, and helped newspapers sell advertisements as well.

There are several standard patterns for placing advertisements on a page. A designer should adopt one of them and use it throughout the publication—except, perhaps, for special sections. But even there, if the section includes more than one page the pattern should be the same for all the pages.

Fig. 16-25 Another inside page arrangement that is gaining popularity is a full page of advertisements. This will free other pages for more extensive display matter. This page is from the *Hastings* (Minnesota) *Star Gazette*, Michael J. O'Connor, publisher.

The most common basic patterns include:

- *Pyramid:* Here the advertisements are built like varying sizes of building blocks from the base of the largest ad in the lower right-hand corner of the page. This leaves the hot spot or area of initial eye contact—the upper-left corner—open for editorial matter. This is actually a half-pyramid, and it is often called that.
- *Magazine:* Here some columns are filled with advertisements and others are completely open for editorial display. This arrangement works well with modular layout and pagination.
- *Well or double pyramid:* Here advertisements are placed on the inside and outside columns plus across the bottom of the page. This forms a sort of well that can be filled with editorial material.

Fig. 16-26 The magazine approach to placement of advertisements was used here. Such an arrangement leaves full columns open for display of editorial matter. It also gives an orderly appearance to the page. (Courtesy *San Francisco Daily Journal*)

Fig. 16-27 Often design efforts are concentrated on the front and section pages. This inside page illustrates effective design for a page containing large advertisements. The advertisements are in a half pyramid arrangement built from the lower right corner of the page. This leaves the upper left "hot spot" of the page open for editorial display. Note the three multicolumn heads and a photograph placed in three orderly rectangles. All the columns are covered by heads that do not tombstone, and white space is used effectively. (Courtesy *Sun News*, Myrtle Beach, S.C.)

• *Modular:* Here advertisements are clustered to form rectangular modules on the page, another arrangement that works well with the pagination equipment coming on line in many newspapers.

Regardless of the pattern adopted, there are a few points to remember when planning inside pages for maximum effectiveness. The important top-of-the-page areas should be kept open for the display of editorial material as much as possible. Advertisements should not be placed so high in the columns that little space is left for reading matter. There should be enough space above advertisements for a headline and arm of body type at least as deep as the headline. If the space left is smaller, the advertisements should be rearranged if at all possible.

There should be at least one "stopper" on each inside page. The attention-getter could be an illustration or a story with a strong headline. Some editors try to have an illustration on every inside page. When editorial art is placed on the page, however, it should not compete with advertising art. The two should not be placed side by side.

Care should especially be taken to harmonize editorial and advertising content. A story about an airline crash, for instance, should not be placed on the same page with an airline advertisement.

The best designed inside pages are pages that have no tombstones; no long, unbroken columns of body type; and no "naked" columns (tops of columns of body type without headlines, art, or rules).

▼ The Editorial and Section Pages

The design of editorial and section pages should reflect the interests and content of that particular unit while preserving the flavor of the entire publication. Some ways this can be accomplished include using page logos that incorporate the publication's nameplate, using the same type family for headlines but in a different posture, and using the same headline form throughout the newspaper. For example, if downstyle, flush left is used in the main section, unity can be preserved if that form is used in all sections.

The Design of the Editorial Page

The editorial page should be distinctly different from the other pages. The readers should understand clearly that this is the page of opinion, the page where ideas clash. But while the page should be different, it should invite the reader by being lively and bright.

There is a trend toward eliminating the editorial page. However, those who believe taking positions is an important function of the media maintain that editorial pages should be improved, not eliminated. They point out that the key to developing a good editorial page is presenting well-written editorials that take strong stands on issues that are important to the readers.

The design of the editorial page can help make it a vital part of the publication. Some things that can be done to make it graphically different and interesting include:

- Eliminate column rules and use more white space between columns.
- Set body type in a larger size.
- Use larger size heads but in a slightly lighter weight than the main headlines.
- Design fewer columns to the page, with a wide measure, but keep within the readability range.

Fig. 16-28 There is a trend among some newspapers to neglect the editorial function. Others help preserve this function and attract readers to this serious aspect of journalism by designing the editorial page with care. (Courtesy *Washington Times*)

• Use graphics to add interest to the page. These could include cartoons, editorial pictures, charts, maps, or diagrams.

The editorial page can help a publication develop its distinctive personality, and it can be an attractive graphic feature that can increase reader interest.

The Design of Section Pages

Each section presents a specific challenge to the editor and designer. The design of each section should reflect its purpose and personality and be in harmony with the complete newspaper. Some sections that most frequently appear in newspapers are "Family Living," "Food," "Sports," and "Business."

The "Family Living" or "Life-Style" section is a recent development. It has taken the place of what was called the "Society" or "Women's" section. The scope and content of this section has been broadened and changed. Traditionally this section was given what was thought to be a "feminine" design treatment. Type styles were softened, and photo treatment was formal and subdued. The trend now is toward the use of magazine design techniques while maintaining the general design philosophy of the entire newspaper.

Food pages often were incorporated with the women's section in the past. Now many men are cooking enthusiasts, and feminine typographic treatment is no longer used. The designer should remember that these are working pages, that recipes and instructions should be presented in

Fig. 16-29 Ornamentation with a purpose: Section logos should identify the section and help unify the entire newspaper package. These section logos preserve unity but give distinct identity to each section with appropriate graphics. (Courtesy *Press Democrat*)

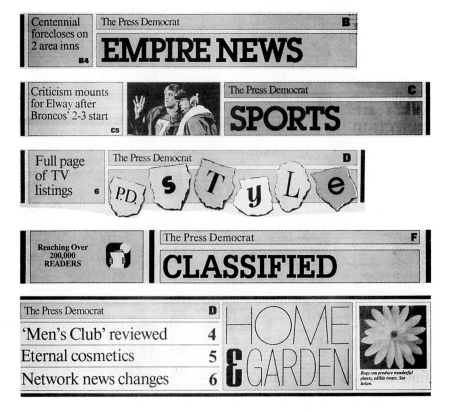

a clear and accurate style. It is also helpful if the arrangement makes the material easy to clip and file.

Sports pages are worlds unto themselves, and the sports staffs have operated as virtual separate entities on many newspapers. They have often had their own head schedules set in type of their own choosing regardless of whether it harmonized with the rest of the newspaper. In

Fig. 16-30 Section pages present specific challenges to their editors and designers. This food section page illustrates the skill and care needed to produce an outstanding, attention-commanding page. (Courtesy, *Register-Guard*, Eugene, Oregon)

recent years a change has taken place, and the appeal of the sports page has been expanded as more sports, such as soccer, have become popular, more attention has been paid to women's sports, and more people have participated in recreational sports.

The typography of the sports pages should reflect the vibrant action-packed activities they record. At the same time, they should be in harmony

Fig. 16-31 Communicators are turning to graphics to help tell the story. Here the skills of the artist and designer, as well as the writer and editor, were used to explain a complicated situation in an attractive and interesting manner. Note the formal balance of this feature section page. (Courtesy *Morning Call,* Allentown, Pennsylvania)

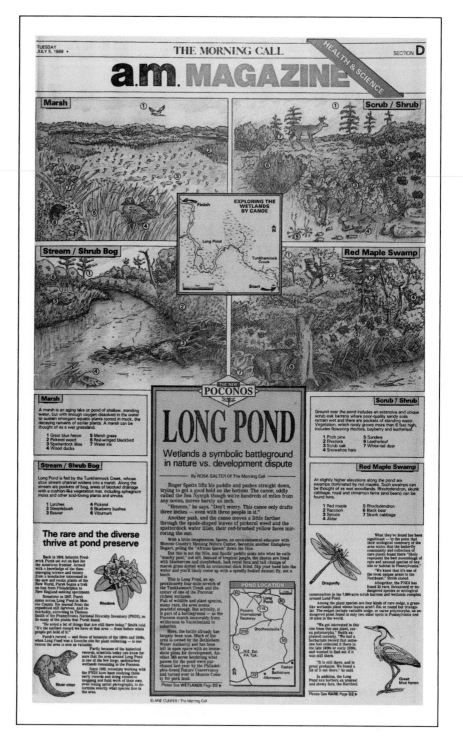

with the rest of the newspaper. A bolder posture of the same type as used in the entire newspaper might give more life to the sports pages. If the newspaper head schedule is upper- and lowercase or if it is downstyle, the sports page heads should use the same style.

Sports pages use more photographs than most other sections. The caption plan for these illustrations should employ the same arrangement as used in the rest of the newspaper.

Sports news includes much statistical matter, box scores, standings, and summaries. The trend is to set this material in agate (5½ point) type and group it for all sports in a separate page or part of a page. This practice helps the orderly organization of sports information and enables the editor to set the type in the optimum line width for agate. The typographic appeal of the statistical matter can be improved with distinctive headings for each topic. Small sketches to identify each sport can also be incorporated with the headings.

The *Wall Street Journal* has the largest daily circulation in the United States. The success of the *Journal* reflects the growing interest in business and economic news.

There was a time when business news was presented in staid, conservative—and even dull—writing and layouts. As the interest in business news has grown, business pages have become more lively. Designers now recognize that the techniques of good typography and graphics should equally apply to the business pages as to the other sections of the newspaper.

▼ A Case Study—The Guthrie Story

Guthrie, a city of 10,312, was Oklahoma's first capital. It has embarked on a major historical renovation program. Regional media have begun to refer to it as "the Williamsburg of the West."

Guthrie is served by the *Guthrie Daily Leader*. Larry Adkisson, the general manager, suggested that the *Leader* consider a redesign to carry out the basic principle of newspaper design—a newspaper's appearance should reflect the character of the community it serves.

The new *Leader* adopted some design techniques that are considered outdated but that turned out to be received enthusiastically by readers and staff. The newspaper adopted a "new" turn-of-the-century style and vertical layout.

Column rules were put back in the pages, and multideck all-capital headlines and bordered photographs were used. The basic design was changed from horizontal to vertical.

How did the staff at Guthrie, assisted by technicians from the Donrey Media Group, go about the redesign?

Goals were set first. They included adoption of a "Territorial-era format" aimed at maintaining the flavor of that period, hopefully without its faults, and at achieving the appearance of a Territorial-era newspaper without sacrificing the integrity and consistency of a modern newspaper.

Then very specific guidelines were adopted for attaining the goals. Because of their thoroughness, these guidelines are useful for anyone contemplating a design or redesign project, although the specifics may be different for different situations.

News Notes

Grassy body shelters car

KANSAS CITY, Mo (UPI) — Artist Bill Harding won over the unconvinced and proved you don't need paint to cover a car, all it takes is grass seed and time

Harding Friday parked his 1966 Buick LeSabre — covered with live grass — at a downtown intersection to show off for the lunchtime crowds

"Hey, it is real grass" exclaimed a believer

Harding glued 30 pounds grass seed to the car's body Sept 1 By Sept 4 his motorized lawn had sprouted

"Nature is our hope People living in the city get confused about their priorities If they look to nature they can see where they came from," said Harding, who showed up covered with grass himself

"I'll tell you, it takes all kinds," a businessman muttered before walking away from the green machine

'Mad world' recalled

MILWAUKEE (UPI) — The plan of whoever tucked a letter in the floorboards of a furniture store 68 years ago worked out just fine

A worker helping tear down the Waldheim furniture store found an envelope Thursday that read, "When this Building is Demolished the enclosed may be of interest"

The letter read "Mrs President died in Washington D C at 5 p.m. today War in Europe raging Germans storming Liege Belgium with great loss of life New plate glass front to this store just completed Weather fair Temperatures 80 Wheat 93 — Corn 72 — Oats 38 cents."

It was dated Aug 6, 1914, and signed "Lee G Smith"

"I've been wrecking for 14 years and this is the first thing like this we've ever come across," Richard Walters, president of Walters Excavating and Wrecking Inc, the firm doing the demolition, said Friday

At the bottom of the letter are the words, "A F Smith Attest"

(See NEWS NOTES—Page 12)

Sloop carries Poles to Newark

Immigration detains four

NEWARK, N.J. (UPI) — Federal officials are considering a request for asylum by four Poles who sailed across the Atlantic and into the shadow of the Statue of Liberty aboard a 38-foot sloop

Immigration and Naturalization Service officials questioned the four refugees in Newark Friday, after a Polish interpreter was found

"The four have asked to remain in the United States and their requests are being considered," said Clifford Landsman, an immigration supervisor in the Newark office

"Until a decision is reached in their case, they will be detained by INS in accordance with immigration service policy," Landsman said, reading from a prepared statement

Landsman would not reveal where the Poles, who sailed into New York Harbor Thursday, were being held

Officials did not identify the four, but The Daily Journal of Elizabeth reported it interviewed one of the men before they were detained

Jarek Neczaj-Hru... ...ics 38, of Lublin, said he and the ...her three men were members of ... outlawed Solidarity union

Neczaj-Hruzewics, the only member of the crew to speak English, identified the other men as Andrzej Plewik, 37, the captain, Andrzej Bienkowski, 34, the first mate, and Stanley Kozak, 38, the second mate, the Journal reported

Instead of returning to their homeland, he said the men headed to the United States, making stops in Africa, France, Spain and Bermuda on the way

Neczaj-Hruzewics was quoted as saying the group received permission four months ago to fly to Athens, where the sloop was moored, to sail it back to Poland

They chose to go to Elizabeth because they have friends there, Neczaj-Hruzewics reportedly said

The men docked in Elizabeth early Thursday and reportedly went to Bernie's Polish Bar, a block away, where they drank a few beers and were allowed to take a shower

Elizabeth police, acting on a tip, arrived and ordered the men to stay on the boat until the Coast Guard came and escorted the sloop to Governor's Island

They spent the night on the boat at Governor's Island, near Liberty Island, before being taken to Newark for their interviews Friday

Guthrie Daily Leader

Historical Capital of Oklahoma

SUNDAY, SEPTEMBER 19, 1982   90th Year No. 367   DAILY 20c SUNDAY 25c

Statehood Day Planning

Preparations for the Guthrie Statehood Day celebration continued with a media gathering Sept. 16 at the Calla Restaurant at Penn Place in Oklahoma City LEFT, Fred Olds talks to saddle maker Mr. Barr and Lt. Gov. Spencer Bernard. Olds is director of the Territorial Museum and a member of the Guthrie Statehood Day committee. ABOVE, Guests enter the buffet line. (Leader Staff Photos by Larry Adkisson.)

Redheads have pride gathering

LAGUNA HILLS, Calif (UPI) — Red-haired people have reason to keep their carrot tops held high.

Steve Douglas, a piano player who sold his $4,000 baby grand and other musical equipment to found Redheads International, is setting out on a campaign to promote red-haired pride.

Douglas has ordered redhead bumper stickers and membership cards along with T-shirts bearing the Redheads International logo.

He rented a cubbyhole office in Laguna Hills, in Orange County, and is gearing up for an Oct. 23 gathering of redheads.

Douglas, who quit his band to launch the club, said more than 1,000 redheads have responded to ads in several national and local publications, paying $10 each to join the club He said several chapters have been formed.

Tacked on the bulletin board is a letter to "Dear Abby" from a man who complained he couldn't find a girl because of "the terrible curse of being a redhead." He asked if there was a club for redheads he could join.

Abby didn't know of any clubs, prompting Douglas to write a letter saying it wasn't a curse to be a redhead and told her about Redheads International.

(See REDHEADS—Page 12)

Commissioner run-off foes differ

Crawford finds voters uneasy

Time is running out on the run-off Democratic campaign for the Logan Co District II commissioner's slot on the November ballot.

Candidates Jim Ferrell and Ralph Taylor found themselves in a run-off when neither got a majority of votes in the August Democratic primary. Five Democrats ran in the August eliminations

To the winner on Sept. 21 goes the right to face Republican candidate Diana Crawford in November

The three candidates all bring a construction background to the race Although the commissioners must sit on 26 different boards and commissions, Ferrell says that the county roads issue is the one "that touches all voters."

Ferrell, who ran for Congress unsuccessfully two years ago, had told voters that he will be a "full-time" commissioner if elected.

He resigned in August as estimator and superintendent for Alexander Plastering and Dry-Wall Co in Yukon. He says this will prevent any possible conflict "with outside business interests."

Even though she has the Republican nomination for District II commissioner already won, Diana Crawford's door-to-door campaign to meet the voters has not slackened

However, she is glad not to have the run-ff pressures of run-off Democrats Ferrell and Taylor

"It's going to be a tough run-off, to be geared up like that to almost a fever pitch is exhausting," she says. She and her husband Joe live near Coyle

"I have been unemployed since after I filed for office. "I quit my job," he says. "If you were able to do it, you would be better able to serve the people"

County commissioners across the state have gotten considerable spotlight across the state in recent times Several have received federal and state indictments for bid-rigging, accepting kickbacks and other illegal

Oklahoma City says that she has encountered a mix of enthusiasm and suspicion from Logan Co residents on the campaign trail

"The county commissioners have been the butt of jokes in the past year. Since I've been meeting people out here they seem to be really interested in county government," she says.

Recently, two Logan Co commissioners and a retired county official pleaded guilty to charges of accepting kickbacks. They became part of a state-wide crackdown on

management activities

Ferrell says that recent state legislation in the form of House Bill 1578 will bring a number of changes to county government

Among these, it will establish a county purchasing agent and county road and bridge engineer positions. Ferrell says that smaller counties will hopefully be able to share an engineer rather than having to each pay their own

abuses in the office.

"People are suspicious of county government. They say, 'what's the ulterior motive here.' when candidates show up

The three candidates got quite a reception at a town gathering earlier this week in Woodward, she says.

"Woodward was the first time all three candidates had been together The community thought it was fantastic," she says

"The act looks to me like it is geared more to the larger counties which can afford more what the act calls for," he says

Ralph Taylor, also wary of previous commissioner pitfalls, has kept his campaign promises to a minimum He says he is making just one

"I'll do the best job I can if they want me for their commissioner That's all I have to say," he notes.

Taylor, along with Ferrell, showed

up at the Langston Co. Fair on Thursday night Republican Diana Crawford was there also.

Taylor recounted his day's travels. "I've been meeting with the people over in Langston-Coyle. I'm weak over there, I may be weaker now," he says with a smile

He knows the campaign is tight, and refuses to make predictions. "We'll just wait until the votes are counted. I just don't know what to think about it," he says

Taylor says he is not impressed with Ferrell's platform of being a "full-time" commissioner

"A full-time commissioner is one who works at it for 24 hours a day, 365 days a year," says Taylor. He plans to keep his rental houses and his construction business.

"I've worked since I was six years old, and I don't plan to stop now. I've worked for what I've got," he says

As a final push before the campaign, Taylo and supporters are sponsoring barbecue Monday night at the American Legion Hall. He is planning for around 700 people.

"I'd rather have too much rather"

(See COMMISSIONER—Page 12)

Ambulance system rides on EMS levy

If the Emergency Medical Service levy does not pass on Sept. 21, the Guthrie fire chief sees the City Council as having three main options for continuing ambulance service.

The council can raise water bills $2.50 a month, quadruple ambulance rates or discontinue ambulance service.

Guthrie's city ambulance service, which the fire department operates, did not receive state recertification in July. The ambulance service has until Dec. 31 to come up to standard or "go out of business."

Chief Bill Ward says that training and equipment status are two areas of deficiency. And as an added problem,

the ambulance is projected by city estimates to lose over $100,000 this year

Authority to establish EMS districts comes from a 1976 amendment to the oklahoma constitution. The amendment allows local government bodies to establish levies of up to three mills per thousand dollars of assessed valuation to fund the districts.

For a $40,000 home assessed at $4,000 with a $1,000 homestead exemption, the city estimates the levy would cost the taxpayer $9 a year.

Crescent is awaiting the results of the vote on the Guthrie School District EMS levy. The town has had a district, but is counting on Guthrie to provide

training for the county, says Ward

Presently Guthrie has four paramedics certified, who also serve as firemen. The city needs four more to meet state rules. None, however, are certified as EMS instructors

Currently, the paramedics also serve as firemen. Ward would like to see them do either one or the other Presently, said Ward, no funds are budgeted for EMS training.

Also, none of the city's three ambulances are up to certification. The city's most serviceable vehicle is a rented one.

It costs about $30,000 to $38,000 to purchase a high-top van and outfit it an ambulance And, the vehicles

generally are retired after 60,000 miles

Leonard Anderson, director of the Emergency Medical Services Division of the Oklahoma Department of Health, says that the district levy would create an entire EMS system for the area

Many areas establish 911 emergency numbers answered by a central dispatcher The dispatcher then sends either police, fire or emergency medical care personnel to the emergency site.

Victims are then taken to the nearest medical facility with the appropriate critical care team for their condition. Four state helicopters and the Military

Assistance for Safety and Traffic (MAST) unit from Fort Sill augment the local districts.

So far, 16 counties and seven school districts have established EMS districts.

Anderson became a firm believer in quick, on-the-site medical care when his twin-engine Army Mohawk observation plane was forced down near the Cambodian border during the Vietnam war.

"I had the misfortune of ejecting in Vietnam, but I also had the fortune to be picked up by a copter. I only had a sprained ankle and a few other problems, but I sure was glad to see them show up," he said.

Today

Today

BALLOTS AVAILABLE

Lorray Dyson, Secretary of the Logan County Election Board stated today the absentee ballots for the Mulhall-Orlando special school bond election are now available at the election board office, 311 East Harrison, Guthrie. Absentee application must be made no later than 5 p.m. September 22

AFS WELCOME SOCIAL

A welcome ice cream social in honor of the two new foreign students at Guthrie Highschool sponsored by the American Field Service will be held Sunday at 4 p.m. at the home of Mr. and Mrs. Norman Jacobs, 810 East Cleveland. The event is open to members of the AFS and would-be members. The students to be welcomed are Suni Fabricius, West Germany, and Adrian Stutz, Switzerland.

ROTARY SPEAKER

Bill Wagoner, superintendent of Guthrie schools, will be the speaker at Rotary Club luncheon Monday noon at the American Legion Building. Marley Smith is program chairman.

BEEKEEPERS TO MEET

Logan County Beekeepers Assn. will meet at the Fairgrounds Tuesday, Sept. 21, from 7 to 9 p.m. Mike Vandeventer, apiary specialist, will present the program.

SATURDAY DECA CARWASH

Guthrie's chapter of the Distributive Education Clubs of America will have a carwash from 9 a.m. to 5 p.m. today at the school bus barn, 802 E. Oklahoma. Cost is $2 per car or pick-up, others

---

Fig. 16-32 The *Guthrie Daily Leader* before redesign. It was in typical, but undistinguished, optimum format. Note the floating flag.

Guidelines for the Redesign of the *Leader*

Basic Layout

1. The paper will be six columns per page throughout, with the exception of the classified and comic pages. Column rules will be in hairline. Eight-column pages will have no column rules.
2. Type size for news will be 9 point on a 10-point base for the present time.
3. Photos will be boxed with a fine line border at the edge of the image and decorative corners. The exception will be on the sports pages where decorative corners will be omitted.
4. Makeup will be vertical, with subheads (centered boldface lines) being used to break up the grayness of the page, using at least two per story where needed, each line centered in a two-line space.
5. Cutoff rules will be used above all headlines and photos where there is type above. The cutoff rules will be the same width as the type for one column and will extend across the column if multicolumn; the column rule will be cut off to meet the cutoff rule.
6. The upper-right-hand corner of the front page is where the dominant story should be placed. This will allow smaller stories of importance to achieve dominance.
7. The flag will be at the top of the page. No story will appear above it.
8. Stories will be two columns wide or less. Three-column stories and headlines may be used on special occasions, as will five- or six-column banners with one- or two-column decks.

General Style

1. On any given story, all headlines and decks will be of the same face and weight. Head and decks should alternate slant: Roman, italic, Roman or italic; Roman, italic.
2. The head will always be flush left, with the exception of banner heads, which will be centered. Heads for boxed stories will be centered.
3. The first deck will be of an alternate slant, two sizes smaller than the head (for example, 36-point head, 24-point first deck). The second deck will be one type size smaller than the first deck (for example, first deck 24 point, second deck 18 point).
4. Heads and decks will be separated by jim dashes. The jim dashes will be 5 picas wide for each column the head covers. They will be situated 18 points from the head or deck above, and 12 points above the lower deck.
5. Heads will always be all caps.

Page 1

1. Headlines directly below the flag will contain one headline and two decks. Stories in this position with sidebars (related stories) may use the same head, and each story will have two decks. Stories above the fold but not attached to the flag will have the head and one deck. Stories below the fold will have the headline only.
2. Tombstoning can help create a territorial look. Tombstoning is defined at the *Leader* as two or three (usually three) one-column heads of the same size (but alternating weight) side by side. Tombstoning may also be done with a two-column or one-column story, alternating weights.
3. Shorts (heads for short stories) may be either 14 or 18 point (18 preferred) but must be in boldface.

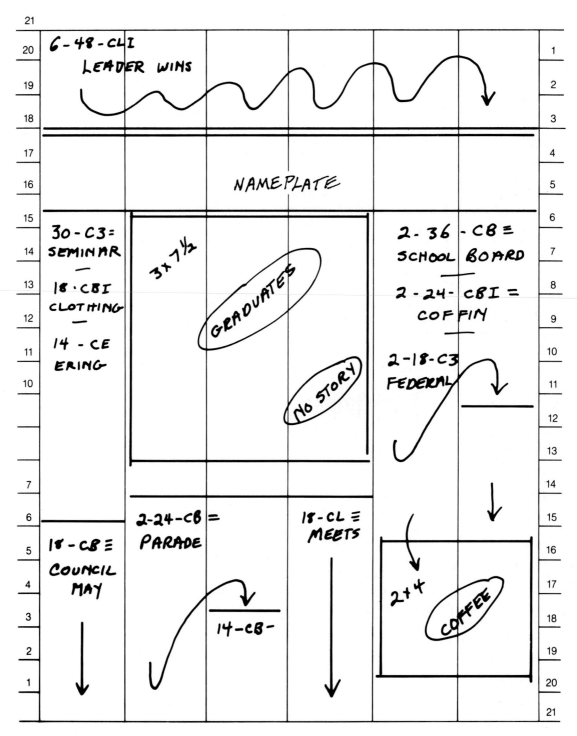

Fig. 16-33 Dummy for prototype page of redesigned *Guthrie Daily Leader* as prepared by the staff. Several prototypes were dummied and composed during the project.

Society Page

1. All heads will be in lightface, but slant may vary.
2. Top heads will be head and one deck; below the top will be headline only.
3. Banner heads will have no decks.

Fig. 16-34 The completed prototype page for the *Guthrie Daily Leader* redesign project.

Before the redesign was put into production, various techniques were tested, several prototypes constructed, and final specifications developed. Adkisson reports that the change has been well received by both readers and advertisers.

Although the changes made in the design of the *Guthrie Daily Leader* might not be right for most communities, they carry out the basic principle of effective newspaper design—the appearance of a newspaper should reflect its editorial philosophy and the audience it seeks to serve.

▼ Effective Design Checklist

- Make sure that all essential elements are included in folio lines and masthead.
- Avoid setting type in unusually narrow measures. Studies show that 9.5 and 10 pica widths simply are not read.
- Keep basic headline types within the same family. If another family is used, it should be used sparingly and for contrast.
- Avoid jumping stories from one page to another. If stories are jumped from page 1, try to continue them all on a convenient page, such as the back page of the front section or page 2. Studies indicate that 80 percent of the readership is lost on most jumped stories.
- Arrange advertisements so there is effective editorial display. Square off columns of advertising, and if the well or pyramid style is used, keep ads as low on the page as possible.
- Avoid pictures (usually mug shots, called "pork chops") that are less than one column wide. Try to use mug shots that show what a person is like, rather than what that person looks like.
- Don't run headlines too wide for the type size. A good rule is to keep headlines to thirty-two characters or less regardless of type size.
- Try to have a "stopper"—art or a dominant head—on each page. Never separate a picture from the story it accompanies.
- Eliminate barriers. But if long copy is broken with art or large subhead arrangements, make sure they do not confuse the reader.
- After a page is designed, try to judge it from the reader's point of view. Is it easy to follow, interesting, attractive?

▼ Graphics in Action

1. A major project in newspaper design could be redesigning your hometown newspaper. Examine current copies of the newspaper and then outline a complete plan for redesign. Make prototype pages by cutting heads, art, and body copy from newspapers that use the styles you would like to adopt. Write justifications for the changes you make.
2. If newspapers did not exist and you decided that a daily printed medium of news and advertisements was needed, how would you design it? What form would it take? What would be its page size? How would the contents be presented? Explain your answers.
3. Obtain a copy of a newspaper. Assume you have inherited the ownership of this newspaper. Since you are interested in graphics, you take a long, hard look at the paper. How would you change the front page? Don't forget that this is a profitable business and you do not want to take a chance on the new design adversely affecting reader acceptance and thus circulation and advertising revenue.

4. Redesign an inside page of a newspaper. Examine the pattern of advertisement placement. If the pyramid or well is used, redesign to magazine or modular. Redesign the editorial content to make its display more effective.

5. If possible, find a nearby community that does not now have a newspaper. Study the community and create a design plan for a newspaper that would reflect the character of the community and thus attract readership. Or choose a community and write to its chamber of commerce to obtain the community's characteristics and demographics to use in this activity.

Notes

[1] *Design,* the Journal of the Society of Newspaper Design, no. 23, p. 4.

[2] Ibid.

[3] *Design,* the Journal of the Society of Newspaper Design, no. 19, p. 7.

[4] Interview with the author, May 24, 1989.

[5] Don March, *Editor & Publisher,* October 16, 1980, p. 15.

Newsletters

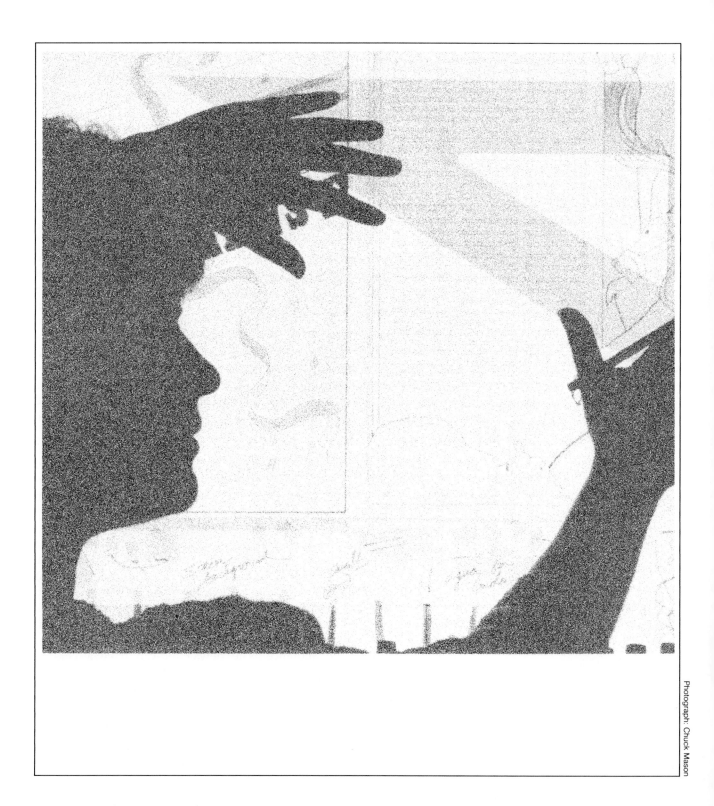

Photograph: Chuck Mason

MERCVRIVS
AULICUS,
A
DIURNALL,
Communicating the intelligence and
affaires of the Court to the reft of
the KINGDOME.

OXFORD,
Printed by H Hall, for W. Webb.
Anu. Dom. M.DC. XLII

Fig. 17-1 Newsletters were popular before newspapers came into existence. Many "newsletters" or news pamphlets were sold in Europe in the 1500s and 1600s.

Laser beams, word processors, editing terminals, cable television, videotext—these are all electronic miracles, and they are causing revolutionary changes in our communications systems. But no matter what the new technology may bring during the coming decades, one ancient form of communication is sure to survive—and prosper. The newsletter seems certain to keep its place in the communications mix.

Communicators agree that the success of an information program often rests on producing a medium for the information on a regular basis and supporting it with auxiliary tools to reinforce and repeat the message. For many organizations and businesses, the ideal tool for accomplishing this basic, regular communication is a *newsletter.*

The newsletter is not a new communications tool. It has been in existence for a long time. Researchers report that the Han dynasty in China published a daily newsletter in 200 B.C. The forerunners of the modern newspapers were leaflets and pamphlets—newsletters—which described an event or happenings from some other place. These were called diurnals, curantos, and mercuries, and they were printed and sold in the streets. And the first successful newspaper in American was called the *Boston News-Letter.*

The modern American commercial newsletter can be traced back to 1923 when Willard Kiplinger brought out the first issue of the *Kiplinger Washington Letter.* Today the newsletter is one of the fastest growing segments of the printing industry in the country. There are more than 100,000 newsletters being produced and distributed on a regular basis. They range from the small mimeographed parish or club sheet to elaborately designed and printed publications that are more closely related to magapapers or in-house magazines.

It is estimated that 3,000 to 5,000 of the 100,000 published newsletters are sent to paid subscribers who pay from one dollar to several thousand dollars a year to receive them. The newsletter with the largest circulation is the *Kiplinger Letter.* It has more than 600,000 paid subscribers.

The newsletter is so popular that many magazines use its format for special interest and updated information pages. *U.S. News & World Report* includes five separate newsletter pages in each issue. Business and organizational publications have found that a page of upbeat information in newsletter style has high readership.

Why are newsletters so popular?

They are liked by communicators and readers. Communicators find the newsletter an ideal communications link with various audiences. A special audience can be targeted easily and reached on a continuing, regular basis. Since the newsletter is brief and to the point, it can be aligned easily with the interests of the target audience. Identification with the interests of an audience is one of the criteria for effective communication.

Messages in a newsletter can be tailor-made for the situation, the time, the location, and the audience. The newsletter can have a personality, and it can come closer to one-on-one personal communication than most other forms of mass communication. Its chatty style can resemble a personal letter.

In addition, the newsletter can accomplish its goals at a low cost because it can be produced very economically. It would be difficult to find a more cost-efficient method of effective printed communication.

Audiences like newsletters because they contain specialized information that cannot be found elsewhere. The newsletter usually condenses information from many sources. Often newsletter readers do not have access to all these sources or the time to peruse them. People like to receive information in brief, to-the-point writing.

▼ A Public Relations Tool

The newsletter can be an effective public relations tool. It can be used to target specific publics with specific interests in an organization. Minne-

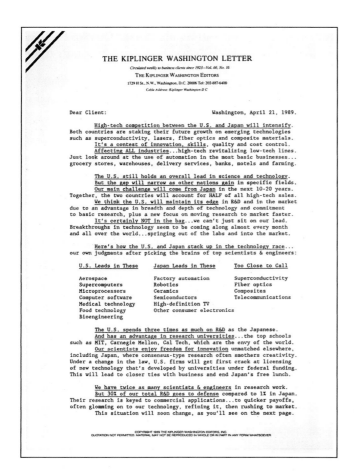

Fig. 17-2 The pioneer commercial newsletter, the *Kiplinger Washington Letter,* was founded in 1923. It set the design style that has been known as the "classic" newsletter format. The fourth page of the classic newsletter concludes with an ending similar to a personal letter. (Reproduced by special permission of the Kiplinger Washington Editors, Inc.)

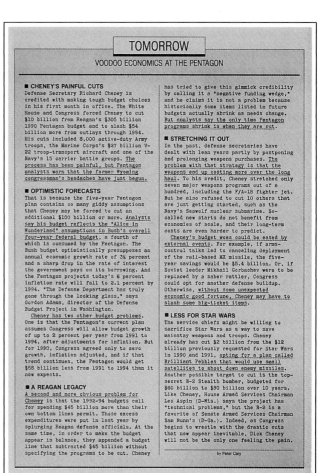

Fig. 17-3 Newsletters are so popular that many magazines adopt the format for informational pages. For example, *U.S. News & World Report* publishes five "newsletters" in each issue of the magazine. They are in the traditional or classic format. (Copyright 1989, U.S. News & World Report, Inc.)

gasco is a public utility headquartered in Minneapolis which serves customers in Minnesota, Nebraska, and South Dakota. It has won a number of awards for its excellent public relations programs. It publishes three newsletters for three different publics.

Minnegasco News, in tabloid format, is the company newspaper that goes to employees and retirees. *In-Touch*, in a 6½ by 7-inch format, is included with bills that are sent to customers. *Minnegasco Week*, is a weekly newsletter that was started in January 1987. It is circulated to all employees at their work locations.

"Following a communications survey of employees taken in 1986, we decided to change from sending out the *News* twice a month to once a month," reports Harlan Johnson, editor of employee publications.[1] The *News* is now zoned so that the content of one or two pages is different in the two editions that are published. In one edition, the zoned pages contain news relating to Minnesota area readers, and in the other content focuses on news of interest to Nebraska and South Dakota readers.

The *Week* is produced to supplement the *News,* which now concentrates more on corporate topics and personal and timely items. The *Week* is sent out each Monday to all employees at their work stations through inter-office mail.

Fig. 17-4 A magapaper in tabloid format, *Minnegasco News* is produced in zoned editions, like many metropolitan daily newspapers.

Fig. 17-5 A zone page of information from *Minnegasco News.* This page is included in the edition for readers in Nebraska and South Dakota.

The results of a survey of employees and retirees printed in the *News*,
April 1989, revealed that 88 percent of those receiving the *News* read half
or more of each issue, and 91 percent of those receiving the *Week* read
more than half of each issue.[2] Responses to questions about content preferences indicated that employees like to read about the company's future
plans, employee benefit programs, personnel policies, and human interest
profiles.

▼ Designing the Newsletter

The newsletter is a rather uncomplicated printed communication. It doesn't
seem to offer much challenge to the editor or designer. But therein lies
the problem. It is such a basic form of printed communication that it can
be put together too easily. As a result, too many lose effectiveness because
they look inept and amateurish. But they don't have to. Application of
basic typographic and design principles can change an unattractive, ineffective newsletter into one that is an asset to its publishers.

Fig. 17-6   Another newsletter used by Minnegasco in its award-
winning information program is *Minnegasco Week*, an 8½ ×
11 newsletter that is distributed to employees at their
workstations every Monday.

Fig. 17-7   The page size of *In Touch* is only 6½ by 7, but a lot
of design work goes into its layout to attract the customers of
Minnegasco in a three-state area.

Why Washington?

The largest publishing house in the world is located in Washington, D.C. It is the government printing office.

Washington has the largest concentration of journalists in the world, representing more than 2,600 newspapers and radio and television stations. There are 3,000 journalists who correspond for periodicals.

Yet Washington is called the cradle of newsletters. Why are so many newsletters published in a place that is saturated with information professionals?

One Washington newsletter publisher offers an answer: "My newsletter will be a profitable business as long as the government doesn't learn to write in plain English. I doubt if that will happen in my lifetime, if ever."

As always, before the designer or editor begins to make decisions concerning the format and graphics of a newsletter, it is necessary to consider its purpose. A thorough understanding of the reasons the newsletter is being produced will help make its appearance appropriate to the subject matter and the audience.

A description of the planned contents and a definition of the audience should be written as a basic policy guide for editing and designing. Potential readers should be described according to such characteristics as their professions, age category, education, sex, beliefs, attitudes, interests, hobbies, and family situation, or whatever else is relevant.

Out of this preliminary research should come some ideas of how to develop the newsletter's personality. Each newsletter should have its own personality, look, and style. This helps it establish its niche in the communications spectrum. The goal should be to design a newsletter that will be welcomed as a letter from a friend each time it appears.

One way to spark ideas about a newsletter's contents and appearance might be to have a brainstorming session in which everyone concerned in the newsletter's production writes down single words that form the basis for a list of goals. Such words as attractive, important, reliable, significant, impressive, lively, and so on can trigger discussions about the goals, format, and style of the newsletter.

Consistency of style is important. A consistent style means each issue will look basically the same as all the others. Since the design elements of a newsletter are rather limited compared with many other printed communications, we may worry about producing a monotonous product. But the page that we spend one or several hours on will only be seen for a few minutes by the reader. And many newsletters have been produced in the same basic design for years with great success.

This does not mean that the simple newsletter format cannot have sparkle and variety. Although a basic, consistent style is important in newsletter design, we should keep in mind that even the smallest change will become immediately apparent to the reader. A different size head type or a boxed item will stand out in a newsletter when it might be virtually unnoticed in a more complicated layout.

If an item is boxed or a few words underlined or set in italic type, the design change will be so apparent to the reader that it will grab attention and send the signal that something of unusual importance is being presented. Changes in design elements should thus be made very carefully. If change is made continually, nothing will stand out and the newsletter will become a confusing design hodgepodge.

Remember, the design of the newsletter sends out signals to the reader. The design tells the audience the attitude of the publication, its approach to the subject matter, and which items are especially important. A feature that is always boxed, for instance, or set in the same typeface every issue tells the reader something about the contents of the feature. Readers will get in the habit of seeking this element for certain information.

The First Step: Selecting a Size

The first step in designing a newsletter is to settle on a size. Just as certain sizes have developed into the standard for newspapers and magazines, so has the 8½ by 11 page become the standard for newsletters. There are

Each newsletter should have a personality. Consistency of style.

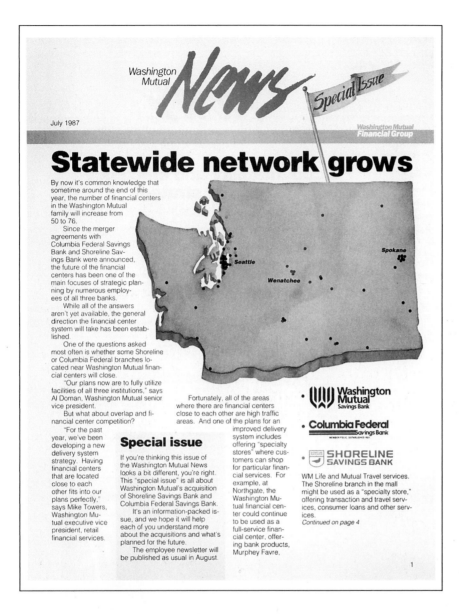

Fig. 17-8 *Washington Mutual News* was awarded a four-star rating by evaluators for *Newsletter Design,* a publication of the Newsletter Clearinghouse. Shown here in an 8½ × 11 format, it was changed to a tabloid after extensive research.

a number of reasons for this. The 8½ by 11 page is the same size as the standard business letter. This size is easy to file or to punch and put into a binder. It folds easily to fit a number 10 (business size) envelope.

In addition, a four-page 8½ by 11 newsletter can be printed on an 11 by 17 sheet, two pages at a time, in most "quick print" or in-house printing facilities. The 11 by 17 (or 17 by 22) sheet is stocked by most shops and so is readily available. It is the standard sheet out of which business letterheads are cut.

Sometimes the method of printing or the equipment available will determine, or at least affect, the page size. If office or in-house duplicating equipment is used, the page size might be limited to a legal-size sheet that is 8½ by 14 inches. The legal-size sheet can be folded in half to create a four-page 7 by 8½-inch newsletter. A small organization or business might find this a workable size. It could also help the newsletter stand out from the more standard-size communications.

Fig. 17-9 Now in tabloid format, *Washington Mutual News* has had a good reception from its readers. Employees like the size and more frequent publication. Tami Peterson, communications specialist and editor, found that a weekly, rather than monthly, gave employees more up-to-date information. The newsletter has been written on floppy disks and taken to an outside designer. Currently, plans are to go to desktop and use the Page-maker program to create camera-ready mechanicals in house.

A newsletter produced on an office duplicator could be an 8½ by 11 sheet folded in half. This would make four 5½ by 8½ pages.

Another possibility would be to fold an 8½ by 11 sheet to make six 3⅔ by 8½ pages. An accordion or gate fold could be used. If such a format is selected, many of the design techniques used in producing brochures could be employed as was discussed in Chapter 11.

Probably the best solution if printing capabilities are limited to 8½ by 11 or 8½ by 14 is to stick to one 8½ by 11 sheet printed on both sides. If more than one sheet is used, a method of binding the sheets together will be needed. Sometimes sheets are held together with a single staple in the upper-left corner. This should be avoided as it is a flimsy device and the pages come apart easily. A solution is to leave a wider margin on the binding side of the page than on the outside and staple the newsletter in sidewire fashion, like a magazine.

Newsletters can be found in all shapes and sizes. Some, called *magaletters*, are actually more closely related to magazines or tabloid newspapers in format. Some, called *magapapers*, are a mix of newsletters, magazines, and tabloid newspapers. The designer of these hybrid publications can employ some of the design techniques of all three in planning such a publication.

Designing the "Classic" Newsletter

Let's begin our discussion of newsletter design by considering the "classic" or traditional newsletter format. This is an 11 by 17 sheet folded in half to produce four 8½ by 11 pages. More pages can be added, of course, but four 8½ by 11 pages seem an ideal size for the content and design of one issue of a newsletter. If there is so much material that it will not fit into a four-page issue, it may be time to consider more frequent publication.

Characteristics of the classic newsletter format, developed for the 8½ by 11 page size, include:

- Typewriter composition (or the use of typewriter-style type if set on equipment other than a typewriter) with one column for each page.
- A short, punchy writing style in which obvious words are often left out and key sentences, phrases, and names are underlined.
- A limited number of graphic elements designed with care and used consistently. Simplicity is stressed.
- One style of type for the content; sometimes italic or boldface is used for limited emphasis.
- The feel of a personal communication.
- Avoidance of a magazine look.

This classic newsletter format was developed by Kiplinger, and many newsletters follow it faithfully.

An example of effective use of the classic, or traditional, newsletter format is *Rotunda Review*, published by the Nebraska Farm Bureau Federation. The *Rotunda Review* was developed in January 1984 by its editor, Susan Larson Rodenburg. Her goal was to keep Farm Bureau leaders up-to-date and informed of the latest activities involving their organization and the Nebraska legislature.

The newsletter presents information in an easy-to-read and easy-to-understand style. All reporting is focused on the one goal of explaining

Fig. 17-10 An effective newsletter in the classic format published by the Nebraska Farm Bureau Federation to inform its members about actions of the legislature. The logo incorporates the bronze statue on top of Nebraska's state capitol. The "rotunda" of the title refers to the area where most of the lobbying takes place. The logo was designed by Susan Larson Rodenburg, editor and assistant director of information and public affairs.

OTUNDA REVIEW

A Report of Nebraska Legislative News

Vol. 6, No. 9   May 1, 1989

Senators Debate Budget Bills

Senators will be back at work Tuesday to debate budget bills, which will continue to occupy most of the morning hours of the session until all are passed. Currently on the agenda is LB 813, the mainline budget proposal which calls for spending $1.015 billion in fiscal 1989-90 and $1.049 billion in 1990-91. Priority bills are taking up most of the afternoon debates. Speaker Bill Barrett is keeping senators on track because there's still much to do before May 24, the scheduled adjournment date.

Meanwhile...the Legislature has given final approval two important ag land valuation bills and Farm Bureau now is turning its attention to property tax relief and other priority issues. But first a word on LB 361 and LR 2...

Senators Approve Ag Land Bills

NFBF Vice President/Public Affairs Trent Nowka said County Farm Bureau leaders deserve much of the credit for the passage of LB 361 and LR 2, called by many "the two most important bills affecting agriculture considered by the lawmakers this year."

The bills represent a two-pronged effort to deal with constitutional problems the State Supreme Court has found with Nebraska's current method of valuing agricultural land for tax purposes on its ability to produce income.

Nowka said Farm Bureau will be extremely involved in the campaign to convince voters to approve LR 2 in 1990. The resolution would allow ag land to be valued differently than other types of property.

But in the short-term, Farm Bureau members should "pay close attention to their county boards because of the valuation increases mandated by LB 361. Any increase in valuation should correspond to a reduction in the mill levy to compensate for the higher valuations on ag land," he said. Nowka offered Farm Bureau's assistance to any County Farm Bureau on this matter.

Property Tax Relief Bills Need FB Action

Farm Bureau leaders are being asked to "take ten minutes today" to help insure property tax relief for farmers and ranchers. The projected state surplus is burning a hole in the pockets of state senators, who are determined to offer some relief to property owners this year.

Continued...

Published by the Nebraska Farm Bureau Federation

how legislation will affect the readers—farmers and ranchers.

The logo of *Rotunda Review* incorporates the bronze statue located on top of Nebraska's state capitol in Lincoln. As readers are aware, the *rotunda* is the area, just outside the legislative chambers, where most lobbying takes place.

Rotunda Review has established itself as a successful tool in providing Farm Bureau members with important information about legislation that affects agriculture and rural living. "We've received many favorable comments from our readers since its inception," reports Rodenburg.[3]

A traditional-format newsletter might be considered when there is a need for short-term communication such as during a campaign or event. Or it can be effective as a regular publication for a small organization or one with limited facilities.

The logo and masthead information can be printed in a large quantity, perhaps a full year's supply, and current information included and the required number printed for each issue. This method of production can help reduce costs.

Some designers have criticized the single-column page, saying that the column width, usually about sixty characters for pica typewriter copy, is too wide for easy reading. However, since this is a standard for type-written letters, people are used to it. If you prefer narrower line widths, an alternative is indenting the columns and placing the heads for items in the left-hand margin.

Even with one column, using wider margins and more space between paragraphs can help readability. If the margins are, say, 6 picas, the line width can be held to about forty-five characters.

There are two ways to handle a two-column page format. One is to vary the widths of the columns. The left-hand column can be 14 to 16 picas wide, for example, and the right-hand column 24 or 25 to 30 picas wide. The other, of course, is to plan two columns of equal size.

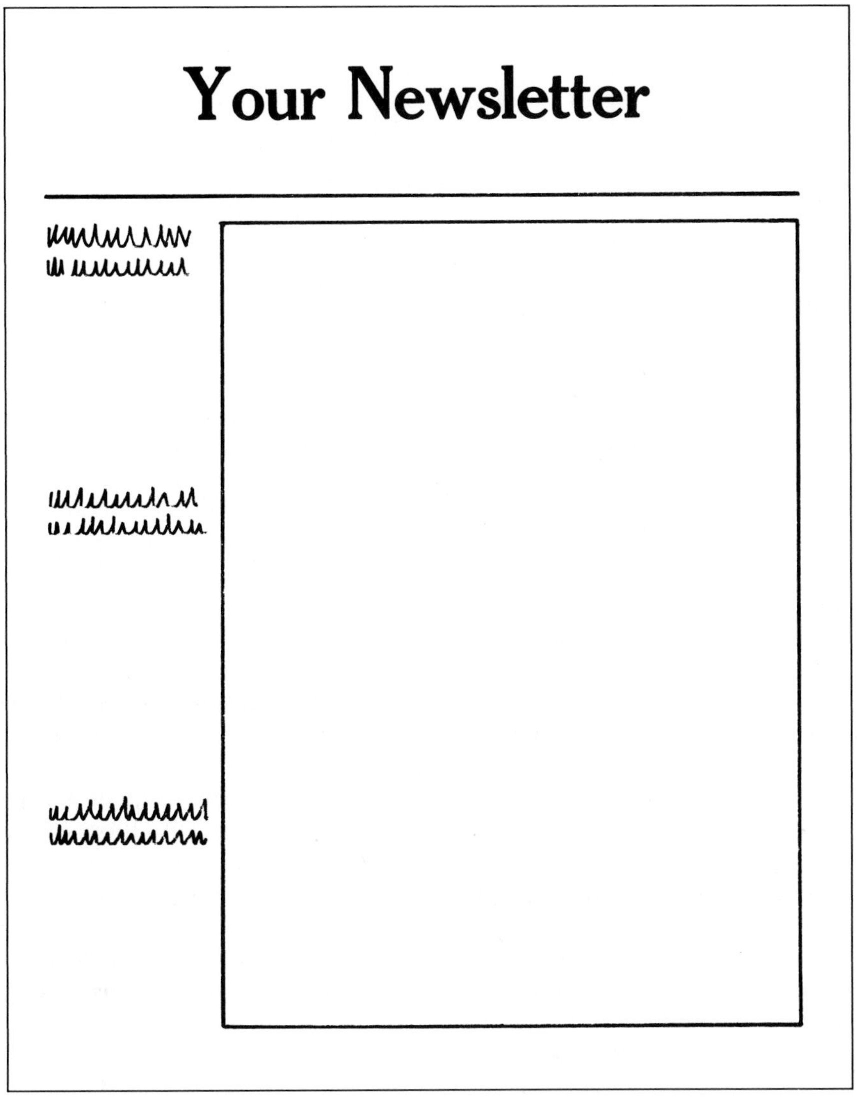

Your Newsletter

Fig. 17-11 The single-column newsletter in an 8½ by 11 format can have a narrower and more readable column width if heads are placed on the side of the reading matter.

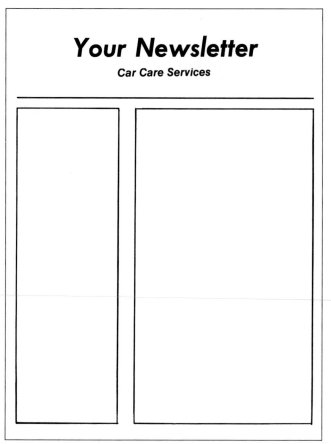

Fig. 17-12 Two possibilities for a two-column newsletter format. Two equal-width columns or one narrow and one wide can be used. Often an identifying slogan can add impact and memorability to the logo.

The two-column format has a number of advantages. It produces a more readable line width, and more words can be accommodated on a page compared with a single-column page. There is greater opportunity for creative design, and more variation and interest can be worked into the layouts. It is easier to use graphics such as charts and illustrations.

If a three-column format is adopted, the newsletter moves toward a magazine or miniature newspaper in appearance. It is necessary to have the body copy typeset for it to look attractive.

Since the three-column format tends to look like a magazine in the eyes of readers, it might be judged as such. Since most newsletters are not produced with the extensive talent and mechanical resources of a large magazine, the newsletter will suffer by comparison.

On the other hand, the three-column page gives the designer an opportunity to blend the best qualities of magazine design with those of the newsletter.

Along with deciding on the number of columns per page, we need to *determine margin size*. Page margins should be at least 3 picas all around. If the pages are to be punched for a binder, the margins should be wide enough to prevent the punch holes from obliterating some of the reading matter. As mentioned, if the newsletter is typewriter-set in a single column, margins of 6 picas all around are not too much.

White space between columns should not be more than 3 picas if copy is typewriter-set or 2 picas if typeset. Too much white space between columns can destroy unity by creating a wide white alley in the middle of the page. But white space should be at least 1½ picas wide for typewriter-set copy and 1 pica for typeset copy if it is to do a neat job of column separation.

Some newsletters are designed with *rules for framing* the type spaces and separating the columns. Rules can be effective design devices, but they should be added to a page only if they serve a useful purpose. A rule across the top and bottom of each page, bled to the edge, can help unify the whole newsletter and be an identifying device. Thin lines under headings can add individuality to the newsletter.

Reading matter should be set ragged right if there is a chance that justifying the lines will create awkward spacing between words. Some computer-generated copy using programs that are not sophisticated page layout or word-processing programs produce unacceptable word spacing if the lines are justified. If the copy is set ragged right, the page will often look more organized if vertical hairline column rules are used. These rules should be centered between columns in at least 1 pica of white space. Be careful in selecting thicker rules between columns. If they are too thick they can create disunity and detract from the appearance of the page.

Fig. 17-13 (Above, left) The newsletter with a three-column page approaches the magazine format. Magazine design techniques can be used. (Above, right) Some newsletters achieve orderly format and memorability plus quick identity with ruled borders around the pages and a symbol added to the logo.

Fig. 17-14 Once the number of columns have been determined, layout of a newsletter can be simplified by constructing grids. They can be done either on lightweight poster board or on the computer for repeated use. Here are typical grids for two-, three-, and four-column layouts.

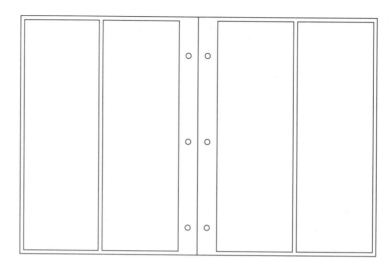

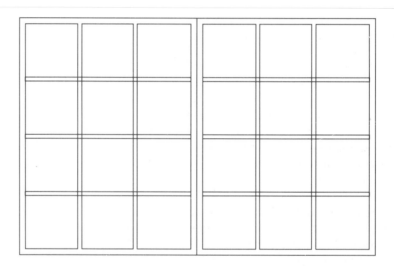

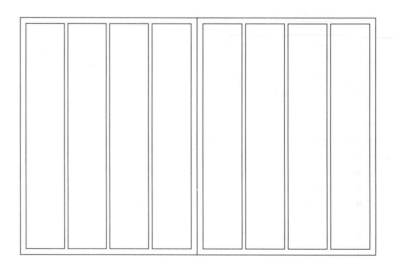

Rules and borders that are too heavy and too complex should be avoided. Such ornamentation usually succeeds in attracting attention to itself rather than enhancing the whole layout. Wide borders, if used, can be toned down by screening.

▼ Newsletter Design Decisions

Along with the basic format for the newsletter, a number of graphic decisions must be made. These include the design of the nameplate or logo, headline treatment, constants such as masthead and folio lines, and heads for regular features, as well as the design of the body type.

The Logo

The newsletter logo deserves serious consideration. In this discussion the term *nameplate* refers to the actual typeset name of the publication and *logo* to the whole treatment of the name plus any symbols or slogans. The logo creates the first impression, and it should identify the newsletter and its scope quickly but not overpower the other elements on the page. It should not occupy more than 20 percent of the page, in most situations.

The newsletter name should be imaginative—it should distinguish the newsletter from the thousands of others that are in circulation. Designers suggest avoiding using a name that is dull and that cannot be expressed with some imaginative graphics. Such names as the initials of the organization with the word *newsletter* or *bulletin* tacked on the end can be dull.

The logo design should help translate the title, and the type style used should be appropriate for the purpose of the newsletter. It should provide contrast with the headline type, but it should not dominate the page to the point that it overshadows the content. It should be distinctive so it will not be confused with the headline types.

Often a symbol or design can be incorporated with the name to create a distinctive logo. This logo can be made in different sizes to be used in the masthead, house ads, letterheads, envelopes, and so on. The more the logo is used, the better it can help create identity and recognition.

The logo should not be so large or strong that it interferes with the content of the newsletter. Designers say the logo plus the folio lines should not take up more than 2 inches at the top of page 1. However, as with most "rules," there might be situations where the exception is the most effective design device.

When Kiplinger created the name for the *Kiplinger Washington Letter*, he deliberately left the word *news* out because his plan was to write about the news rather than report it. Many newsletter publishers have adopted this personalized style. There is the *Granville Report* on stocks, and the *Lundborg Letter* on oil, for instance.

A touch of distinction that can set a newsletter out of the ordinary is to print the logo in color. It is possible to do this inexpensively if arrangements are made with the printer to print a supply of blank paper for the newsletter when color is on the press. Then the body of the newsletter

Headers

"The little headings on the top of the page, called running 'titles' by some. Headers are a great way to establish the identity of an article for those perverse readers who start at the back and flip toward the front."

Large Text

"Let it out! Try a full page of 24-point text for a real feeling of liberation. Or, start out with big text—to make it easy for the reader and to create a vigorous graphic texture."

Mug Shot

"No editorial layout would be complete without a tightly cropped head shot."

Slugs

"Subheads between paragraphs that help organize, or at least break up, long text. They can be big or small—they don't even have to take their own line—they can start on the first line of a paragraph, in type that's bigger, bolder, both."

Caption

"It is the caption that is important here, copy often read before anything else. Editors always wait until the last minute to write them, and the result is terse, if not cryptic."

White Space

"For readers it is a godsend. It's the needed 'oxygen' in a magazine. A page packed with type is forbidding, it looks hard to read, or dull, or confusing. Judiciously used, white space can invite readers into the story."

Quotations

"Skillfully extracted from the piece, these devices allow a reader to get an idea of what a story is about before taking the cold bath of actually reading it."

Rules

"Column rules made a comeback some time ago, but now magazines are replete with Oxford rules (a thick and thin), Scotch rule (thick, with a thin on either side), and even dotted leader rules. My rule is: Take them all out and see if the layout is any worse for it."

The Right Type

"I seem to always start a layout with the typeface. It is often a matter of association. The subject will bring to mind a type design from a place or time that fits. . . . Of course, few readers are aware of these associations. At best, they may notice that the layout is 'appropriate.' But that's all I ask. Typographical design is almost all in the subconscious."

Fig. 17-15 A page composed on a computer to illustrate the elements used in designing a newsletter in magapaper format. (Copyright 1989 Adobe Systems Incorporated. All rights reserved.)

can be printed in black, or whatever, at a later date. Some newsletter producers have a year's supply of colored logos printed at one time at a considerable savings.

Other ways to add color and brighten the newsletter is to print a screened tint block across the top of the first page and surprint the logo over it. Or a rule that extends across the top or both the top and bottom of all pages can be printed in color. The logo can be printed from a reverse plate in color, too.

Some newsletter names are printed in all lowercase letters. But this design technique can cause problems. One newsletter is titled *communications briefings*. There is no problem when the title appears in the content of that newsletter because it is always set in italics. But if it were to appear in another publication, it could be confusing. Then there was the headline that appeared in a newspaper trade journal announcing "presstime staff changes." Readers had to know that "presstime" was the name of an association publication that elected to use a lowercase *p* for the first letter in its name. Readers should not have to stop and figure out an ambiguous or puzzling communication device.

Sometimes a slogan can be useful. It might be incorporated with the nameplate and, perhaps, a symbol to create an effective logo. Such phrases as "all about airplanes" or "your financial adviser" can help identify the newsletter and create memorability.

For a more detailed discussion of the planning that goes into the creation of a logo and ideas for its design, see Chapter 11.

Fig. 17-16 Two logo approaches to avoid. All-capital letters in Black Letter, Script, or Miscellaneous typefaces most likely will be unattractive and unreadable. A logo in all-lowercase letters can cause confusion when used within reading matter or if it appears in articles in other publications.

The Folio Line

Newsletter design should make it easy for the reader to find out by whom, where, and when the newsletter is produced. A good practice is to include the volume number, issue number, and date in a folio line just below the logo. The name of the originator, address, telephone, and copyright information can be added in small type at the bottom of page 1. This information, plus staff members, subscription date, and so on, can be incorporated in a masthead at the bottom of page 4. This plan is standard practice for many four-page newsletters.

PUBLISHED FOR OCHSNER EMPLOYEES   INSIDE OCHSNER • VOL. 13, NO. 3 • MAY, 1989

INSIDE *Ochsner*

New Nursing Scholarship Program Offers Tuition, Books, Fees, Stipend

An Ochsner Nurse Scholarship Program, which provides full tuition, books, lab fees as well as a monthly stipend of $200 to help with living expenses, will be offered by the Alton Ochsner Medical Foundation in September.

"The scholarship program reflects the Ochsner commitment to attracting highly qualified and motivated individuals into the nursing profession," says Linda Sims Matessino, director of nursing and associate director of the Ochsner Foundation Hospital in New Orleans.

Any student interested in pursuing a nursing career can apply for the scholarship. Students will be selected on the basis of grades, college entrance scores, letters of recommendation and motivation towards a career in nursing. The scholarship may be used at any state approved and National League of Nursing (NLN) accredited nursing program whether it is a baccalaureate (BSN) degree, an associate (AD) or diploma nursing program.

Application deadline for the academic year 1989-90 is August 1, 1989. Scholarships will be awarded in September. Scholarship finalists will be required to complete a personal interview with an Ochsner representative. Individuals will also be re-

quired to complete a medical examination. Applicants must be full-time nursing students and submit a written education plan identifying courses and a projected graduation date.

For more information contact the Ochsner Nurse Scholarship Program, Alton Ochsner Medical Foundation, 1516 Jefferson Highway, New Orleans, La. 70121 or call 504-838-3600.

Ochsner Has A New Chef

A chef has to be a lot of things — not only a cook. He should also be a business man, an artist and most important of all, a diplomat.

Michael Ward, the new chef at Ochsner, hopes to fit into all of these catagories while he supervises the serving of 1100 to 2000 "tasty and nutritious" meals a day.

"Everything I've learned about food, I've learned right here in New Orleans during the last 14 years," Ward says. He comes to Ochsner after 11 years at Christian's restaurant which specializes in classical French and Creole food.

Ward is a native of Albuquerque, N.M., where he went to high school and graduated from the University of New Mexico where he "concentrated on the history of the American Indian, anthropology and economics". After graduation he came to New Orleans in 1975 and "started an 18-month apprenticeship program under executive Chef Willy Coln at the Royal Sonesta Hotel. I was moved to different areas of the kitchen every three months" he relates. He also worked for a year at LeRuth's restaurant on the west bank before coming to Christian's, "where I coordinated the purchasing of seafood and

continued on page 3

Fig. 17-17 This award-winning newsletter uses a reverse for its folio line and a hairline rule to box pages plus 6-point cutoff rules. Although there is liberal use of rules, they do not interfere with readability. (Courtesy *Inside Ochsner,* Ochsner Medical Institution, New Orleans. Editor: Luba B. Glade; designer: P. Douglas Manger)

MAYO CLINIC NUTRITION LETTER

Information on Nutrition and Fitness for a Healthier Life

Vol. 2, No. 4   mayo   April 1989

Discover these tasty seasonings while you reduce sodium

Herbs & Spices

You've heard the recommendations to cut down on sodium. The National Academy of Sciences estimates that an acceptable intake is 1,100 to 3,300 milligrams of sodium a day, but many Americans consume much more.

Sodium is not the same as salt. Salt is 40 percent sodium. One teaspoon contains more than 2,000 milligrams of sodium.

Sodium is one factor associated with high blood pressure (hypertension). More than 60 million Americans have high blood pressure. Their risks of heart disease, stroke and other serious illnesses are heightened.

Small amounts of sodium can raise blood pressure in some people. Others can consume much higher levels of sodium with no apparent effect on blood pressure. There is no foolproof way to predict who can lower their blood pressure by reducing sodium intake. The recommendation of 1,100 to 3,300 milligrams of sodium a day is safe and adequate for most adults.

An effective way to begin reducing your sodium intake is to limit the amount of salt you use in cooking and at the table. Taste your food *before* you add salt. Liberal use of the saltshaker can account for about one-third of the sodium you consume. If your physician recommends a low-sodium diet, carefully select processed foods to avoid those containing large amounts of sodium.

But if you cut back on salt, won't food taste dull? Not necessarily. A cornucopia of herbs and spices, ranging from the delicate to the pungent, is available to tantalize your taste buds. Some common examples of those with dominant flavors are bay, cardamom, curry, ginger, hot peppers, mustard, black pepper, rosemary and sage. Delicate-flavored herbs include chervil, chives and parsley.

Herbs and spices contain almost no sodium. In some recipes, they also can help you reduce the amount of fat and sugar — other traditional flavor enhancers whose excess consumption is linked to various health problems.

We don't recommend eliminating salt, sugar and fat altogether, but regardless of health concerns, herbs and spices provide a welcome alternative.

Spices grow in tropical climates. They come from the flowers, seeds, buds, fruits, roots or bark of plants. Spices have a sharp, distinctive taste. Most spices are available in dried or powdered form.

Herbs can be grown locally. You might consider starting a backyard garden; the appearance and aroma of such a cultivation are delightful. Herbs are soft-stemmed plants. The leaves contain the seasoning ingredients.

Shop carefully

When purchasing herbal blends, watch out for products that mention sodium, salt or monosodium glutamate near the top of the ingredient list. Ingredients are listed in descending order of concentration. Some products claim to be "salt-free" while using sea salt (instead of salt mined inland). Sea salt, however,

Fig. 17-18 The *Mayo Clinic Nutrition Letter* was a first-place winner in a Newsletter Association competition. It was established in 1988 and in one year attained a circulation of 75,000. Note the unity created by placing the herb container so it extends into the logo area.

Creative Communication

"Fools rush in"—creativity can be foiled and the process can take longer if preparation and analysis of the problem are inadequate. Creative scientists who seek to solve problems spend considerable time analyzing the situation. For example, Albert Einstein spent seven years on intensive study and fact gathering. But then it only took him five weeks to write the resulting revolutionary paper on relativity. He was working full time as a clerk in a Swiss patent office as well.

Thorough preparation often can shorten the period spent on actual creativity. Thorough preparation might include assembling pertinent facts, asking questions and finding answers, and seeking out leads for possible solutions.

Repeating the name of the publication, page number, and date on every page is not necessary in a four-page newsletter. However, if the publication is filed or stored for future reference and a semiannual or annual index is issued, it might be worthwhile to consider a page-numbering system.

One such system follows the style used by the printers of "mercuries" in England in the late 1500s and early 1600s. The mercuries were series of news pamphlets of continuing accounts of affairs. The first pamphlet of a series might contain pages 1 through 4. The second issue would have pages 5 through 8, and so on until the series was completed. *National Geographic* follows this page-numbering system today.

In addition to their own system of filing, some newsletter publishers obtain an International Standard Serial Number (ISSN) and include it with the page 1 or masthead information. Anyone can obtain an International Serial Number at no cost. This number goes into a worldwide computer data bank, and libraries refer to it when subscribing to publications. Information concerning the number can be obtained from the National Serials Data Program (Library of Congress, Washington, DC 20540).

Headlines

Headline type for newsletters should harmonize but stand out from the body type. Since newsletter design stresses simplicity, an uncomplicated Sans Serif or clean modern Roman type will work well for headlines. The headlines should be kept small; usually 12 to 18 points are adequate. If the newsletter is typewriter-set in typical typewriter Roman, a Sans Serif of the same size but in all capital letters can be effective.

Some newsletter designers simply set the heads in all capitals of the same type as the body or set the first few words in the first paragraph of an item in all capitals. Others do not use heads or capitals but simply underline the first few words of the first paragraph.

Regardless of the style selected for headline treatment, it should be consistent. If some headlines are set flush left, all should be flush left. If some are centered, all should be centered. The flush left is the simplest and quickest to use, and it can be given distinction by indenting it an en or em instead of lining it up with the body type. This will let more light into the page. But no matter what style is adopted, allow ample white space around the heads to help them stand out.

Subheads

Since the newsletter thrives because it gives information quickly, most articles are short. When a long article is included, *subheads* should be used to break it up into short takes to enhance the punchy appearance of the newsletter. Subheads look best when set in the same type style, though in a smaller size, as the main heads.

Some designers do use a different type style for subheads, but one that harmonizes well with the main head type.

For example, they say there is nothing wrong with using a Sans Serif subhead with a Roman main head. But they do suggest that an old style Roman of one family will not mix well with an old style Roman of another

Fig. 17-19 The *Getty Newsletter* is a six-page 8½ by 11 format with two parallel folds. This arrangement could be effective in smaller page sizes as well. Note how unity is achieved with a rule running across all pages.

family. The same basic principles of good type selection apply to newsletters just as they do to all other printed communications.

Italics or lightface of the same family as used for the main head work well for subheads. One rule of thumb concerning subhead size is to make the subhead about half the size of the main head. If this practice is used, the subhead should be checked to ensure that it is at least the same size as the body type. Subheads should be placed so they do not interfere with the story line of the article. They should be in natural breaks. Also, they should not confuse the reader by appearing to signal the start of an entirely different article.

Punch can also be added by using typographic devices to emphasize points in the body copy. But such use should be very limited.

Standing Heads

Newsletters often contain features that continue from one issue to the next. Many times these regular features are titled with heads such as "A Chat with the President" or "Front Office Notes" that never change. These standing heads can make a newsletter seem dull and static. We should plan a method of handling them so the material they identify comes alive.

A standing head can be supplemented with subheads, or, better yet, the identifying head can be used as a *kicker* (a head above the main head, usually underlined) so the subject head can be changed with each issue. An arrangement such as this can help:

A Chat with the President
Things Are Looking Up;
Membership Is Increasing

Fig. 17-20 Note the design elements that were used to make this effective inside page of *Write Up*. The uneven column endings harmonize with the contour of the art, and a drop-in ruled quote is placed effectively to give added weight to the upper half of the page and break the gray of the reading matter. (Courtesy Marie L. Lerch, Booz-Allen & Hamilton, Inc.)

Researchers study food poisoning causes

Continued from p. 5

The incidence of parasites in raw fish and shellfish found in Hawaiian waters is another task on the agenda. With the rising popularity of sushi and sashimi in this country, parasites that would normally be killed in the cooking process are kept alive in the raw fish. These may pose potential dangers. Research in this area is being conducted at the University of Hawaii.

The risk of botulism poisoning in improperly smoked fish is being studied at Virginia Tech. Following botulism outbreaks from contaminated smoked fish in 1960 and 1963, the FDA proposed Good Manufacturing Practice (GMP) regulations for smoked fish which, due to high processing temperatures, would insure destruction of the Clostridium botulinum organism. However, fish produced under this GMP were unacceptable to the consumer. As a result, a Federal court ruled against the FDA on enforcement of the GMP and smoked fish processors here ignored its recommendations. The purpose of the current research sponsored by FDA is to find procedures which ensure the safety of the product while providing a more palatable smoked fish.

> "...a deeper understanding of the molecular biology of bacteria and toxins will be gained..."

The findings from these and other tasks may be used by the Bureau of Foods to develop new Good Manufacturing Practice requirements to assist industry in processing safe, quality foods. The objective is to develop better food processing and packaging procedures while gaining understanding of diseases caused by contaminated foods.

Booz·Allen's involvement in the FDA-sponsored research program will continue at current levels for another two years. Some of the direct benefits will be the development of new methods for differentiating disease-producing from benign bacteria in foods, and finding new techniques for detecting contaminated or decomposed foods. In the area of food processing, study of bacteria and parasites will help identify food processing techniques (such as heat, freezing, preservatives) that provide the best protection. A deeper understanding of the molecular biology of bacteria and their toxins will also be gained.

After verifying the results and final reports from each task, the firm will assist the investigators and the FDA in disseminating the results.□

Researchers are studying parasites in raw fish and shellfish

Quite often the organization of the newsletter can be improved by boxing constants, copy that does not change from issue to issue such as the masthead information. Another way to handle the constants might be to surprint this information over a screen.

Things to Avoid

One newsletter designer summed up the shortcomings of some newsletters: small, difficult-to-read typefaces for reading matter; crowded pages; dull and static layouts with little accent or variety; too much gingerbread in the designs. Other shortcomings included tiny photos that should have been enlarged for more impact, group shots in which faces were hardly recognizable, and inconsistency in design.

When big rigs crash on urban highways, the daily rush hour can turn into a nightmare for commuters and local authorities.

Truck Crashes and Congestion: A Growing Concern Across the U.S.

In a check of cities and towns around the country, *Status Report* finds truck crashes are Page One news in just about every community that generates a rush hour. In a one-month period late last summer, scores of newspapers had Page One stories reporting 125 local truck crashes that involved 232 vehicles.

The crashes resulted in 61 immediate fatalities and at least 150 injuries. Traffic delays were extensive. In some of the more complex collisions, delays of eight and nine hours were reported. Fire equipment was often involved. And cleanups, seldom simple because of the huge cranes needed to right and haul away disabled tractor trailers, were more difficult when crashed trucks spilled their cargoes, caught on fire, or hazardous materials were aboard.

Of the 162 heavy trucks involved in the 125 incidents, 17 reportedly were hauling hazardous materials, which pose special problems for local officials.

From Atlanta to Seattle, transportation officials agree, truck crashes are a major contributor to traffic congestion, and they are looking for new ways to solve the urban impasse.

In Washington, D.C., following a series of dramatic heavy truck crashes, Federal Highway Administration (FHWA) Executive

(Cont'd on Page 6)

SPECIAL ISSUE: TRUCK CRASH CONGESTION

Fig. 17-21 *Status Report* is a consistent award winner in newsletter contests. It is produced by the Insurance Institute for Highway Safety and goes to 15,000 readers who are members of target audiences concerned with highway safety. James H. Mooney is editor. Heads and body type are a condensed Cheltenham, which is used throughout.

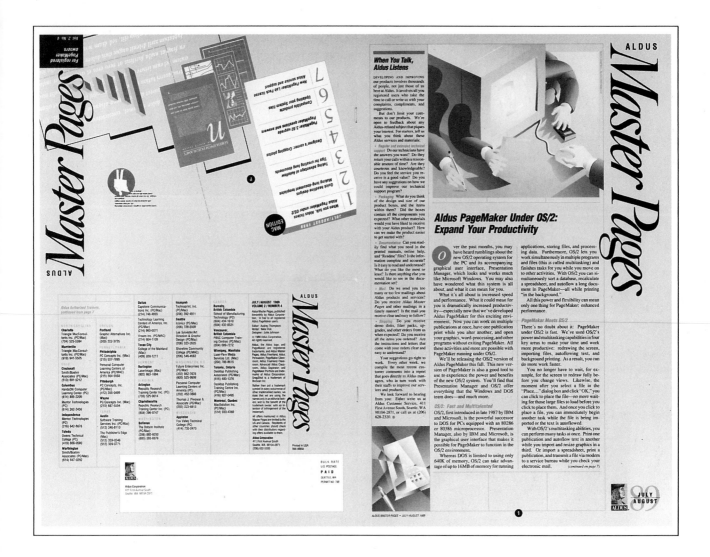

Fig. 17-22 *Aldus Master Pages*, in tabloid format, is folded to create two 8½ by 11 pages for the front and back cover pages. The size of these pages is thereby reduced to make the publication a convenient size for mailing and filing. Note the handling of teasers on the cover and the address label area on page 8. (Courtesy Aldus Corporation)

▼ Award-Winning Newsletters

Since 1972 the *Newsletter on Newsletters*, published by the Newsletter Clearinghouse, has conducted a competition held in conjunction with the International Newsletter Conference. Awards for overall excellence are made to newsletters that exhibit design, typographic, photographic, and printing quality. Judges for the competition are selected from the communications and graphics industries.

Comments made by judges in recent contests indicate that they do not like odd, unusual layouts. They pay particular attention to the designs of front pages and nameplates. (After all, the nameplate is the major design element of many newsletters.)

They applaud the use of color. They are impressed by newsletters that are printed on colored paper stock rather than white paper. They point

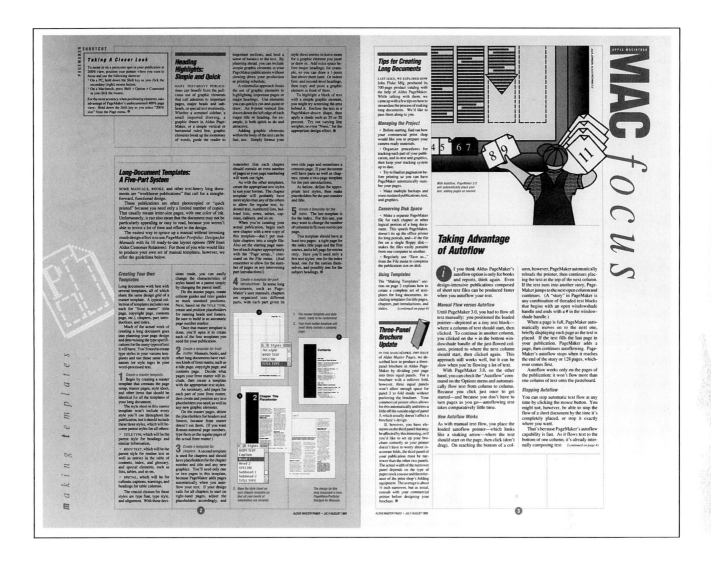

Fig. 17-23 The inside of *Aldus Master Pages* combines effective use of color and graphics. Note the use of white space, hairline rules, and a cool background color that recedes and thus does not hamper legibility. The unique initial letter, the placement of elements so they do not compete, and the colorful graphics demonstrate the skills of the designer. (Courtesy Aldus Corporation, Audrey Thompson, editor; Julie Johnson, designer)

out that color can be an important element in creating identification for a newsletter.

On the negative side, the judges criticize the use of nondescriptive names and initials for newsletter names. Such tags make it difficult to identify the newsletter quickly. If initials are used for a newsletter name, they should be followed quickly by a descriptive phrase.

The judges also criticize many newsletters for using small, difficult-to-read type. (We have recommended that a newsletter not be set in less than 10-point type.) They found that type packed into a newsletter without heads for items or breathing space was difficult to read. White space should be used to get air into a newsletter page just as it is used for all types of printed communication. Packed pages are unappealing and tend to turn readers away.

An effective newsletter should be attractive and neat. Its design should be uncomplicated and consistent from issue to issue.

▼ Effective Design Checklist

- Consider using an initial letter at the start of each article. However, don't use initial letters if subheads are being used, as too many elements in a newsletter can create a typographic mishmash.
- Use a different type style for cutlines, or a different size or weight of the same type family as the body type, to brighten the page. The cutlines should all be in the same style and used in the same form, however.
- Use a distinctive style for handling bylines to add a touch of contrast. But use one style for all the bylines.
- Set off a list of specific points in an article by numbers or appropriate small ornaments. Or indent the points an em or so to bring in additional white space to brighten the page.
- Since so much printed material is produced on white paper, consider using colored stock and/or ink to make your newsletter stand out from the rest.
- Typewritten body copy will save space and look more like professional typesetting if reduced 10 percent.
- Once the basic format is designed, stick with it and do not make format changes from issue to issue. Such changes can be costly and time consuming.
- Study the effects of various screen percentages on the ink used. Often the effect of a second color can be obtained without the cost.

▼ Graphics in Action

1. The prototype illustrated in figure 17-24 was created by a student in a graphics class. Evaluate it on the basis of good design, appropriate type styles, and balance, unity, harmony, proportion, contrast, rhythm. Make recommendations for revisions.
2. Plan the format for a four-page 8½ by 11 newsletter about your favorite hobby. Begin by outlining the research you would do, including evaluation of the target audience. Then design the logo, folio lines, number of columns per page, and all the other graphic and typographic elements. Explain what you did, and why.
3. Design and paste up a prototype for the front page of a newsletter. Select a size for the page after investigating production possibilities (for instance, if a 10 by 15 or 11 by 17 press is available in your shop, the prototype might be designed so it could be printed with one or the other). Find headline types, body-type blocks, illustrations, and other design elements in magazines. Clip and use them in the project. Use elements that resemble as closely as possible what you would specify in reality—such as ragged-right body type in the size you would select, if the newsletter is to be set ragged right.
4. Find a four-page newsletter that has a one-column page and redesign it to a two-column format. Be creative and consider using such

Your Newsletter

Number 00   April 0, 1900   Volume 00

Three-year drive garners $1,050,000

Five properties here get 4-diamond rating

John Deere Life Insurance Company, while still pioneering new markets, also emphasizes the establishment of long-term relationships with its customers. The life agent is now assisting customers in non-insurance areas such as financial planning which requires an ongoing relationship.

All of the efforts of the two divisions have been directed at developing a distribution system that will enable them to develop these relationships on the most efficient and dependable basis. The company's long-established reputation for value, service and integrity is a great asset in marketing the services provided by the Insurance Group.

A key strategy in making sure that the John Deere name is prominent in the customer relationship was to develop a sales

organization which essentially represents the John Deere insurance product lines exclusively. In the property/casualty division, sales personnel, whether employee or independent contractor, are required to sell only John Deere Insurance products unless specific exceptions have been made. Life insurance producers have not yet developed an exclusive relationship with the company, but efforts are aimed at that objective. Altogether, we now have nearly 200 exclusive John Deere Insurance agents and more than 700 independent John Deere Life agents.

Personal lines such as automobile and homeowner insurance, are emphasized in Minnesota, Iowa, Illinois and Wisconsin. Exclusive John Deere Insurance contract agents also sell life insurance and business

insurance to John Deere dealers, the last surance industry. The relationship between the company, the agent, and the customer is somewhat unique, however. We want this relationship to be long-term and stable. We regard the close relationship of John Deere Insurance with its customer base as the cornerstone of our entire insurance operation.

The next five years will provide the Insurance Group with even greater challenges as we strive to become a billion tinues to emphasize expansion in rural and small town areas where the John Deere name is most recognized. We have now begun to go after the Midwestern metropolitan areas, however, and we're finding the John Deere name is readily recognized and conveys a marketing advantage.

Aspen Tree going out of busin

The year 1986 was an especially challenging and rewarding year for the John Deere Insurance Group. Both sales and earnings hit record highs. The property casualty division is meeting the increased insurance needs of existing and new customers in an extremely chaotic market. the same time, the life division has been undergoing some radical changes in product lines and distribution system.

John Deere Life Insurance Company reached an important milestone on October 24, 1986, when it reached $1 billion ordinary life insurance in force. Of the 2, life insurance companies, John Deere ranks 286 in life insurance volume.

The year 1986 was also especially gratifying for the property/casualty side of business because we began getting dorsements from important associatic

emphasize its core business of provid insurance to John Deere dealers, the surance industry. The relationship tween the company, the agent, and customer is somewhat unique, howev We want this relationship to be long-te and stable. We regard the close relations of John Deere Insurance with its custor base as the cornerstone of our entire ins ance operation.

The next five years will provide Insurance Group with even greater cl lenges as we strive to become a bill tinues to emphasize expansion in rural a small town areas where the John De name is most recognized. We have n begun to go after the Midwestern met politan areas, however, and we're find the John Deere name is readily recogniz and conveys a marketing advantage.

Edgar Sanchez

Fig. 17-24 This prototype newsletter was created by a student using materials clipped from publications. Although the text bears no relation to the heads, the design can be used to evaluate possible arrangement of a newsletter page. What are the design and typography strengths and weaknesses of this prototype?

innovations as pages boxed with rules, tint blocks for surprinting the logo, screens of various designs for illustrations.

5. Select a section from a metropolitan Sunday newspaper (sports, business, travel, and so on) and plan a newsletter devoted to that area of interest. Write a prospectus for such a newsletter including the research that should be done and a description of the format and graphic elements the newsletter would contain.

Notes

[1] Letter from Harlan Johnson, Minnegasco editor, to the author, May 3, 1989.
[2] *Minnegasco News*, April 1989, p. 4.
[3] Letter from Susan Larson Rodenberg, *Rotunda Review* editor, to the author, May 25, 1989.

CHAPTER EIGHTEEN

Designing in the Twenty-First Century

Photograph: Mark Jenkinson

An executive, who is general manager of a branch plant of a major manufacturer, and her friend were inspecting the display of electronic equipment in Macy's department store during their lunch break. The executive pointed to a hand-held calculator the size of a business card. It had an $8.95 price tag.

"I can remember when I paid $150 for a calculator that could not perform all the functions that little thing is capable of," she remarked.

So it is with graphic design today. The electronic equipment is available to convince even the most unbending skeptic that a designer at a computer work station can turn out graphics and create layouts far more efficiently and of far higher quality than many produced in the days of hot type and the linotype and, more recently, with airbrush, pen, and pastepot.

▼ Electronics and Designers

Several companies are producing electronic layout and design stations that cost $100,000 or more. These stations can speed up the creation of design concepts, comprehensive layouts, and final mechanicals. They shorten the publication cycle. They eliminate intricate mechanical operations like cutting, scaling, and pasting stats of pictures and galleys of type. They create grids that can be used in laying out pages for current use or stored in the computer for retrieval. They eliminate process camera work for enlargement and reduction of art.

Color pictures, line drawings, graphs, charts, headlines, subheads, and text can be brought together in layouts on the console. Text can be brought in from a number of other work stations. Type can be seen in black, or in reverse, or in any color. Letters can be kerned, and copy can be wrapped around art, or superimposed on art.

Pictures can be stored and called up to be resized, scaled, shaped, cropped, flopped, reproportioned, and repeated throughout a layout. And, of course, borders, various shapes, and tint blocks can be created precisely and quickly.

The final proofs thus created can be used as mechanicals for conventional printing.

An example of this new, highly technical equipment is the Scitex Response console, which is being used at *Time* magazine.

Although such expensive consoles are beyond the reach of many in the design and publishing business, each year affordable equipment to produce quality printed communications is becoming available to all shops, large and small. A new magazine, *Smart*, was conceived, designed, and produced with desktop publishing equipment. Many believe a quality product cannot be made in this manner. Terry McDonell, the president and editor-in-chief, doesn't agree.

"Those people are not very well informed, are they? The large, extremely sensitive systems put in by such magazines as *Time* and *Newsweek* over the past four or five years can now be duplicated by a $40,000 desktop system," McDonell said in 1989.[1]

Fig. 18-1 A desktop publishing system for the 1990s. All functions from typesetting to scaling and screening art to page layout can be done at a desk to prepare mechanicals and send them to the laser printer. This complete design and composition workstation produces high-quality professional materials when operated by someone with a graphic and typographic background.

The equipment used to produce the magazine was two Macintosh Plus computers, a Macintosh SE computer, and an Apple LaserWriter Plus printer. One of the computers was equipped with a full-page display monitor.

Most of the writers wrote their articles on personal computers. They sent their disks to *Smart* or transmitted them by modem to the magazine's electronic mail box. The editors then used Microsoft Word 3.01 software to edit the submitted stories and produce final copy.

Smart's designer used QuarkXPress software to design and lay out all the pages. The headlines and graphic elements were created, combined with final copy, then pasted up electronically on the Macintosh screen. A logo for the magazine was drawn on the Macintosh using Adobe Illustrator software.

The Laser Writer Plus printer proofed pages as they were created. The finished pages were printed directly from the Macintosh to a Linotronic 300 laser imagesetter.

While the technological revolution has opened up doors of opportunity for editors, designers, and all who create printed communications, they

Fig. 18-2   (Right and on the facing page) A national magazine produced on the desktop. SMART's designer used Quark-XPress software to design and lay out all pages. Macintosh computers and Apple laserwriters were used to prepare mechanicals. The Adobe Illustrator program was used to create the logo.

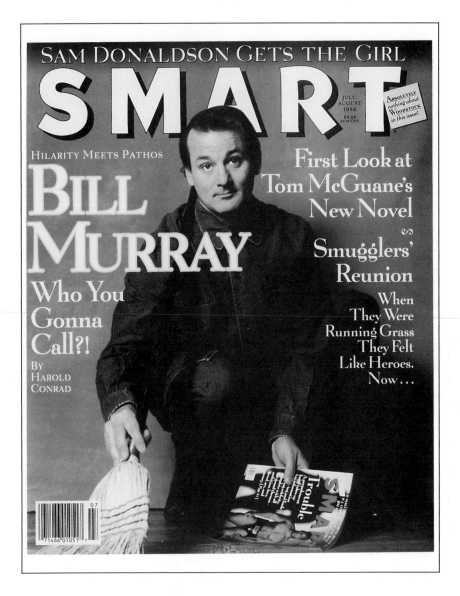

have engendered some concerns as well. Printing, once a craft that required long and intricate training before a person could produce a quality product, has been placed in the hands of anyone who can afford the relatively inexpensive equipment. This has become a problem for the trained graphic designer.

"There's an art to designing printed matter, and I don't see it having a very long life if everyone who writes something is his own publisher," commented Kerry Bierman, the American Medical Association director of marketing.[2]

But many people without a background in graphic design who have attempted to be their own publishers do realize they need help. And their need has created a new business opportunity for designers—the service bureau.

A service bureau is a small shop set up as a designer-operated desktop publishing center where someone who has produced a publication on a computer can go for help. It can offer design services and can create mechanicals for printing. It can provide the expertise and equipment that the desktop publisher may lack.

▼ Robots and Satellites

Many of the technological advances of this information age are being adopted by the graphics arts industry. Robots and satellites are playing their part.

Although robots have been used in manufacturing for many years, they are just beginning to invade the graphic arts. These are not the robots we visualize as resembling humans with arms and legs but with antennae added that run around doing simple tasks. They are machines that are programmable to perform quite complex tasks and to relieve humans from the monotony of assembly-line work.

Robots have been used to manufacture machinery, to load, handle parts, weld, drill, and die cast.

In graphic arts they are being used in large printing plants for handling material from the press to the bindery. They load signatures of books and magazines from the bindery onto skids. They watch over perfect-binding machines and wrapping equipment as well as printed pieces as they pass through the bindery.

It is anticipated that in the future robots will be working in newspaper and magazine publishing plants handling paper, changing press plates, and sorting the printed product for distribution on delivery vehicles.

Satellites have been used for some time by such publications as *USA Today*, the *Wall Street Journal*, *Time*, and *Newsweek*, but the use of satellite technology by monthly magazines is relatively new and increasing.

For instance, in September 1988 *Vanity Fair* assigned a reporter to look into the background of Senator Dan Quayle after he was picked by Republican presidential nominee George Bush to be his running mate. On October 3 the magazine distributed its November issue with an in-depth article on Quayle. Monthly magazines usually take months to develop articles from inception to press. How did *Vanity Fair* do it so fast? The answer is with satellite technology.

Every month since March 1988 a late section of up to ninety-six pages has found its way into *Vanity Fair*'s production cycle. The layouts are digitized and sent via satellite from New Jersey to a printing plant in Illinois. There the digital information is received, decompressed, and made into film for plate making.

"In the past, we could get caught very easily by a political event," noted Pamela Maffei McCarthy, *Vanity Fair*'s managing editor, "in that no man's land, that four to six weeks between the time a monthly goes to press and is distributed. And we absolutely cannot afford that. Our readers, and I think most readers today, were raised in an era of instant communication."[3]

Satellite transmission has given monthly magazines the tool to compete with the more frequently published weekly magazines.

Satellites and telephone lines are being used in a number of ways in addition to shortening the production cycle and making publishing more timely. Advertising agencies have the capability to transmit advertisements on film directly to a publication. Electronic mail (E-Mail), which was devised from the technologies of fax, computers, videotext, sound, and graphics, allows phone calls to become greeting cards, a keyboarded message to be digitally voiced, and single messages to be edited and distributed to many receivers. In 1989 E-Mail was a $415 million service industry; by 1991 it is expected to surpass $2.2 billion. Predictions are that 60 billion messages will be buzzing around the world via E-Mail by the year 2000.[4]

Cocooning, the practice of working at home with the aid of electronic communications equipment, has come to the graphics and design world. A California designer has set up a desktop publishing and typesetting service in his home. Customers send their work to him by fax over a toll-

Fig. 18-3 *Vanity Fair,* a monthly magazine, uses modern technology, including satellites, to speed the production cycle. Each issue contains a late section that includes material prepared just days before the publication goes to press. A case in point is the cover story for the November, 1989 issue. Barishnikov announced during the week of September 28th that he was leaving the American Ballet Theater. *Vanity Fair,* with the help of satellite transmission, was able to turn this news into its cover story, complete with photographs, and still get the issue out and to its readers in under two weeks' time. (Courtesy *Vanity Fair.* © 1989 by The Conde Nast Publications Inc.)

free number. The customer is promised type set in twenty-four hours and proofs are faxed back to the customer.

When the work is approved, the customer receives reproduction proofs by overnight mail. The owner of this cocoon business says he has not met most of his clients in person, just by mail, phone, fax, and modem.[5]

▼ Competition from the Electronic Media

Any discussion of the future of graphic design cannot overlook the effect of the electronic media on printed communications. The outlook, as reported by the *American Printer,* is both bright and threatening. It is estimated that by 1996 electronic substitutes will replace about 2 percent of print production. By the year 2001 electronic media will take over 5 percent of the print production, and by 2006 they will have captured about 7 percent.

But at the same time, the output of the printing industry will grow dramatically. Even though printed communications will have lost as much as 5 to 10 percent of **potential** growth to electronic alternatives, the printing industry will still produce nearly $300 billion in printed products compared with the $70 billion it was producing twenty years earlier.[6]

And now comes the paperless magazine! The owner of Digital Media Publications, John Wang, created *InfoStacks,* a magazine distributed on a computer floppy disk in 1989. The magazine targets owners of HyperCard, a popular and powerful software information storage program. The program can be arranged by the owner with "cards" of information. These cards, or information bits, can be shuffled and arranged to fit the needs of the owner.

InfoStacks is the first magazine to be produced and distributed on a computer disk. Wang lists these benefits for readers of his magazine:

1. If a reader is searching for something in a conventional magazine, he or she must sort through the table of contents and then if it can't be found must engage in much shuffling of pages. In the magazine on a disk all a reader has to do is type a word or phrase and the computer will search through the disk to find what is sought.
2. *InfoStacks* uses sounds and animation as well as words. As Wang puts it, "Paper magazines are a passive format; they lie there waiting for you to read them. In an electronic magazine articles can have sound and animation. I use passages of music to grab the reader's attention at the start of several sections of the magazine." Sound and animation are available for advertisers, as well.[7]

Issues of a paperless magazine can be put in the computer memory, and if a reader seeks an article that appeared in the magazine a month or a year ago she or he can gain access to it quickly and easily. While *InfoStacks* is produced in black and white now, the publisher anticipates it will be full color in the future.

Designers working in the newspaper industry undoubtedly will see changes during the coming years as well. There is a certain amount of gloom in the industry concerning the future. The population continues to increase, but newspaper circulations have not kept pace. Newspapers have formed a committee to study the situation and help chart future directions.

One newspaper success story might be examined for possible trends. The internationally known French newspaper *LeMonde* was in trouble financially and circulation was lagging but its managing director was able to reverse both trends, increasing circulation and moving it into the black. Some of the changes made that might be considered by other newspapers include:

- Creating a new design to meet readers' requests, as revealed through research, for a newspaper that would be easier to read.
- Using lots of color.

- Planning special supplements aimed at specific target audiences such as a special supplement for doctors.
- Producing more regional and neighborhood editions.
- Storing the entire text of each issue so readers can access it on their personal computers.
- Indexing all book reviews electronically so readers can call up reviews on their personal computers and even order books via their computers.[8]

▼ Trends in Corporate Graphic Design

The graphic designer who works in the corporate world will see increased use of sophisticated printed communications. Everything from letterheads to annual reports are being produced with a greater focus on type and graphics. In a world of information overload, corporations and their public relations professionals are turning more and more to graphic designers to produce printed material that is not only attractive but innovative and memorable as well.

Fig. 18-4 This two-page spread from a General Mills annual report illustrates the increasing reliance of major corporations on graphic design in their communications. Annual reports are more than financial reports, they are valuable communication tools, created by design professionals. Note the arrangement of the products around the art to create unity with the art and the facing page. Note also the placement of the art that crosses the gutter; the gutter crease does not mar the composition. A large initial letter signals the start of the reading matter and helps unify and balance the layout.

Fig. 18-5 A two-page spread from an Adobe Systems Incorporated annual report. The annual report is being used more and more to build the company image. Bolder and more expansive layouts are making such reports more interesting and colorful.

In seeking to achieve the corporation's goals, the designer is planning communications that capture the attention and pull the reader into and through the entire message. Also, the designer seeks to reinforce the image the corporation wishes to create with the use of appropriate visuals.

Current trends in graphic design for corporations include:

- Increased use of the techniques of the graphic journalist to help explain information.
- More use of the tabloid format for employee publications. Many company newsletters are adopting this format; editors and designers like the opportunity for graphic display this larger page provides.
- Using more colorful art and concise copy for brochures.
- Using the annual report to build the company image and bolder and more expansive graphics to make it more interesting and colorful.
- Increased use of strong visual elements such as bold initial letters and brilliant color bars in layouts to give impact to points the company wishes to emphasize.

Rose DeNeve, former managing editor of *Print* and a corporate design specialist, notes, "Today design is an integral part of the communications effort, as inseparable from it as words and thoughts are from content and meaning."[9]

▼ Television Graphics: A Growing Opportunity

The video display terminal is playing an increasingly important role in the work of editors and designers. But anyone who is investigating graphics in communications should consider it from another aspect—its function as a conveyor of information to an audience—its role in television.

Television graphics is a growing area of opportunity for designers. The medium offers challenges quite different from those of the print media. The more senses a message can stimulate, the more impact it can have. Television offers the opportunity to combine motion, sound, and color to create presentations that stimulate several senses.

While the basic principles of design, type selection, and the use of illustrations still hold for television, there are unique differences in their application.

First of all, graphics for television are viewed in a rectangular area that is in the ratio of 4:3. All design elements and art must be prepared with this in mind. Television screen sizes are measured on a diagonal line from one corner to another. Thus the actual viewing area of a 22-inch television set is 17.6 by 13.2 inches. A 26-inch set has a viewing area of 20.8 by 15.6 inches.

The designer concerned with preparing visuals for television also needs to know how these visuals are transmitted. A television picture is made by a series of fine horizontal lines. The image is scanned much like a person's eyes scan a printed page while reading. An electronic beam traces the visual in a 525-line zigzag thirty times each second. As a result, it is difficult for television to transmit delicately drawn lines or fine shading. Typefaces with very thin strokes or serifs can be distorted or become virtually illegible when seen on the television screen. Spacing is also an important factor in type placement for television transmission. Lines and letters that are too close together can merge and become blurs.

The amount of the total picture that leaves the station to be seen on various receivers can vary depending on the types and sizes of the receivers used by the audience. The designer must be aware of this and allow for a "safety margin." This is a margin or frame around the visual to compensate for the loss of the full picture by some receivers. If a visual is made on a 12 by 9 card, for example, the actual design area should be only about 10 by 7 inches.

The designer must be aware of viewing conditions on the receiving end of the transmission, too. If the program is viewed in the home by a few people sitting 6 to 8 feet from the screen, the design elements should be quite different than if the program is educational and intended for viewing by a large audience in a classroom.

As a general guideline, designers believe that the minimum size for type in a graphic for television should be at least ⅟₂₅ of the total picture height. The type size used in a 12 by 9 layout should be at least 36 point if this guideline is followed.

The designer of graphics for television has to become skilled at working with a number of techniques that do not apply to design for the print media. Probably the most complex is the whole area of animation. Others are such devices as inlay, overlay, back projection, color separation overlay, and split-screen effects, to mention a few.

Inlay is a technique by which part of a television picture is cut out electronically and replaced by visuals from another source. Overlay is accomplished in a similar way. Foreground material from one camera is projected on top of background originating from another source.

Back projection involves showing an image of a photographic transparency on a translucent screen. Then a person or object in front of the screen is recorded by the television camera along with the image. Color separation overlay is another method of projecting graphics and art from

Fig. 18-6 This image was created by Joseph Dettmore for NBC Nightly News. It is a fine example of illustrating a difficult subject in a creative way. This work would take hours to produce using the traditional tools. The Quantel Paintbox provides the speed & flexibility needed to create graphics within tight deadline situations.

two sources at the same time. Split-screen effects also enable the producer to project material from more than one source simultaneously.

All of these techniques require close coordination between the designer and the other staff members engaged in producing a television program. It is obviously helpful for the designer of graphics for the video screen to become familiar with all phases of television production.

The road to a successful career in television graphics leads the beginning designer through many complicated processes that require much study and practice. The starting place, however, is an introductory course in graphics that covers the same basic information and techniques needed for work in the print media. Ralph Famiglietta, head of NBC news graphics, says that basic design skills are of paramount importance in broadcast design, as they would be anywhere else. "The first thing I look for in a portfolio are good drawing skills, the ability to storyboard and, above all, good conceptualization and communications skills." He adds, "I always look for concept work in the portfolio—that's very important." Indeed, the ability to take a complicated news story, extract the essential facts, and render those facts visually in the period of time that NBC's deadlines allow is a talent few artists possess. What's important in Famiglietta's department is not the computer, but making sure that the person who sits in front of it is a first-rate designer.[10]

▼ A New Emphasis on Creativity

Now that the new technology is in place and is being continually refined, designers are contemplating their future in this electronic age. Some observers predict less of a separation between art directors, designers, editors, and production personnel. The designer will become more involved in the work of all those who create and produce communications and the others will become more involved in the work of the designer.

The graphic designer in the twenty-first century will not only design but will also be a problem solver. The designer will seek the correct visual solution to communication problems. This will require developing the skills needed to analyze problems and find solutions. The designer will need to be able to plan the implementation of these solutions and present them without the interference of semantic or channel noise.

The designer who does not have a broad background that extends well beyond technical design skills will find it difficult, if not impossible, to fit into the age of information overload. And, in this age it may become more difficult for the designer to develop creative skills, especially if the designer falls into what could be called the *information bits trap.*

We increasingly receive our information in bits and pieces. In the newspaper world, for instance, the trend is to fill the publication with bits and pieces of information rather than depth accounts that probe for answers. The popularity of *USA Today* testifies to this.

Television is the information medium of choice for increasing numbers of people. The average television image lasts only 3.5 seconds, and it is possible for the average person to view more than 7,200 scenes each day on television alone.

A recent study indicates that four out of five Americans can remember no more than three stories aired during a half-hour newscast the night before.[11] The average 30-minute newscast offers thirteen stories. That means less than a quarter of what is broadcast is remembered.

This glut of information, much of it useless, can hamper our ability to create. We run the risk of being conditioned to form decisions on bits and pieces of information. We might be tempted to base our creative efforts on information that is too flimsy.

As we have discussed in earlier chapters, creativity occurs when a person synthesizes existing information or knowledge and uses it in a new and useful way. To be creative in this world of information overload, graphic designers might need to develop skills that enable them to:

- Eliminate useless information, or at least recognize the shallow nature of much of the information they receive.
- Look behind the headlines, and beyond material designed to entertain, to obtain knowledge from solid sources.

Fig. 18-7 (Right and below) Although predesigned templates for page layouts and stock art available for desktop publishing are said by critics to stifle creativity, programs are available that open the door to almost limitless possibilities. These examples of creative design were produced on a computer. (Courtesy Computer Support Corporation).

- Seek out information from as many sources as possible and not just from sources that reinforce preconceived beliefs.
- Read widely, and not only read but also experience and participate in actions that relate to the graphic problem for which a solution is sought.

▼ Preparation for Today's Graphics

Where does all this leave someone planning a career in communication or the graphic arts? How should a person prepare to work and advance in the world of graphic communications today?

"There will be no careers in the field that won't be touched by computers. The design field will rely more and more on computers," comments David B. Gray, managing editor graphics of the *Providence Journal*.[11]

Gray suggests that people planning careers in communications should have a strong liberal arts background. In addition, he believes design schools should spend more time teaching writing and journalism schools should spend more time teaching design.

Other professionals suggest that students who wish to function effectively with design and graphics in the electronic era learn as much as possible about the principles and philosophies of the great movements or schools of design. These include classic revival, art nouveau, art deco, Bauhaus, Swiss design, and international architectural style.

In addition, students should learn about the principles, philosophy, and richness of typographical design and develop an interest in all the graphic arts, especially photography. Students should study layouts in newspapers, magazines, and other publications. The idea is to know not only the rudiments of design but, perhaps more important, what the designer is trying to accomplish.

Study of the principles of statistical and informational graphics is a must. And students are urged to learn to think of design and typography as absolutely integrated and interrelated with the processes of communication, not as decorative appendages. The goal of design and journalism schools today is to teach students to think, to explore, to question, to wonder, and to approach old problems with new solutions.

And, it is important to remember that, even with the new technology, printed communications are still "consumed" with a device that has not changed in the nearly 550 years since Johann Gutenberg cast the first piece of movable type—the human eye. Design should always help, not hinder, the human eye.

The Same, Only More So

When the first edition of this book was being written, designer David B. Gray was asked to make suggestions for those who are planning careers in communications.

At that time he said, "There will be no careers in the publishing field that won't be touched by computers. More especially, the design field in particular will rely more and more on computers."

Gray recommended that those planning careers in communications should have strong liberal arts backgrounds. Designers should spend more time on writing and journalism schools should spend more time on design.

When asked if he would like to revise the recommendations for a new edition of this book, Gray said, "The same thing applies today as when I said it five years ago, *only more so.*"

▼ Career Tips from Professionals

- Computers are here to stay. Learn as much as possible about them, but don't get bogged down in the bits and bytes.

- Broaden your horizons by reading, studying, learning—in the liberal arts.
- Start now and make a habit of reading the leading trade periodicals in the world of printed communications.
- Study publications and advertisements on a regular basis to monitor trends in graphics and typography.
- Read the popular books that are concerned with the directions society is taking as we move into the information age.
- Writers and editors should learn as much as they can about design, and designers should learn as much as they can about writing and editing.

▼ Graphics in Action

1. Collect samples of unusual type arrangements and see if you can determine if minus letterspacing or minus leading was used. Analyze the samples to determine if the arrangements helped or hindered readability.
2. Arrange a field trip to a newspaper, graphic arts firm, or printing plant that is using electronic editing and design equipment. Try to get some hands-on practice.
3. Prepare a research paper on electronic equipment, how it operates, and what it is capable of doing in your particular area of interest— advertising, newspapers, public relations, magazines, and so on.
4. Devise a model program of courses of study you believe would be suitable for someone planning a career in print or electronic graphics in the information age with emphasis on your area of special interest.
5. Devise a "printed communications IQ test" that a person could take to determine if he or she were acquiring the attributes that would lead to a successful career. Include a scoring system weighted according to the relative importance of the items you include.

Notes

[1] "How 'Smart' did it," *Pre-*, April 1989, p. 36.
[2] "New Roles for Graphic Designers," *Electronic Publishing & Printing*, December 1988, pp 28–30.
[3] "The Satellite Connection," *Magazine Design & Production*, February 1989, pp. 22–25.
[4] "The Future of Print," *American Printer*, August 1986, p. 44.
[5] "Cocooning," *Electronic Publishing & Printing*, January/February 1989, p. 4.
[6] "Forecast '89," *American Printer*, December 1988, p. 52.
[7] "The Paperless Magazine," *Pre-*, April 1989, pp. 16–17.
[8] "The View From the Danube," *Graphic Arts Monthly*, October 1987, p. 114.
[9] "How to Choose and Use a Graphic Designer," *Public Relations Journal*, July 1985, p. 33.
[10] "Making the News," *How*, November/December 1988, p. 93.
[11] "Film at eleven . . . ," *OSU Quest*, Winter 1986, pp. 12–13.
[12] David B. Gray, personal communications with the author, November 26, 1983, and May 5, 1989.

A Basic Communications Design Library

It would be impossible and impractical to include a truly comprehensive list of publications in graphic design. However, the books and periodicals listed have been chosen to form a good, basic library for the graphics communicator.

▼ Books

Adams, James L. *Conceptual Blockbusting*. Stanford, California: Stanford Alumni Association, 1974.
A concise, readable approach to creative problem solving with thought-provoking ideas of value to all who write or design.

Arnold, Edmund C. *Designing the Total Newspaper*. New York: Harper & Row, 1981.
A comprehensive examination of all aspects of newspaper design by America's leading newspaper designer for many years.

Beach, Mark. *Editing Your Newsletter*. Portland, Oregon: Coast-to-Coast Books, 1982.
A guide to the writing, design, and production of newsletters. Excellent for beginners in newsletter production.

Beach, Mark and Ken Russon. *Papers for Printing: How to Choose the Right Paper at the Right Price for Any Printing Job*. Portland, Oregon: Coast-to-Coast Books, 1989.
This no-nonsense book describes classifications of paper, features and benefits of various stocks, selecting and specifying, and business considerations. Includes 40 samples of the most common papers specified by printing buyers.

Beaumont, Michael. *Type Design, Color, Character and Use*. Cincinnati: North Light Books. 1987.
A delightful, colorful, and profusely illustrated collection of typography from around the world.

Birren, Faber. *Creative Color*. New York: Van Nostrand Reinhold, 1961.
An introduction to color principles and the application of color to design; it includes simple to complex exercises to help one understand and use color.

Bonura, Larry S. *Desktop Publisher's Dictionary*. Plano, Texas: Wordware Publishing, Inc., 1989.
A valuable addition to the professional library of anyone involved in graphics in this age of electronic publishing.

Bove, Tony. *The Art of Desktop Publishing*. New York: Bantam Books, 1987.
Probably the best introduction to desktop publishing for the novice. Very easy to understand.

Craig, James. *Designing with Type, Revised Edition*. New York: Watson-Guptill, 1983.
A new edition of a basic introductory workbook/text on typography. It includes complete alphabets and simple how-to information.

Garcia, Mario R. *Contemporary Newspaper Design*. Englewood Cliffs, N.J.: Prentice-Hall, 1984.
An outline of the basics of newspaper design, very contemporary, with many illustrations plus examples of redesign projects completed by the author.

Gottschall, Edward M. *Typographic Communications Today*. Cambridge, Massachusetts: MIT Press, 1989.
A handsomely produced critical review of 20th century typographic design. Over 900 images (half are in color) show influences on design and type, including examples from Japan, Finland, Brazil, Canada, and Czechoslovakia. A worthwhile reference.

Graham, Walter. *Complete Guide to Pasteup*. Philadelphia: North American, 1975.
A comprehensive, fundamental guide aimed at the "quick print" and newsletter producer. It is an unsophisticated introductory book, ideal for the neophyte.

Heller, Steven and Seymour Chwast. *Sourcebook of Visual Ideas*. New York: Van Nostrand Reinhold, 1989.
A collection of graphic solutions organized by category. It provides models, showing how others have solved common problems in fresh ways, while underscoring how personal style increases the effectiveness of an idea.

Hudson, Howard Penn. *Publishing Newsletters*. New York: Scribner's 1982.
This is considered the most complete guide to all aspects of newsletters, including editing and design.

Hurlburt, Allen F. *The Grid, A Modular System for the Design and Production of Newspapers, Magazines and Books*. New York: Van Nostrand Reinhold, 1978.
An overview of the use of the grid in publication layout, helping one understand pagination and computer design techniques.

International Paper Company. *Pocket Pal*. New York: International Paper, 1983.
A must for all those working with graphic arts. Write International Paper Company, 220 East 42nd Street, New York, NY 10017.

Kneller, George F. *The Art and Science of Creativity*. New York: Holt, Rinehart and Winston, 1965.
Easy reading, concise, and a good introduction to creativity.

Lem, Dean Phillip. *Graphics Master 4*. Los Angeles: Dean Lem Associates, 1986.
A workbook of planning aids, guides, and graphic tools for the design and production of printing; it also contains a line gauge and ruler, copy-fitting charts, and a photo sizing proportion scale.

Lindegren, Erik. *ABC of Lettering and Printing Typefaces*. New York: Greenwich House, 1982.
A collection of representative typeface alphabets from the various races to use in selecting and tracing types.

Moen, Daryl R. *Newspaper Layout and Design*. Ames: Iowa State University Press, 1984.
This paperback gives a well-balanced overview of newspaper design, with emphasis on how it is done.

Munce, Howard. *Graphics Handbook.* Westport, Conn.: North Light, 1982.
A guide for beginners; a good book to help neophytes design simple printed pieces.

Nelson, Roy Paul. *The Design of Advertising.* Dubuque, Iowa: Wm. C. Brown, 1989.
A comprehensive book on all aspects of advertising design and production with an overview of the basics of typography and graphics.

Nelson, Roy Paul. *Publication Design.* Dubuque, Iowa: Wm. C. Brown Company, 1987.
Emphasizes magazine and periodical design, and touches on newspapers and typography. It was originally published in 1972. Many illustrations.

Nesbitt, Alexander. *The History and Techniques of Lettering.* New York: Dover, 1957.
A classic by one of the leading students of letter forms and typography; it traces the development of letter forms and the basic races of types as well as their uses in typography and graphic design.

Parnau, Jeffrey. *Desktop Publishing: The Awful Truth.* New Berlin, Wis.: Parnau Graphics, 1989.
A delightful, readable book that debunks advertising hype about desktop publishing and offers many valuable suggestions for anyone contemplating becoming a desktop publisher.

Rosen, Ben. *Type and Typography.* New York: Van Nostrand Reinhold, 1976.
A workbook for students of graphic design and all those who work with type. It contains more than 1,500 specimens and alphabets of text and display types.

Sanders, Norman. *Graphic Designer's Production Handbook.* New York: Hastings House, 1982.
A valuable, comprehensive reference work for anyone involved in the creation and production of printed communciations. It explains with illustrations and diagrams such processes as overlays, keying, preparing for printing.

Sausmarez, Maurice de. *Basic Design: The Dynamics of Visual Form.* New York: Van Nostrand Reinhold, 1983.
A classic introduction to the basic elements and dynamics of design.

Silver, Gerald A. *Graphic Layout and Design.* New York: Van Nostrand Reinhold, 1981.
An overview of principles and practices of graphic arts in easy-to-understand language. A good self-teaching book for those desiring to sharpen layout skills.

Stevenson, George A. *Graphic Arts Encyclopedia.* New York: McGraw-Hill, 1979.
The encyclopedia of printing and graphic arts; a valuable reference for any professional's library.

Tschichold, Jan. *Treasury of Alphabets and Lettering.* London: Omega Books, 1985.
An excellent treatise on the proper selection and arrangement of type. Contains many unique and historic alphabets.

V and M Typographical. *The Type Specimen Book.* New York: Van Nostrand Reinhold, 1974.
Complete alphabets of 544 different typefaces in more than 300 sizes.

Van Uchelen, Rod. *Pasteup Production Techniques and New Applications*. New York: Van Nostrand Reinhold, 1976.
A profusely illustrated, easily understood introduction to pasteup; it contains simple projects that can be produced with a minimum of tools.

Watkins, Don. *Newspaper Advertising Handbook*. Wheaton, Ill.: Dynamo, 1980.
Although devoted mainly to designing newspaper advertisements, it contains many valulable tips for all types of graphic design. Available from Dynamo, Inc., P.O. Box 175, Wheaton, IL 60187.

White, Jan V. *Mastering Graphics*. New York: R. R. Bowker, 1983.
A handy reference tool for those who edit or produce internal or external publications for businesses or organizations.

Wildbur, Peter. *Information Graphics*. New York: Van Nostrand Reinhold, 1989.
A survey of typographic, diagrammatic, and cartographic communication. Rather complex and not a beginner's book, but worthwhile as a reference.

▼ Sources for New Publications

In the fast-paced world of the information age, new books are being produced constantly. The following are sources for books on graphics and typography. Lists of current offerings are available.

Dynamic Graphics, Inc.
6000 North Forest Park Drive
P.O. Box 1901
Peoria, IL 61656-1901
1-800-255-8800

Graphic Artist's Book Club
P.O. Box 12526
Cincinnati, OH 45212-0526

Print Book Store
6400 Goldsboro Road
Bethesda, MD 20817

▼ Periodicals

Communication Arts—A bimonthly publication containing detailed reports on designers, illustrators, photographers, and art directors. *Communication Arts*, 410 Sherman Avenue, P.O. Box 10399, Palo Alto, CA 94303.

Communicators, The—A monthly publication of the International Association of Business Communicators that contains many articles and regular features on magazine design. IABC, 879 Market Street, Suite 940, San Francisco, CA 94102.

DESIGN—A quarterly journal of the professional society dedicated to improving newspapers through design; it provides information on newspapers undergoing design overhauls, computer graphics, trends. The Society of Newspaper Design, The Newspaper Center, Box 17290, Dulles International Airport, Washington, DC 20041.

HOW—A bimonthly, profusely illustrated magazine emphasizing ideas and techniques in graphic design. Worthwhile but somewhat advanced for the neophite designer. HOW, Post Office Box 12575, Cincinnati, OH 45212-9927.

Ninth Edition—An annual publication of the Society of Newspaper Design showing the winners in numerous categories of design. Issued yearly as First Edition, Second Edition, and so on. The Society of Newspaper Design, address above.

Newsletter on Newsletters, The—A twice-a-month newsletter for those who edit and produce newsletters; it contains a frequent "Graphics Clinic" section. The Newsletter Clearinghouse, 44 West Market Street, P.O. Box 311, Rhinebeck, NY 12572.

Personal Publishing—A magazine for desktop publishers that monitors new products and emphasizes how to get the most out of your desktop programs. Personal Publishing, 191 South Gary Avenue, Carol Stream, IL 60188.

Print—A bimonthly, profusely illustrated design periodical; a source of new ideas and trends in design and typography. *Print,* 6400 Goldsboro Road NW, Washington, DC 20034.

Publish!—A monthly magazine for desktop publishers that provides professional help in typography, graphics, page design and layout. Practical tips and how-to articles appear in every issue, as well as reviews of the latest technology. *Publish!,* Subscription Department, P.O. Box 51966, Boulder, CO 80321-1966.

Step-by-Step Electronic Design—A monthly newsletter for desktop publishers. *Step-by-Step Electronic Design,* 6000 N. Forest Park Drive, P.O. Box 1901, Peoria, IL 61656-9975.

Step-by-Step Graphics—As the title suggests, this is a practical, how-to magazine in which designers and illustrators share their secrets and demonstrate how they work. Published seven times per year. *Step-by-Step Grapics,* 6000 North Forest Park Drive, P.O. Box 1901, Peoria, IL 61656-9975.

Typographic i—A great little publication for ideas, information, and inspiration for anyone involved in producing printed communications. Typographers International Association, 2262 Hall Place NW, Washington, DC 20007.

U&lc—A quarterly tabloid-size free magazine about typography; an excellent source of ideas. U&lc, International Typeface Corporation, 2 Hammarskjold Plaza, New York, NY 10017.

Type Specimens

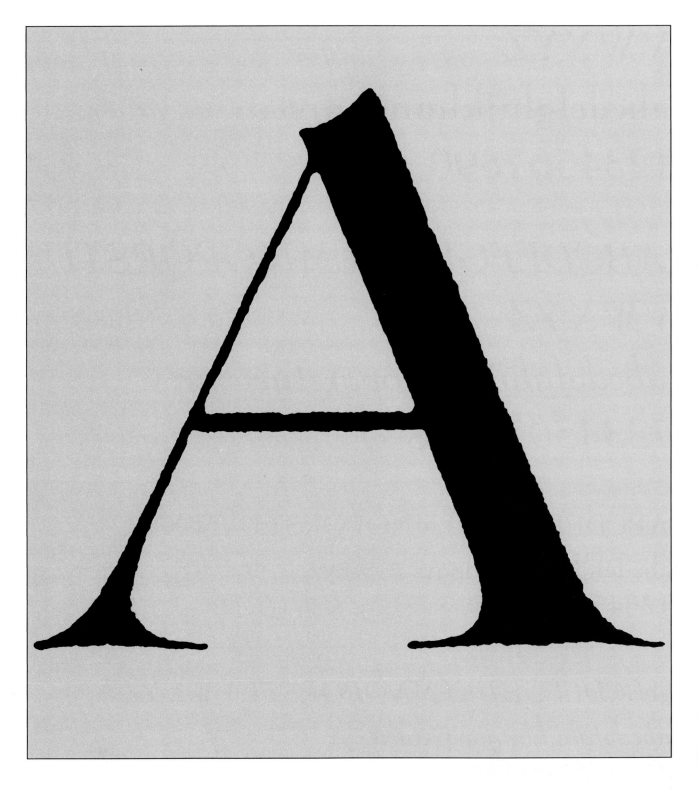

---

36 POINT BERNHARD MODERN BOLD

ABCDEFGHIJKLMNOPQRSTU VWXYZ
abcdefghijklmnopqrstuvwxyz
1234567890

---

36 POINT BERNHARD MODERN BOLD ITALIC

ABCDEFGHIJKLMNOPQRSTU VWXYZ
abcdefghijklmnopqrstuvwxyz
1234567890$

---

24 POINT BERNHARD MODERN BOLD

ABCDEFGHIJKLMNOPQRSTUVWXYZ
abcdefghijklmnopqrstuvwxyz
1234567890$&

---

24 POINT BERNHARD MODERN BOLD ITALIC

ABCDEFGHIJKLMNOPQRSTUVWXYZ
abcdefghijklmnopqrstuvwxyz
1234567890$&

72 POINT BERNHARD MODERN BOLD

ABCDEFGHI
JKLMNOPQR
STUVWXYZ
abcdefghijklm
nopqrstuvwxyz
1234567890

36 POINT BODONI BOLD

ABCDEFGHIJKLMNOPQRSTUV WXYZ
abcdefghijklmnopqrstuvwxyz
1234567890$&

36 POINT BODONI BOLD ITALIC

ABCDEFGHIJKLMNOPQRSTUV WXYZ
abcdefghijklmnopqrstuvwxyz
1234567890$&

24 POINT BODONI BOLD

ABCDEFGHIJKLMNOPQRSTUVWXYZ
abcdefghijklmnopqrstuvwxyz
1234567890$&

24 POINT BODONI BOLD ITALIC

ABCDEFGHIJKLMNOPQRSTUVWXYZ
abcdefghijklmnopqrstuvwxyz
1234567890$&

72 POINT BODONI BOLD

ABCDEFGHIJ
KLMNOPQR
STUVWXYZ
abcdefghijklm
nopqrstuvwxyz
1234567890$

36 POINT CASLON OLD STYLE

ABCDEFGHIJKLMNOPQRSTUV WXYZ
abcdefghijklmnopqrstuvwxyz
1234567890&$

36 POINT CASLON OLD STYLE ITALIC

ABCDEFGHIJKLMNOPQRS TUVWXYZ

abcdefghijklmnopqrstuvwxyz ABCD EGFKLMNPQRT 1234567890

24 POINT CASLON OLD STYLE

ABCDEFGHIJKLMNOPQRSTUVWXYZ
abcdefghijklmnopqrstuvwxyz

1234567890

24 POINT CASLON OLD STYLE ITALIC

*ABCDEFGHIJKLMNOPQRSTUVWXYZ
abcdefghijklmnopqrstuvwxyz ABCDEGFK LMNP RTUY& 1234567890&$*

72 POINT CASLON OLD STYLE

ABCDEFGHIJ
KLMNOPQRS
TUVWXYZ
abcdefghijklmn
opqrstuvwxyz
1234567890&$

36 POINT CENTURY EXPANDED BOLD

ABCDEFGHIJKLMNOPQRST UVWXYZ
abcdefghijklmnopqrstuvwxyz
1234567890

36 POINT CENTURY EXPANDED BOLD ITALIC

ABCDEFGHIJKLMNOPQRST UVWXYZ
abcdefghijklmnopqrstuvwxyz
1234567890

24 POINT CENTURY EXPANDED BOLD

ABCDEFGHIJKLMNOPQRSTUVWXYZ
abcdefghijklmnopqrstuvwxyz
1234567890

24 POINT CENTURY EXPANDED BOLD ITALIC

ABCDEFGHIJKLMNOPQRSTUVWXYZ
abcdefghijklmnopqrstuvwxyz
1234567890

72 POINT CENTURY EXPANDED BOLD

ABCDEFGHIJ
KLMNOPQRS
TUVWXYZ
abcdefghijklmn
opqrstuvwxyz
1234567890

---

36 POINT GARAMOND BOLD

ABCDEFGHIJKLMNOPQRSTUV WXYZ
abcdefghijklmnopqrstuvwxyz
1234567890$&

---

36 POINT GARAMOND BOLD ITALIC

ABCDEFGHIJKLMNOPQRSTUV WXYZ
abcdefghijklmnopqrstuvwxyz
1234567890$&

---

24 POINT GARAMOND BOLD

ABCDEFGHIJKLMNOPQRSTUVWXYZ
abcdefghijklmnopqrstuvwxyz
1234567890$&

---

24 POINT GARAMOND BOLD ITALIC

ABCDEFGHIJKL MNOPQRSTUV WXYZ
abcdefghijklmnopqrstuvwxyz
1234567890$&

72 POINT GARAMOND BOLD

ABCDEFGHIJ
KLMNOPQ
RSTUVWXYZ
abcdefghijklm
nopqrstuvwxyz
1234567890$&

36 POINT GOUDY BOLD

ABCDEFGHIJKLMNOPQRST UVWXYZ
abcdefghijklmnopqrstuvwxyz
1234567890

36 POINT GOUDY BOLD ITALIC

ABCDEFGHIJKLMNOPQRST UVWXYZ
abcdefghijklmnopqrstuvwxyz
1234567890

24 POINT GOUDY BOLD

ABCDEFGHIJKLMNOPQRSTUVWXYZ
abcdefghijklmnopqrstuvwxyz
1234567890

24 POINT GOUDY BOLD ITALIC

ABCDEFGHIJKLMNOPQRSTUVWXYZ
abcdefghijklmnopqrstuvwxyz
1234567890

72 POINT GOUDY BOLD

ABCDEFGHIJK
LMNOPQRSTU
VWXYZ
abcdefghijklmnop
qrstuvwxyz
1234567890

36 POINT PALATINO SEMIBOLD

ABCDEFGHIJKLMNOPQRSTU VWXYZ

abcdefghijklmnopqrstuvwxyz
1234567890   *1234567890*

36 POINT PALATINO SEMIBOLD ITALIC

ABCDEFGHIJKLMNOPQRSTU VWXYZ

*abcdefghijklmnopqrstuvwxyz
1234567890*

24 POINT PALATINO SEMIBOLD

ABCDEFGHIJKLMNOPQRSTUVWXYZ
abcdefghijklmnopqrstuvwxyz
1234567890 *1234567890*

24 POINT PALATINO SEMIBOLD ITALIC

ABCDEFGHIJKLMNOPQRSTUVWXYZ
*abcdefghijklmnopqrstuvwxyz
1234567890*

ABCDEFGHIJK
LMNOPQRST
UVWXYZ
abcdefghijklmn
opqrstuvwxyz
1234567890
1234567890

36 POINT TIMES NEW ROMAN BOLD

ABCDEFGHIJKLMNOPQRST UVWXYZ& abcdefghijklmnopqrstuvwxyz 1234567890$

36 POINT TIMES NEW ROMAN BOLD ITALIC

ABCDEFGHIJKLMNOPQRST UVWXYZ& abcdefghijklmnopqrstuvwxyz 1234567890$

24 POINT TIMES NEW ROMAN BOLD

**ABCDEFGHIJKLMNOPQRSTUVWXYZ&
abcdefghijklmnopqrstuvwxyz
1234567890$**

24 POINT TIMES NEW ROMAN BOLD ITALIC

*ABCDEFGHIJKLMNOPQRSTUVWXYZ&
abcdefghijklmnopqrstuvwxyz
1234567890$*

72 POINT TIMES NEW ROMAN BOLD

ABCDEFGHIJ
KLMNOPQRS
TUVWXYZ&
abcdefghijklmno
pqrstuvwxyz
1234567890$

60 POINT WEISS

ABCDEFGHIJKLM
NOPQRSTUV
WXYZ&
abcdefghijklmnopqr
stuvwxyz
1234567890$

36 POINT WEISS

ABCDEFGHIJKLMNOPQRST
UVWXYZ&
abcdefghijklmnopqrstuvwxyz
1234567890$

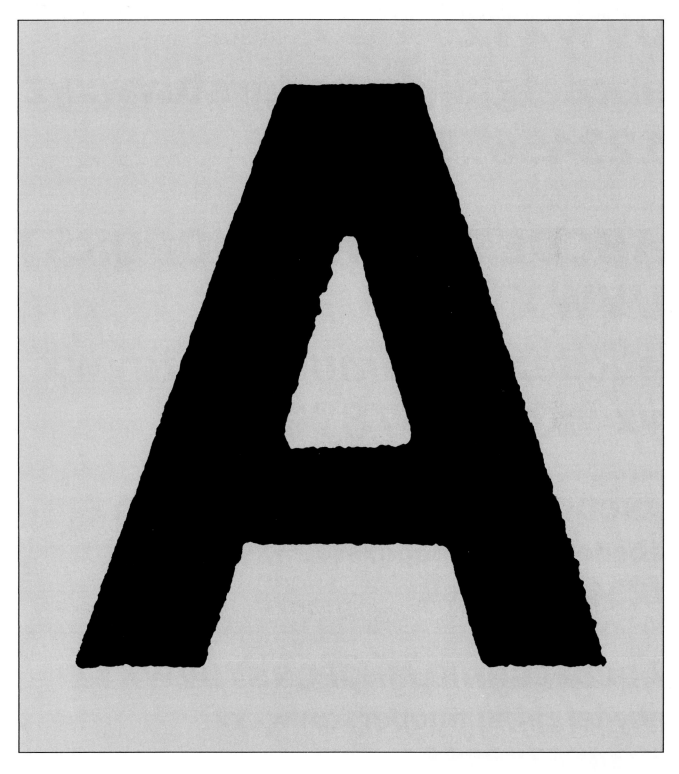

36 POINT FRANKLIN GOTHIC BOLD

ABCDEFGHIJKLMNOPQRST UVWXYZ abcdefghijklmnopqrstuvwxyz 1234567890$&

36 POINT FRANKLIN GOTHIC BOLD ITALIC

ABCDEFGHIJKLMNOPQRST UVWXYZ abcdefghijklmnopqrstuvwx yz 1234567890$&

24 POINT FRANKLIN GOTHIC BOLD

ABCDEFGHIJKLMNOPQRSTUVWXYZ abcdefghijklmnopqrstuvwxyz 1234567890$&

24 POINT FRANKLIN GOTHIC BOLD ITALIC

ABCDEFGHIJKLMNOPQRSTUVWXYZ abcdefghijklmnopqrstuvwxyz 1234567890$&

72 POINT FRANKLIN GOTHIC BOLD

ABCDEFGHIJK
LMNOPQRSTU
VWXYZ
abcdefghijklmno
pqrstuvwxyz
12345

60 POINT FUTURA LIGHT

ABCDEFGHIJKLMNOP QRSTUVWXYZ abcdefghijklmnopqrstuv wxyz& 1234567890

36 POINT FUTURA LIGHT

ABCDEFGHIJKLMNOPQRSTUVWXYZ
abcdefghijklmnopqrstuvwxyz&
1234567890

36 POINT FUTURA LIGHT ITALIC

ABCDEFGHIJKLMNOPQRSTUVWXYZ
abcdefghijklmnopqrstuvwxyz&
1234567890

60 POINT FUTURA DEMIBOLD

ABCDEFGHIJKLMNO
PQRSTUTUVWXYZ
abcdefghijklmnopqrst
uvwxyz 1234567890

36 POINT FUTURA DEMIBOLD

ABCDEFGHIJKLMNOPQRSTUVWXY
Z abcdefghijklmnopqrstuvwxyz&
1234567890

36 POINT FUTURA DEMIBOLD ITALIC

ABCDEFGHIJKLMNOPQRSTUVWXY
Z abcdefghijklmnopqrstuvwxyz&
1234567890

60 POINT HELVETICA LIGHT

ABCDEFGHIJKLMN
OPQRSTUVWXYZ
abcdefghijklmnopqrst
uvwxyz&1234567890

36 POINT HELVETICA LIGHT

ABCDEFGHIJKLMNOPQRST
UVWXYZ
abcdefghijklmnopqrstuvwxyz&
1234567890

36 POINT HELVETICA LIGHT ITALIC

ABCDEFGHIJKLMNOPQRST
UVWXYZ
abcdefghijklmnopqrstuvwxyz&
1234567890

60 POINT HELVETICA

ABCDEFGHIJKLMN
OPQRSTUVWXYZ
abcdefghijklmnopqrst
uvwxyz 1234567890

36 POINT HELVETICA

ABCDEFGHIJKLMNOPQRS
TUVWXYZ
abcdefghijklmnopqrstuvwxyz&
1234567890

36 POINT HELVETICA ITALIC

ABCDEFGHIJKLMNOPQRS
TUVWXYZ
abcdefghijklmnopqrstuvwxyz&
1234567890

---

36 POINT HELVETICA BOLD

ABCDEFGHIJKLMNOPQRSTUV WXYZ
abcdefghijklmnopqrstuvwxyz& 1234567890

---

36 POINT HELVETICA BOLD ITALIC

ABCDEFGHIJKLMNOPQRSTUV WXYZ
abcdefghijklmnopqrstuvwxyz& 1234567890

---

24 POINT HELVETICA BOLD

**ABCDEFGHIJKLMNOPQRSTUVWXYZ
abcdefghijklmnopqrstuvwxyz&
1234567890**

---

24 POINT HELVETICA BOLD ITALIC

***ABCDEFGHIJKLMNOPQRSTUVWXYZ
abcdefghijklmnopqrstuvwxyz&
1234567890***

72 POINT HELVETICA BOLD

ABCDEFGHIJKL
MNOPQRSTUV
WXYZ
abcdefghijklmno
pqrstuvwxyz&
1234567890

---

72 POINT HELVETICA EXTRA BOLD

ABCDEFGHIJ
KLMNOPQRS
TUVWXYZ
abcdefghijklm
nopqrstuvwx
yz1234567890

---

36 POINT HELVETICA EXTRA BOLD

ABCDEFGHIJKLMNOPQ
RSTUVWXYZ
abcdefghijklmnopqrstuv
wxyz& 1234567890

60 POINT OPTIMA

ABCDEFGHIJKLMN
OPQRSTUVWXYZ
abcdefghijklmnopqrs
tuvwxyz&1234567890

36 POINT OPTIMA

ABCDEFGHIJKLMNOPQRST
UVWXYZ
abcdefghijklmnopqrstuvwxyz&
1234567890

36 POINT OPTIMA ITALIC

*ABCDEFGHIJKLMNOPQRST
VWXYZ
abcdefghijklmnopqrstuvwxyz&
1234567890*

60 POINT UNIVERS 45

ABCDEFGHIJKLMN
OPQRSTUVWXYZ
abcdefghijklmnopqrst
uvwxyz&
1234567890

36 POINT UNIVERS 45

ABCDEFGHIJKLMNOPQRSTUV
WXYZ
abcdefghijklmnopqrstuvwxyz&
1234567890

36 POINT UNIVERS 46

ABCDEFGHIJKLMNOPQRSTUV
WXYZ
abcdefghijklmnopqrstuvwxyz&
1234567890

60 POINT UNIVERS 55

ABCDEFGHIJKLMN OPQRSTUVWXYZ abcdefghijklmnopqrs tuvwxyz& 1234567890

36 POINT UNIVERS 55

ABCDEFGHIJKLMNOPQRSTU VWXYZ abcdefghijklmnopqrstuvwxyz& 1234567890

36 POINT UNIVERS 56

ABCDEFGHIJKLMNOPQRSTU VWXYZ abcdefghijklmnopqrstuvwxyz& 1234567890

72 POINT UNIVERS 75

ABCDEFGHIJK
LMNOPQRS
TUVWXYZ
abcdefghijklmn
opqrstuvwxyz
1234567890

36 POINT UNIVERS 75

ABCDEFGHIJKLMNOPQR
STUVWXYZ
abcdefghijklmnopqrstuvwxyz&
1234567890

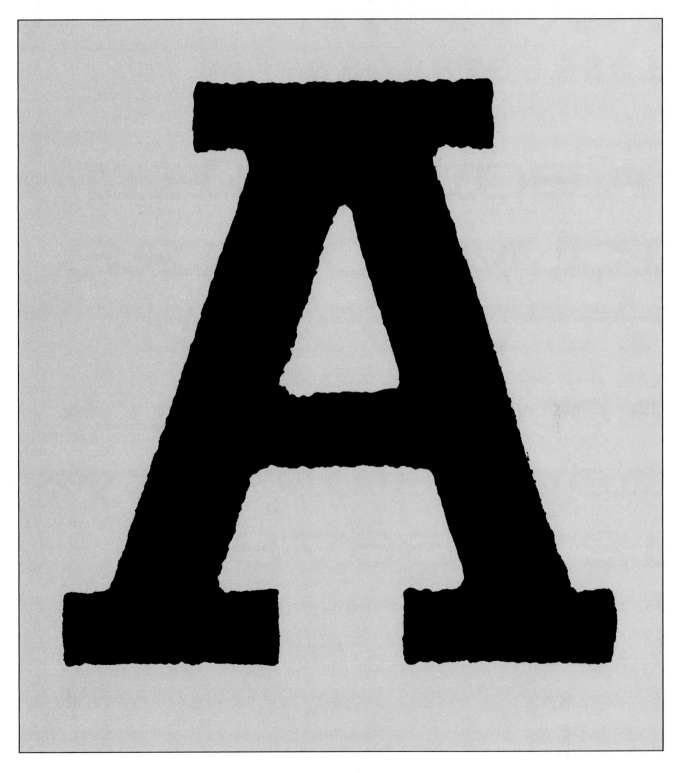

24 POINT GOLD RUSH

ABCDEFGHIJKLMNOPQ
RSTUVWXYZ
1234567890 0$&

36 POINT HELLENIC WIDE

ABCDEFGHIJ
KLMNOPQRS
TUVWXYZ
abcdefghijklm
nopqrstuvwxyz
1234567890

72 POINT P. T. BARNUM

ABCDEFGHIJKLM
NOPQRSTUVW
XYZ abcdefghijk
lmnopqrstuvwxyz
1234567890$&

48 POINT P. T. BARNUM

ABCDEFGHIJKLMNOPQRST
UVWXYZ
abcdefghijklmnopqrstuvw
xyz 1234567890$&

36 POINT LUBALIN GRAPH LIGHT

ABCDEFGHIJKLMNOPQRSTU
VWXYZ
abcdefghijklmnopqrstuvwxyz
1234567890

36 POINT LUBALIN GRAPH BOOK

ABCDEFGHIJKLMNOPQRSTU
VWXYZ
abcdefghijklmnopqrstuvwxyz
1234567890

36 POINT LUBALIN GRAPH BOLD

ABCDEFGHIJKLMNOPQRS
TUVWXYZ
abcdefghijklmnopqrstuvw
vwxyz&
1234567890

---

36 POINT STYMIE LIGHT

ABCDEFGHIJKLMNOPQRST
UVWXYZ
abcdefghijklmnopqrstuvwxyz&
1234567890

---

36 POINT STYMIE LIGHT ITALIC

ABCDEFGHIJKLMNOPQRST
UVWXYZ
abcdefghijklmnopqrstuvwxyz&
1234567890

---

24 POINT STYMIE MEDIUM

ABCDEFGHIJKLMNOPQRSTUVWXYZ
abcdefghijklmnopqrstuvwxyz&
1234567890

---

24 POINT STYMIE MEDIUM ITALIC

ABCDEFGHIJKLMNOPQRSTUVWXYZ
abcdefghijklmnopqrstuvwxyz&
1234567890

36 POINT STYMIE BOLD

ABCDEFGHIJKLMNOPQRST UVWXYZ
abcdefghijklmnopqrstuvwxyz
1234567890

36 POINT STYMIE BOLD ITALIC

ABCDEFGHIJKLMNOPQRST UVWXYZ
abcdefghijklmnopqrstuvwxyz
1234567890

24 POINT STYMIE BOLD

ABCDEFGHIJKLMNOPQRSTUVWXYZ
abcdefghijklmnopqrstuvwxyz&
1234567890

24 POINT STYMIE BOLD ITALIC

ABCDEFGHIJKLMNOPQRSTUVWXYZ
abcdefghijklmnopqrstuvwxyz&
1234567890

72 POINT STYMIE BOLD

ABCDEFGHIJK
LMNOPQRST
UVWXYZ
abcdefghijklm
nopqrstuvwxyz
1234567890

48 POINT STYMIE EXTRA BOLD

AABCDEFGHIJKL MNOPQRSTUVW XYZ

aabcdefghijklmnop qrstuvwxyz 1234567890

36 POINT STYMIE EXTRA BOLD

AABCDEFGHIJKLMNOP QRSTUVWXYZ

aabcdefghijklmnopqrstu vwxyz

1234567890

48 POINT STYMIE OPEN

ABCDEFGHIJKL
MNOPQRSTUV
WXYZ&
1234567890$

36 POINT STYMIE OPEN

ABCDEFGHIJKLMNOP
QRSTUVWXYZ&
1234567890$

24 POINT STYMIE OPEN

ABCDEFGHIJKLMNOPQRSTUV
WXYZ&
1234567890$

---

48 POINT TOWER

ABCDEFGHIJKLMNOPQRSTUV WXYZ

abcdefghijklmnopqrstuvwxyz

1234567890

---

36 POINT TOWER

ABCDEFGHIJKLMNOPQRSTUVWXYZ

abcdefghijklmnopqrstuvwxyz

1234567890

---

36 POINT TRYLON

ABCDEFGHIJKLMNOPQRSTUVWXYZ&

abcdefghijklmnopqrstuvwxyz

1234567890$

48 POINT BERNHARD TANGO

ABCDEFGHIJKLMNOPQ
RSTUUVWXYZ
abcdefghijklmnopqrstuvwxyz
1234567890

24 POINT BERNHARD TANGO

ABCDEFGHIJKLMNOPQRSTUUVWXYZ
abcdefghijklmnopqrstuvwxyz   1234567890

48 POINT CORONET BOLD

ABCDEFGHIJKLMNO
PQRSTUVWXYZ&
abcdefghijklmnopqrstuvwxyz
1234567890$

24 POINT CORONET BOLD

ABCDEFGHIJKLMNOPQRSTUVWXYZ&
abcdefghijklmnopqrstuvwxyz   1234567890$

72 POINT BRUSH

ABCDEFGHIJ
KLMNOP2R
STUVWXYZ
abcdefghijklmnopqrst
uvwxyz1234567890

36 POINT BRUSH

ABCDEFGHIJKLMNOP2
RSTUVWXYZ
abcdefghijklmnopqrstuvwxyz
1234567890$&

48 POINT LEGEND

ABCDEFGHIJKLMNOP
QRSTUVWXYZ
abcdefghijklmnopqrstuvwxyz

1234567890

36 POINT LIBERTY

ABCDEFGHIJKLMNOPQ
RSTUVWXYZ

abcdefghijklmnopqrstuvwxyz

1234567890

60 POINT LYDIAN CURSIVE

ABCDEFGHIJKLMN
OPQRSTUVWXYZ
abcdeefghijklmnopqrstuv
wxyz
1234567890

36 POINT LYDIAN CURSIVE

ABCDEFGHIJKLMNOPQRSTUVW
XYZ
abcdeefghijklmnopqrstuvwxyz
1234567890

48 POINT MISTRAL

ABCDEFGHIJKLMNOPQRSTU
VWXYZ&
abcdefghijklmnopqrstuvwxyz
1234567890$

48 POINT STRADIVARIUS

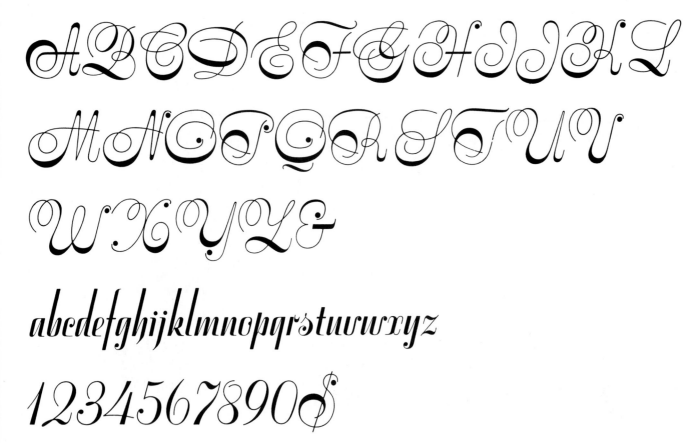

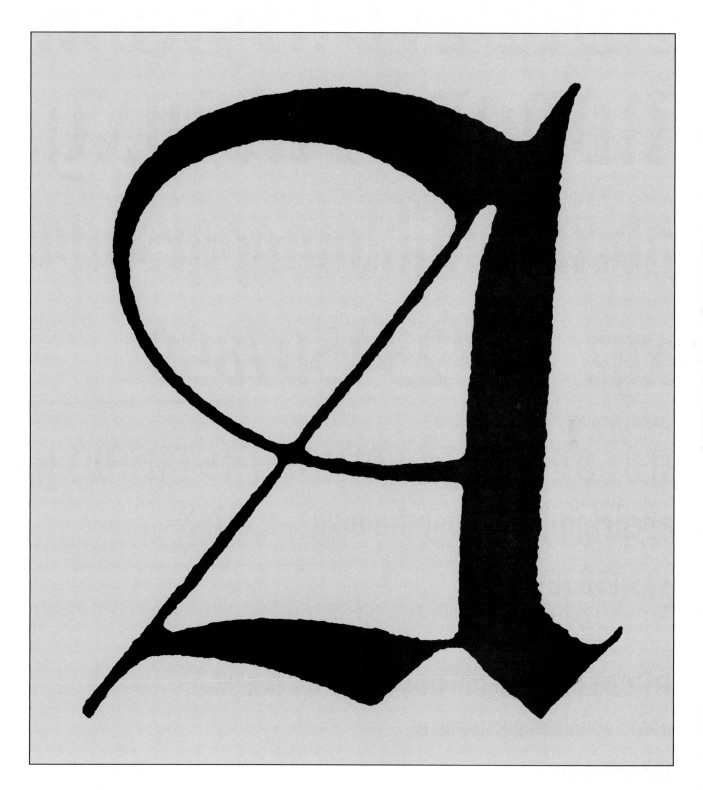

---

72 POINT AMERICAN TEXT

ABCDEFGHIJKLM
NOPQRSTUVWXYZ
abcdefghijklmnopqrstuvw
xyz 1234567890

---

36 POINT AMERICAN TEXT

ABCDEFGHIJKLMNOPQRSTUVWXYZ
abcdefghijklmnopqrstuvwxyz.-:;, ˇ'!?$&
1234567890

---

24 POINT AMERICAN TEXT

ABCDEFGHIJKLMNOPQRSTUVWXYZ
abcdefghijklmnopqrstuvwxyz
1234567890

60 POINT CLOISTER BLACK

ABCDEFGHIJK
LMNOPQ
RSTUVWXYZ&
abcdefghijklmnopqrstu
vwxyz 1234567890$

36 POINT CLOISTER BLACK

ABCDEFGHIJKLMNOPQ
RSTUVWXYZ&
abcdefghijklmnopqrstuvwxyz
1234567890$

72 POINT GOUDY TEXT

ABCDEFGHI
JKLMNOPQ
RSTUVWXYZ
abcdefghijklmnopqrstu
vwxyz 1234567890$

36 POINT GOUDY TEXT

ABCDEFGHI JKLMNOPQR
STUVWXYZ&
abcdefghijklmnopqrstuvwxyz
1234567890$

36 POINT AMELIA

ABCDEFGHIJKKLMNOPQRSTUVWXYZ
abcdef ghijklmnopqrstuvwxyz
1234567890&!?$

36 POINT GALLIA

ABCDEFGHIJKLMN
OPQRSTUVWXYZ&
1234567890$
A E R S T

36 POINT JIM CROW

ABCDEFGHIJKLMNOPQRSTUV
WXYZ&   1234567890

36 POINT KISMET

ABCDEFGHIJKLMNOPQRSTUVW
XYZ
abcdefghijklmnopqrstuvluwxxyz
1234567890&$?

60 POINT POSTER ROMAN BOLD

ABCDEFGHIJKL MNOPQRSTU VWXYZ&
abcdefghijklmnopq
rstuvwxyz
1234567890$

36 POINT POSTER ROMAN BOLD

ABCDEFGHIJKLMNOPQRST
UVWXYZ&
abcdefghijklmnopqrstuvwxyz
1234567890$

36 POINT PRISMA

ABCDEFGHIJKLMNOPQR
STUVWXYZ&
1234567890$

36 POINT ROMANTIQUE

ABCDEFGHIJKLMN
OPQRSTUVWXYZ&
1234567890

36 POINT SMOKE

ABCDEFGHIJKLMNOPQRSTUVWXYZ
abcdefghijklmnopqrstuvwxyz
1234567890&!?$

36 POINT STENCIL

ABCDEFGHIJKLMN
OPQRSTUVWXYZ&
1234567890$

48 POINT STUDIO

ABCDEFGHIJKLMNOPQ
RSTUVWXYZ&
abcdefghijklmnopqrstu
vwxyz
1234567890$

36 POINT STUDIO

ABCDEFGHIJKLMNOPQRSTU
VWXYZ&
abcdefghijklmnopqrstuvwxyz
1234567890$

Glossary

A

Access To retrieve information from a storage device (internal memory, disk, tape). Access time is the time it takes to retrieve the stored data.

Accordian fold Two or more parallel folds with adjacent folds in opposite directions.

Achromatic The absence of color; black, gray, or white.

Agate This is 5½-point type. There are fourteen agate lines in 1 inch. The term is used to measure advertisements and can be used to designate tabular and classified matter in newspapers.

Alignment The positioning of letters so all have a common baseline; it also refers to the even placement of lines of type or art.

Alphabet length The width of lowercase (usually) characters when lined up *a* through *z*.

Ampersand The symbol used for *and* (&).

Antique A coarse and uneven paper finish.

Ascender. The letter stroke that extends above the *x* height of a lowercase character.

B

Bank One line of a multiline headline.

Banner A large multicolumn headline, usually extending across the top of page 1 in a newspaper.

Base alignment The positioning of characters so the bottom of the *x* height lines up evenly on a horizontal line; in phototypesetting this alignment is used for the even positioning of different type styles on a common line.

Ben Day The regular pattern of dots or lines used to add tonal variation to line art.

bf The designation for setting type in boldface.

Bidirectional printer A printing device that speeds hard copy production by printing left to right and then right to left and so on until the printout is completed.

Bit BInary digiT. This is the single digit of a binary number; 10 is composed of two bits.

Bit-map graphic A graphic image document formed by a series of dots, with a specific number of dots per inch. Also called a "paint-type" graphic.

Black Letter A race or group of type characterized by its resemblance to medieval northern European manuscript characters.

Bleed To run a photograph to the edge of the page.

Blind embossing Embossing (see below) without printing.

Block A group of words, characters, or digits forming a single unit in a computerized system.

Block letter A letterform without serifs (the finishing stroke at the end of a letter), in the Sans Serif type group.

Blueline Copy composed of blue lines on a white background.

Blurb Copy written with a sales angle, usually in brief paragraphs.

Body type "Reading matter" type as differentiated from display or headline type.

Boldface Characters of normal form but heavier strokes.

Bond paper Paper with a hard, smooth finish for ruling, typing, and pen writing.

Boot Getting your computer going; getting it started up and into the program you're going to use.

Border A frame around the type, art, or complete layout in either plane lines or an ornamental design.

Bowl the interior part of a letter in a circle form such as in a *b*, *c*, *d*, or *o*.

Box A border or rule that frames type.

Bracketed serif A serif (see below) connected to the character stem with a curved area at the connecting angle.

Break of the book The allocation of space for articles, features, and all material printed in the magazine.

Broadsheet A standard-size newspaper page as contrasted to the small tabloid size.

Brownline A brown-line image on a white background.

Bullet A round, solid ornament resembling a large period: ●.

Byte A number of binary digits, or bits, needed to encode one character such as a letter, punctuation mark, number, or symbol.

C

Calender The process in papermaking that creates the amount of smoothness in the paper.

C&lc The symbols for setting type in which the first character of each word is in capitals.

Camera ready The completed pasteup from which the printing plate is made.

Canned format The specifications for composition and/or makeup of type kept on magnetic or paper tape for repeated use to command a typesetter.

Capitals Large characters, the original form of Latin characters.

Caption The term used in magazine layout for the explanatory matter accompanying art; usually called a *cutline* in newspaper editing and layout.

Card A printed circuit board; computer systems are made up of these boards.

Cast off To determine the space a typewritten manuscript will occupy when set in type.

Catchline A line of display type between a picture and cutline.

Cathode ray tube (CRT) An electronic tube used to project images on a screen; it is also called a *visual display unit;* a television picture tube.

Clipboard A temporary holding place for material, in the computer. You can store text, graphics, or a group selection on a clipboard for later use.

Colophon The data about design, type styles, and production of a book; usually found at the end of the book.

Combination plate Halftone and line art on a single printing plate.

Composing room The area of a printing plant where type is set and arranged for plate making or printing.

Comprehensive A completed, detailed layout ready for making a plate; also called a *comp.*

Computer graphics Any charts, diagrams, drawings, and/or art composed on a computer.

Condensed type A vertically compressed character.

Constants The typographic and graphic elements in a publication that don't change from issue to issue.

Continuous tone Any art, such as a photograph or painting, which contains black and white and the variations of grays between the two.

Copy Information to be printed or reproduced.

Copy block A segment of body type or reading matter in a layout.

Copy fitting Determining the area a

certain amount of copy will occupy when set in type.

Crop To eliminate unwanted material or change the dimensions of art.

Cropping L The two right angles used to frame art to determine where it should be cropped.

Cursive A form of type that resembles handwriting.

Cursor A spot of light on a video screen that the user manipulates to indicate where changes in copy are to be made.

Cut A piece of art ready for printing; originally referred to as a mounted engraving used in letterpress printing.

Cutline The descriptive or identifying information printed with art—a caption.

Cutoff rule The dividing rule between elements, usually used in newspaper format.

Cyan A vivid blue color used in process (full color) color printing.

Cylinder press A printing press in which the form to be printed is flat and the impression is made on paper clamped on a cylinder that is rolled over the form.

D

Daisy wheel A metal or plastic disk with typewriter characters on spokes radiating from its center. It is about 3 inches in diameter. Hard copy printers can be equipped with more than one daisy wheel to mix faces. On typewriters, the wheel can produce type with differential spacing.

Dash A small horizontal rule in layouts; also a punctuation mark.

Data base A collection of information that is organized and stored so that an application program can access individual items.

Debugging Correcting errors in programs.

Deck One unit of a headline set in a single type size and style.

Descender A stroke of a lowercase letter that extends below the x height.

Digital computer A device used to manipulate data and perform calculations; most work on the binary number system (the number system based on powers of 2 rather than powers of 10).

Digitized type A form of type produced photographically by computer instructions created by patterns of black and white spots similar to the way television images are produced.

Dingbat A typographic ornament.

Diskette A flexible plastic recording medium, also called floppy disks or flexible disks.

Double pyramid The placement of advertisements on a page or facing pages to form a center "well" for editorial material.

Double truck A single advertisement that occupies two facing pages.

Download To transfer data from one electronic device to another. You could download information from one computer to another with a modem, for instance, or you could download information from a hard disk to a floppy disk.

Downloadable fonts Fonts that you can buy separately and install so as to expand the variety of fonts available on your printer.

Downstyle A form of headline in which only the first word and proper nouns are capitalized.

dpi Dots per inch.

DTP Desktop publishing.

Dummy A "blueprint" or pattern, usually half size for newspaper pages, used as a guide in making finished layout or pasteup.

Duotone A technique for color printing in which two plates are made from a black and white photo and printed in different colors to produce a single image.

E

Ear The editorial matter alongside the nameplate (the name of the publication) on page 1.

Em A unit of space equal to the square of the type size being used.

Embossing The process of impressing an image in relief to achieve a raised surface over printing. Embossing on blank paper is called *blind embossing*.

En A unit of space that is the vertical half of an em.

Extended A form of type in which the normal character structure is widened.

F

Face The style of a type, such as boldface.

Family A major division of typefaces.

File A collection of stored information with matching formats, the computer version of filing cabinets.

Finder The file that manages all the other files; the finder is like an index; it saves, names, renames and deletes things in a file.

Flag Synonymous with *nameplate*; the name of the publication in a distinctive design.

Flexography Printing from relief plates usually made of rubber as in letterpress but using a water-base ink rather than paste ink.

Floating flag A flag set in narrow width and displayed in a position other than the top of a page.

Flush left Type set even on the left margin and uneven on the right margin.

Flush right Type set with uneven lines on the left and even lines on the right margin.

Folio lines Originally the page numbers, but usually now the line giving date, volume, and number; or page number, name of publication, and date in small type on the inside pages.

Font All the characters and punctuation marks of one size and style of type.

Footer One or more lines of text that appear at the bottom of every page, similar to folio lines or running feet.

Form All the typographic elements used in a particular printed piece arranged to be placed on a printing press.

Format The general appearance of a printed piece, including the page size and number of columns per page.

Foundry type Printing type made of individual characters cast from molten metal.

Frame A newspaper makeup pattern in which the left and right outside columns are each filled with a single story.

Function code A computer code that controls the machine's operations other than the output of typographic characters.

Functional typography A philosophy of design in which every element used does an efficient and necessary job.

G

Galley A three-sided metal tray used to hold type; the term also refers to long strips of printed photographic or cold type ready to be proofread and used to make pasteups.

Galley proof The impression of type used for making corrections.

Gothic A group, or race, of monotonal types that have no serifs; also called *Sans Serif*.

Gravure An intaglio (see below) printing method that uses recessed plates.

Grid In graphic design, a pattern of horizontal and vertical guidelines for making layouts or dummies; in typesetting, an image carrier (a piece of film containing the characters) for a font of type for phototypesetting.

Grotesk The European name for Gothic type.

Gutter   The margin of the page at the point of binding, or the inside page margin.

H

Hairline   The thinnest rule used in printing, or the thinnest stroke in a letter form.

Halftone   A printing plate made by photographing an image through a screen so that the image is reproduced in dots.

Hanging indent   A headline style in which the first line is full width and succeeding lines are indented the same amount from the left margin.

Hard copy   Printed or typed copy, usually the printout from a computer or word processor or similar device.

Hardware   The actual equipment that makes up a computer system (see also *software*).

Head   An abbreviation for headline.

Header   Same as footer, but at the top of each page; like a running head.

Headletter   The type used for headlines.

Headline   The title of an article or news story.

Headline schedule   A chart showing all the styles and sizes of headlines used by a publication.

Highlight   The lightest portion of a halftone photograph; the area having the smallest dots or no dots at all.

Horizontal makeup   An arrangement of story units across columns rather than vertically.

Hot metal   Type, borders, and rules made of molten metal cast in molds.

Hue   A color, or the quality that distinguishes colors in the visible spectrum.

H&J   Hyphenation and justification; there are programs that will do the hyphenation of text for you following a standard dictionary.

I

Icon   A small graphic image that identifies a tool, file, or command displayed on a computer screen.

Imposition   An arrangement of pages for printing so they will appear in proper order for folding.

Impression cylinder   A printing press unit that presses paper on an inked form to make the print.

Initial   The first letter in a word set in a larger or more decorative face, usually used at the beginning of an article, section, or paragraph.

Ink-jet printing   A method of placing

characters on paper by spraying a mist of ink through tiny holes in the patterns of the characters.

Inline   A style of type in which a white line runs down the main stroke of the letter.

Insert   Reference lines inserted in the body of an article; also called a *refer* or *sandwich*.

Intaglio   A printing method in which the image is carved into, or recessed, in the plate; also called *gravure*.

Inverted pyramid   A headline style in which each centered line is narrower than its predecessor; term is also used for a newswriting form.

Italic   A form of Roman type design that slants to the right.

J

Jim dash   A small rule—usually used to separate decks in a headline or title.

Jump head   A headline on the part of a story continued from another page.

Justify   Setting type so the left and right margins are even.

K

Kerning   Placing two adjacent characters so that part of one is positioned within the space of the other; kerning may be controlled by the keyboard operator or programmed into the computer.

Keylining   A process of using an overlay in a layout to indicate color separations, reverses, outlines, or other special effects.

Kicker   A small headline, usually underlined, above a main headline.

Kilobyte   1,024 bytes or 1K; a 3½ inch floppy disk holds 800 kilobytes (800K) or about 400 double-spaced typewritten pages (a rough estimate).

L

Layout   A diagram or plan, drawing, or sketch used as a guide in arranging elements for printing.

lc   The abbreviation for lowercase.

lca   The abbreviation for lowercase alphabet.

Lead   The space between lines of type, usually 2 points (pronounced "led").

Leaders   The dots or dashes often used in tabular matter.

Letterpress   A printing method that uses raised images.

Letterspace   The space added to the nor-

mal spacing between the letters in a word.

Ligature   Two or more characters joined to make a single unit.

Lightface   Characters with strokes that have less weight than normal.

Light pen   An electronic stylus used to position elements or indicate changes in copy on a CRT.

Line art   A piece of art or a plate in black and white, not continuous tones.

Line conversion   A line printing plate made from a continuous tone original by eliminating the halftone screen.

Linen paper   A paper made from linen or having a finish resembling linen cloth.

Lithography   A flat-surface printing method based on the principle that oil and water are mutually repellent.

Logo   Abbreviation for logotype.

Logotype   A distinctive type arrangement used for the name of a publication, business, or organization.

Lowercase   The small letters of the alphabet.

M

Magenta   Also called *process red*; it is a purplish red color and is used in process (or full) color printing.

Magnetic ink   An ink that contains ferrous (iron) material that can be sensed magnetically.

Makeup   The art of arranging elements on a page for printing.

Markup   The process of writing instructions on a layout for the size and styles of types and the other elements desired by the designer.

Masthead   The area in a publication that lists the staff, date of publication, and other pertinent information.

Matrix (mat)   The brass mold from which type is cast (or molded) in the hot metal process.

Measure   The width of the lines being set.

Mechanical   A pasteup ready for plate making.

Mechanical separation   Copy prepared by a designer with each individual color in a separate section.

Menu   A list of commands that appears when you point to and press the menu title in the menu bar.

Menu bar   The area at the top of the publication window that lists the menus.

Mezzotint   A screen used for creating a crayon drawing effect on a printing plate.

Minimum line length   The shortest width of lines of type of acceptable readability.

Minuscules Small characters.

Minus leading The elimination of space between lines, a technique possible with photographic or electronic typesetting.

Minus letterspacing Reducing the normal space between characters; a technique possible with photographic or electronic typesetting.

Miter To cut a rule or border at a 45-degree angle for making corners on a box.

Mixing Combining more than one style or size of type on the same line.

Mnemonics Ancient memory-aiding devices; also used to refer to abbreviations of complex terms used in encoding computer instructions.

Mock-up A full-size, experimental layout for study and evaluation.

Modem A telecommunications device that translates computer signals into electronic signals that can be sent over a telephone line; a way to get information from one computer to another or from your computer to a print shop and so on.

Modular makeup The arrangement of elements in rectangular units on a page, also called *Mondrian*.

Moiré A distracting pattern that results when a previously screened halftone is screened again and printed.

Mondrian makeup The arrangement of elements into rectangles of various sizes and shapes, also called *modular*.

Monotonal Typefaces with strokes of equal thickness.

Montage A composite picture, usually made of two or more combined photographs.

Mortise An area cut out of a piece of art for the insertion of type or other art.

N

Nameplate The name of a publication set in a distinctive type form, also called *flag*.

Newsprint A low-quality paper mainly used for printing handbills and newspapers.

Notch mortise A rectangle cut from a corner of a rectangular illustration.

Novelty A category of type that is usually ornamental in design and does not display any strong characteristics of one of the basic races or species.

O

Object-oriented graphic An illustration created in an object-oriented, or draw-type application. An object-oriented graphic is created with geometric elements. Also called a "draw-type" graphic.

Oblique Letters that slant to the right, usually Sans Serif or Square Serif; Roman slanted letters are called *italics*.

OCR (optical character recognition) A device that electronically reads and encodes printed or typewritten material.

Offset A printing process in which the image is transferred from a printing plate to a rubber blanket to paper.

Old style A Roman typeface subdivision characterized by bracketed serifs and little difference between the thick and thin strokes.

Opaque Something that blocks light; in paper a lack of show-through.

Optical center A point about 10 percent above the mathematical center of a page or area.

Optimum format A format in which the width of the type columns is within the range of maximum readability.

Optimum line length The line width at which reading is easiest and fastest.

Ornament A decorative typographic device.

Outline A type design in which the letter is traced or outlined by lines on the outside of the strokes and the inside of the letter is blank.

Overlay A sheet of transparent plastic or paper placed over a piece of art or a layout on which instructions are written and areas to be printed in color are drawn.

Overline A display type heading placed above a picture.

Overprint To print over an area that has already been printed.

Overset Body matter that exceeds the allotted space.

Oxford rule Parallel heavy and light lines.

P

Pagination The process of arranging pages for printing; a computer-generated page layout; and/or the numbering of the pages of a book.

Parallel fold Two or more folds in the same direction.

Pasteup The process of fixing type and other elements on a grid for plate making; a *mechanical*.

Pebbling Embossing paper in its manufacture to create a ripple effect.

Perfect binding A method of binding that uses flexible glue rather than stitching.

Photocomposition Phototypesetting by film or paper.

Photoengraving A printing plate with a raised surface.

Photolithography An offset printing process that uses a plate made by a photographic process.

Phototypesetter A device that sets type by a photographic process; letter images are recorded on light-sensitive film or photographic paper that is then developed and printed.

Pi To mix type; individual metal characters that have been mixed up by accident.

Pica A 12-point unit of measurement.

Pixel (picture element) The smallest part of a graphic that can be controlled through a program. You could think of it as a building block used to construct type and images. The resolution of text and graphics on your screen depends on the density of your screen's pixels.

Planography A printing process that uses a plate with a flat surface; offset lithography is a planographic process.

Plate A printing surface.

Platen press A machine in which paper is held on a flat surface and pressed against the form for printing.

PMT (photo mechanical transfer) A positive print that is ready for pasteup.

Point A unit of measurement approximately 1/72 of an inch.

Pork chop A small head shot of a person, usually half a column wide.

Poster makeup An arrangement of a newspaper's front page that usually consists of large art and a few headlines to attract attention.

Prescreen A halftone positive print that can be combined with line copy in paste-up, thus eliminating the need to strip in a screened negative with a line copy negative.

Primary color The colors red, yellow, and blue, which combine to make all other hues.

Primary optical center The spot where a reader's eye usually first lights on a page, the upper left quadrant.

Process color Printing the three primary colors in combination to produce all colors; full color.

Production department The mechanical department of a printing plant.

Proof A preliminary print of set type or a comprehensive, used to detect errors before the final printing.

Prototype A mock-up; a model that is the pattern for the final product.

Pyramid An arrangement of advertisements on a page to form a stepped half pyramid.

Q

Quad   A unit of space in setting type.

Quad left, quad middle, quad right   Commands that instruct a typesetting machine to put space in lines.

Quadrant makeup   A plan for a page in which each quarter is given a strong design element.

R

Race   The basic division of type styles, sometimes called *species;* type groups.

Ragged right   Type set unjustified (uneven) at the right margins.

Ragged left   The opposite of ragged right.

RAM   Random access memory; the temporary memory inside the computer that allows you to find stored text and graphics.

Readability   The characteristic of type and/or its arrangement that makes it easy to read.

Readout   The headline or unit of a headline between a banner head and the story; also used to refer to devices for breaking up body matter such as quotes (called quote-outs or pulled quotes) set in display type and embellished with typographic devices.

Recto   In book or pamphlet design, the odd-numbered, right-hand pages.

Register   To line up color printing plates so the multiple impressions will create an accurate reproduction of the original.

Relief printing   Printing from a raised surface; letterpress.

Repro (reproduction proof)   The final image used for pasteup and plate making.

Resolution   The number of dots per inch (dpi) used to represent a character or graphic image. The higher the resolution the more dots per inch and the clearer the image looks.

Reverse   A printing area in which the background is black and the image is white.

Reverse kicker   A headline form in which the kicker is larger than the primary headlines.

Reverse leading   The technique of operating a phototypesetting device so superior figures or mixed display type can be added to the line.

Rivers   Vertical strips of white space in areas of type created by excessive space between words; also called "rivers of white."

Roman   A basic race or species of type in which the characters have serifs and variations in the widths of their strokes.

ROP (run of press)   Color or other matter that isn't given a specific special position in the publication.

Rotary press   A press that uses a curved plate to print on a continuous roll of paper.

Rotogravure   A recessed-image (intaglio) printing process on a continuous roll of paper.

Rough   A sketchy dummy or layout to show the placement of elements.

Rule   An element that prints a continuous line or lines usually used to frame type, art, or a layout.

Run-in head   A headline that is part of the first line of the text.

Running foot   A line at the bottom of a book page that indicates the book, chapter, or section title, and/or the page number.

Running head   A line at the top of a book page that contains the same information as a running foot.

S

Saddle stitch   A binding method for magazines or pamphlets that uses a wire stitch on the centerfold.

Sandwich   A short notice placed within the body of an article.

Sans Serif   A race or species of type without serifs (the finishing strokes on characters) usually containing monotonal letter strokes.

Scale   To size art to certain enlarged or reduced dimensions.

Scanner   A hardware device that reads information from a photograph, image, or text, converting it into a bit-map graphic.

Score   To crease paper on a line to facilitate folding.

Screen   A device, available in various densities used to reduce continuous tone art to a halftone plate.

Script   A typeface that resembles handwriting.

Scroll   To move a story up or down on a video display terminal screen, usually for editing purposes.

Secondary color   A hue (color) produced by mixing two primary colors.

Section logo   A typographic device used to identify a section of a publication.

Series   A basic subdivision of a type family; it has family characteristics but an individual posture; all the sizes of that particular posture of a family.

Serif   The finishing stroke at the end of a primary stroke of a character.

Set solid   Type set with no leading between the lines.

Shade   A darker hue obtained by adding black to a color.

Shaded   Type that gives a gray instead of solid black imprint.

Sideline   An arrangement with short display lines to the left of the cutlines.

Sidestitch (sidewire stitching)   A binding in which the staple is placed on the side rather than on the spine as in saddle stitch.

Signature   A group of pages printed on a single sheet.

Silhouette   Art in which the background has been removed.

Sinkage   A point below the top margin of a page where chapter openings or other material is set.

Skyline   A headline or story at the top of the first page of a publication, above the nameplate.

Slug   A unit of space, usually between lines of type and usually 6 or 12 points thick; a line of cast hot type; an identifying line on copy.

Small cap   A capital letter for a font of type smaller than the regular capital letter.

Software   The instructions or programs that cause a computer to operate and perform desired functions.

Sorts   Characters that are obtainable but not ordinarily included in a font of type, such as mathematical signs and special punctuation and accent marks.

Species   A basic division of type; a race of type.

Spine   The midpoint area between the front and back covers of a book or magazine; the center point of the outside cover.

Spot color   One hue (color) in addition to black, usually in a headline, display line, border, or ornament.

Square Serif   A typeface characterized by monotonal strokes and heavy, squared serifs; also called *Egyptian, Slab Serif.*

Standing head   A headline that remains the same from one issue of a publication to another.

Stepped head   An arrangement of display type in which the top line is flush left, the middle line (if used) is centered, and the third line is flush right, and all the lines are less than full width to create a stair step effect.

Stereotyping   The process of casting a printing form from molten metal by using a mold (called a *mat*) or paper, usually in a curved form, to create a curved printing plate for a rotary press.

Stet   A term meaning "do not change," used in proofreading.

Stick-on letters   Alphabet characters printed on paper, usually self-adhesive, for cold type composition.

Straight matter   Reading matter; body type; the text material in a book.

Streamer   A banner headline.

Strike-on   Type produced by a typewriter or other percussion keyboard device that impacts character forms directly on paper.

Stripping   Combining halftone and line negatives to create a comprehensive negative for printing plates that contain line and screened art.

Subhead   A display line that is auxiliary to the main headline or used to break up masses of body type.

Sunken initial   An initial inset in reading matter.

Surprint   Something printed over art.

Swash   A letter decorated with an elongated stroke, usually decorative.

Symmetrical   An arrangement of elements in formal balance.

T

Tabloid   A newspaper format usually about half the size of a broadsheet or approximately 11 by 15 inches.

Tabular matter   Statistics arranged in table or columnar form, such as stock market reports, financial statements, and so on.

Terminal   A device in a communications network or system where information can be entered, removed, or displayed for viewing and arranging.

Text   The Black Letter race or species of type.

Text   Reading matter

Text wrap   To run text around an illustration on a page layout. Some programs have an automatic text-wrap feature that will shorten lines of text when a graphic is encountered; in other systems you need to change the length of lines to go around a graphic.

Thumbnail   A small preliminary sketch of a possible arrangement of elements; also the term for a portrait that is less than a column wide and inset in reading matter.

Tint   A value of color created by adding white to a hue (color).

Tint block   An area on a printed page produced in a tint, usually with type or art surprinted on it.

Tombstone   Identical side-by-side headlines that compete for attention.

Tone   Shading or tinting a printing element.

Transfer letter   A letter for printing obtained by rubbing from a master sheet onto the layout sheet.

Transitional   A subdivision of the Roman type race with characteristics of both old style and modern Romans.

Type   The characters from which printing is done.

Typeface   The distinctive design of an alphabet of letters and related characters.

Typo   An error in set type.

Typography   The use and arrangement of elements for printing.

U

U&lc   The designation for setting type in capital and lowercase letters where it is appropriate.

Unit   A fraction of an em; in a 36-unit phototypesetting system, for instance, an em would have 36 units; more units allows more latitude in programming space between letters and words in designating character widths.

Unit count   A method of determining whether display type will fit a given area.

Uppercase (uc)   Capital letters.

V

Value   Synonym for tone, or the relative tint or shade of a printing element.

Velox   A black and white print of a halftone photograph.

Video display terminal (VDT)   A device and screen for arranging elements.

Verso   The even-numbered, left-hand pages.

Vertical justification   Automatic adjustment of leading or the space between lines, in very small amounts so columns on a page can all be made the same depth.

Vignette   An illustration in which the margins appear to fade into the background.

W

Watermark   A design, name, and/or logotype impressed on paper during manufacture.

Waxer   A device for coating the back of layout parts with melted wax to attach them to a grid or layout sheet.

Web   A wide strip or roll of paper that travels through a press for printing.

Weight   The comparative thickness of strokes of letters.

Well   The arrangement of advertisements on the right and left sides of a page so editorial matter can be placed between.

Widow   A short line at the top of a column that completes a paragraph from the bottom of the preceding column; also used to refer to a very short final line of a paragraph.

Wrong font (wf)   A type character set in a different family or series from the rest of the specified set matter.

WYSIWYG   An acronym for "what you see is what you get."

X

x height   The distance between the baseline and meanline of type; the height of a lowercase letter excluding the ascender and descender.

Z

Zipatone   A transparent sheet containing dot or line patterns that provides a tonal effect similar to that provided by Ben Day.

Index

C O L O P H O N

The second edition of *Graphic Communications Today* was designed on a Macintosh IIcx computer using Pagemaker 3.02. The text type is Palatino, with display material set in Optima. Type specifications are as follows:

General text: 10/12 Palatino × 27.5 picas, justified. One em paragraph indents.

Headings: first level heads: 14 pt. Optima bold; second level heads: 12 pt. Optima bold.

Figure captions: 8.5/10 Palatino bold for ''Fig.'' and number; caption set 8.5/10 Optima × 13 picas, flush left, ragged right.

Running heads: 8.5 pt. Optima bold, flush with left text margin (27.5 picas).

Folios: 10 pt. Optima bold, flush with outside margins (42 picas).

Chapter Openings: Chapter number is 10 pt. Palatino, all caps, flush left × 42 picas. Chapter title is 36/40 Optima bold, clc, flush left × 42 picas.

The type was set at Parkwood Composition Service, Inc., New Richmond, Wisconsin. The second color used in the text is PMS 286. In the four-color sections, this color is matched with C100, M65 (100% cyan, 65% magenta). The text was printed at Webcrafters in Madison, Wisconsin, using 45 lb. Courtland Matte paper. Prepress work for all illustrations and photographs (color separations, halftone screens, and other camera work) was completed at West Publishing Company.